Annals of Mathematics Studies
Number 191

Classification of Pseudo-reductive Groups

Brian Conrad

Gopal Prasad

PRINCETON UNIVERSITY PRESS

PRINCETON AND OXFORD

2016

Published by Princeton University Press, 41 William Street,
Princeton, New Jersey 08540
In the United Kingdom: Princeton University Press, 6 Oxford Street,
Woodstock, Oxfordshire OX20 1TW

press.princeton.edu

Library of Congress Cataloging-in-Publication Data

Conrad, Brian, 1970-
Classification of Pseudo-reductive Groups / Brian Conrad, Gopal Prasad.
pages cm. – (Annals of mathematics studies; number 191)
Includes bibliographical references and index.
ISBN 978-0-691-16792-3 (hardcover) – ISBN 978-0-691-16793-0
(pbk. : alk. paper) 1. Linear algebraic groups. 2. Group theory. 3. Geometry,
Algebraic. I. Prasad, Gopal. II. Title.
QA179.C665 2016
512'.55–dc23
2015023803

British Library Cataloging-in-Publication Data is available

This book has been composed in LaTeX using MathTime fonts.

The publisher would like to acknowledge the authors of this volume for
providing the camera-ready copy from which this book was printed.

1 3 5 7 9 10 8 6 4 2

Contents

Classification of Pseudo-reductive Groups

1

Introduction

1.1 Motivation

Algebraic and arithmetic geometry in positive characteristic provide important examples of imperfect fields, such as (i) Laurent-series fields over finite fields and (ii) function fields of positive-dimensional varieties (even over an algebraically closed field of constants). Generic fibers of positive-dimensional algebraic families naturally lie over a ground field as in (ii).

For a smooth connected affine group G over a field k, the unipotent radical $\mathscr{R}_u(G_{\overline{k}}) \subset G_{\overline{k}}$ may not arise from a k-subgroup of G when k is imperfect. (Examples of this phenomenon will be given shortly.) Thus, for the maximal smooth connected unipotent normal k-subgroup $\mathscr{R}_{u,k}(G) \subset G$ (the *k-unipotent radical*), the quotient $G/\mathscr{R}_{u,k}(G)$ may not be reductive when k is imperfect.

A *pseudo-reductive group* over a field k is a smooth connected affine k-group G such that $\mathscr{R}_{u,k}(G)$ is trivial. For any smooth connected affine k-group G, the quotient $G/\mathscr{R}_{u,k}(G)$ is pseudo-reductive. A pseudo-reductive k-group G that is perfect (i.e., G equals its derived group $\mathscr{D}(G)$) is called *pseudo-semisimple*. If k is perfect then pseudo-reductive k-groups are connected reductive k-groups by another name. For imperfect k the situation is completely different:

Example 1.1.1. Weil restrictions $G = \mathrm{R}_{k'/k}(G')$ for finite extensions k'/k and connected reductive k'-groups G' are pseudo-reductive [CGP, Prop. 1.1.10]. If G' is nontrivial and k'/k is not separable then such G are *never* reductive [CGP, Ex. 1.6.1]. A solvable pseudo-reductive group is necessarily commutative [CGP, Prop. 1.2.3], but the structure of commutative pseudo-reductive groups appears to be intractable (see [T]). The quotient of a pseudo-reductive k-group by a smooth connected normal k-subgroup or by a central closed k-subgroup scheme can fail to be pseudo-reductive, and a smooth connected normal k-subgroup of a pseudo-semisimple k-group can fail to be perfect; see [CGP, Ex. 1.3.5, 1.6.4] for such examples over any imperfect field k.

A typical situation where the structure theory of pseudo-reductive groups is useful is in the study of smooth affine k-groups about which one has limited information but for which one wishes to prove a general theorem (e.g., cohomological finiteness); examples include the Zariski closure in GL_n of a subgroup of $\mathrm{GL}_n(k)$, and the maximal smooth k-subgroup of a schematic stabilizer (as in local-global problems). For questions not amenable to study over $\overline{k}$ when k is imperfect, this structure theory makes possible what had previously seemed out of reach over such k: to reduce problems for general smooth affine k-groups to the reductive and commutative cases (over finite extensions of k). Such procedures are essential to prove finiteness results for degree-1 Tate-Shafarevich sets of arbitrary affine group schemes of finite type over global function fields, even in the general *smooth* affine case; see [C1, §1] for this and other applications.

A detailed study of pseudo-reductive groups was initiated by Tits; he constructed several instructive examples and his ultimate goal was a classification. The general theory developed in [CGP] by characteristic-free methods includes the open cell, root systems, rational conjugacy theorems, the Bruhat decomposition for rational points, and a structure theory "modulo the commutative case" (summarized in [C1, §2] and [R]). The lack of a concrete description of commutative pseudo-reductive groups is not an obstacle in applications (see [C1]).

In general, if G is a smooth connected affine k-group then $\mathscr{R}_{u,k}(G)_K \subset \mathscr{R}_{u,K}(G_K)$ for any extension field K/k, and this inclusion is an equality when K is separable over k [CGP, Prop. 1.1.9] but generally not otherwise (e.g., equality fails with $K = \overline{k}$ for any imperfect k and non-reductive pseudo-reductive G). Taking $K = k_s$ shows that G is pseudo-reductive if and only if G_{k_s} is pseudo-reductive (and also shows that if k is perfect then pseudo-reductive k-groups are precisely connected reductive k-groups). Hence, any smooth connected normal k-subgroup of a pseudo-reductive k-group is pseudo-reductive.

Every smooth connected affine k-group G is generated by $\mathscr{D}(G)$ and a single Cartan k-subgroup. Since $\mathscr{D}(G)$ is pseudo-semisimple when G is pseudo-reductive [CGP, Prop. 1.2.6], and Cartan k-subgroups of pseudo-reductive k-groups are commutative and pseudo-reductive, the main work in describing pseudo-reductive groups lies in the pseudo-semisimple case. A smooth affine k-group G is *pseudo-simple* (over k) if it is pseudo-semisimple, nontrivial, and has no nontrivial smooth connected proper normal k-subgroup; it is *absolutely pseudo-simple* if G_{k_s} is pseudo-simple. (See [CGP, Def. 3.1.1, Lemma 3.1.2] for equivalent formulations.) A pseudo-reductive k-group G is *pseudo-split* if it contains a split maximal k-torus T, in which case any two such tori are conjugate by an element of $G(k)$ [CGP, Thm. C.2.3]

Remark 1.1.2. If G is a pseudo-semisimple k-group then the set $\{G_i\}$ of its pseudo-simple normal k-subgroups is finite, the G_i's pairwise commute and generate G, and every perfect smooth connected normal k-subgroup of G is generated by the G_i's that it contains (see [CGP, Prop. 3.1.8]). The core of the study of pseudo-reductive groups G is the absolutely pseudo-simple case.

Although [CGP] gives general structural foundations for the study and application of pseudo-reductive groups over any imperfect field k, there are natural topics not addressed in [CGP] whose development requires new ideas, such as:

(i) Are there versions of the Isomorphism and Isogeny Theorems for pseudo-split pseudo-reductive groups and of the Existence Theorem for pseudo-split pseudo-simple groups?

(ii) The *standard construction* (see §2.1) is exhaustive when $p := \mathrm{char}(k) \neq 2, 3$. Incorporating constructions resting on exceptional isogenies [CGP, Ch. 7–8] and birational group laws [CGP, §9.6–§9.8] gives an analogous result when $p = 2, 3$ provided that $[k : k^2] = 2$ if $p = 2$; see [CGP, Thm. 10.2.1, Prop. 10.1.4]. More examples exist if $p = 2$ and $[k : k^2] > 2$ (see §1.3); can we generalize the standard construction for such k?

(iii) Is the automorphism functor of a pseudo-semisimple group representable? (Representability fails in the commutative pseudo-reductive case.) If so, what can be said about the structure of the identity component and component group of its maximal smooth closed subgroup $\mathrm{Aut}^{\mathrm{sm}}_{G/k}$ (thereby defining a notion of "pseudo-inner" k_s/k-form via $(\mathrm{Aut}^{\mathrm{sm}}_{G/k})^0$)?

(iv) What can be said about existence and uniqueness of pseudo-split k_s/k-forms, and of quasi-split pseudo-inner k_s/k-forms? ("Quasi-split" means the existence of a solvable pseudo-parabolic k-subgroup.)

(v) Is there a Tits-style classification in the pseudo-semisimple case recovering the version due to Tits in the semisimple case? (Many ingredients in the semisimple case *break down* for pseudo-semisimple G; e.g., G may have no pseudo-split k_s/k-form, and the quotient G/Z_G of G modulo the scheme-theoretic center Z_G can be a proper k-subgroup of $(\mathrm{Aut}^{\mathrm{sm}}_{G/k})^0$.)

The special challenges of characteristic 2 are reviewed in §1.3–§1.4 and §4.2. Recent work of Gabber on compactification theorems for arbitrary linear algebraic groups uses the structure theory of pseudo-reductive groups over general (imperfect) fields. That work encounters additional complications in characteristic 2 which are overcome via the description of pseudo-reductive groups as central extensions of groups obtained by the "generalized standard" construction given in Chapter 9 of this monograph (see the Structure Theorem in §1.6).

1.2 Root systems and new results

A maximal k-torus T in a pseudo-reductive k-group G is an almost direct product of the maximal central k-torus Z in G and the maximal k-torus $T' := T \cap \mathscr{D}(G)$ in $\mathscr{D}(G)$ [CGP, Lemma 1.2.5]. Suppose T is *split*, so the set $\Phi := \Phi(G,T)$ of nontrivial T-weights on $\mathrm{Lie}(G)$ injects into $\mathrm{X}(T')$ via restriction.

The pair $(\Phi, \mathrm{X}(T')_{\mathbf{Q}})$ is always a root system (coinciding with $\Phi(\mathscr{D}(G), T')$ since $G/\mathscr{D}(G)$ is commutative) [CGP, Thm. 2.3.10], and can be canonically enhanced to a root datum [CGP, §3.2]. In particular, to every pseudo-semisimple k_s-group we may attach a *Dynkin diagram*. However, $(\Phi, \mathrm{X}(T')_{\mathbf{Q}})$ can be non-reduced when k is imperfect of characteristic 2 (the non-multipliable roots are the roots of the maximal geometric reductive quotient $G_{\overline{k}}^{\mathrm{red}}$). A pseudo-split pseudo-semisimple group is (absolutely) pseudo-simple precisely when its root system is irreducible [CGP, Prop. 3.1.6].

This monograph builds on earlier work [CGP] via new techniques and constructions to answer the questions (i)–(v) raised in §1.1. In so doing, we also simplify the proofs of some results in [CGP]. (For instance, the standardness of all pseudo-reductive k-groups if $\mathrm{char}(k) \neq 2, 3$ is recovered here by another method in Theorem 3.4.2.) Among the new results in this monograph are:

- (i) pseudo-reductive versions of the Existence, Isomorphism, and Isogeny Theorems (see Theorems 3.4.1, 6.1.1, and A.1.2),
- (ii) a *structure theorem* over arbitrary imperfect fields k (see §1.5–§1.6),
- (iii) existence of the automorphism scheme $\mathrm{Aut}_{G/k}$ for pseudo-semisimple G, and properties of the identity component and component group of its maximal smooth closed k-subgroup $\mathrm{Aut}_{G/k}^{\mathrm{sm}}$ (see Chapter 6),
- (iv) uniqueness and optimal existence results for pseudo-split and "quasi-split" k_s/k-forms for imperfect k, including examples (in *every* positive characteristic) where existence *fails* (see §1.7),
- (v) a Tits-style classification of pseudo-semisimple k-groups G in terms of both the Dynkin diagram of G_{k_s} with $*$-action of $\mathrm{Gal}(k_s/k)$ on it and the k-isomorphism class of the embedded anisotropic kernel (see §1.7).

We illustrate (v) in Appendix D by using anisotropic quadratic forms over k to construct and classify absolutely pseudo-simple groups of type F_4 with k-rank 2 (which never exist in the semisimple case).

1.3 Exotic groups and degenerate quadratic forms

If $p = 2$ and $[k : k^2] > 2$ then there exist families of *non-standard* absolutely pseudo-simple k-groups of types B_n, C_n, and BC_n (for every $n \geqslant 1$) with no analogue when $[k : k^2] = 2$. Their existence is explained by a construction with certain *degenerate* quadratic spaces over k that exist only if $[k : k^2] > 2$:

Example 1.3.1. Let (V, q) be a quadratic space over a field k with $\mathrm{char}(k) = 2$, $d := \dim V \geqslant 3$, and $q \neq 0$. Let $B_q : (v, w) \mapsto q(v + w) - q(v) - q(w)$ be the associated symmetric bilinear form and $V^\perp$ the defect space consisting of $v \in V$ such that the linear form $B_q(v, \cdot)$ on V vanishes. The restriction $q|_{V^\perp}$ is 2-linear (i.e., additive and $q(cv) = c^2 q(v)$ for $v \in V$, $c \in k$) and $\dim(V/V^\perp) = 2n$ for some $n \geqslant 0$ since B_q induces a non-degenerate symplectic form on $V/V^\perp$.

Assume $0 < \dim V^\perp < \dim V$. Now q is non-degenerate (i.e., the projective hypersurface $(q = 0) \subset \mathbf{P}(V^*)$ is k-smooth) if and only if $\dim V^\perp = 1$, which is to say $d = 2n + 1$. It is well-known that in such cases $\mathrm{SO}(q)$ is an absolutely simple group of type B_n with $\mathrm{O}(q) = \mu_2 \times \mathrm{SO}(q)$, so $\mathrm{SO}(q)$ is the maximal smooth closed k-subgroup of $\mathrm{O}(q)$ since $\mathrm{char}(k) = 2$. Assume also that (V, q) is *regular*; i.e., $\ker(q|_{V^\perp}) = 0$. Regularity is preserved by any separable extension on k (Lemma 7.1.1). For such (possibly degenerate) q, define $\mathrm{SO}(q)$ to be the maximal smooth closed k-subgroup of the k-group scheme $\mathrm{O}(q)$; i.e., $\mathrm{SO}(q)$ is the k-descent of the Zariski closure of $\mathrm{O}(q)(k_s)$ in $\mathrm{O}(q)_{k_s}$. In §7.1–§7.3 we prove: $\mathrm{SO}(q)$ is absolutely pseudo-simple with root system B_n over k_s where $2n = \dim(V/V^\perp)$, the dimension of a root group of $\mathrm{SO}(q)_{k_s}$ is 1 for long roots and $\dim V^\perp$ for short roots, and the minimal field of definition over k for the geometric unipotent radical of $\mathrm{SO}(q)$ is the k-finite subextension $K \subset k^{1/2}$ generated over k by the square roots $(q(v')/q(v))^{1/2}$ for nonzero $v, v' \in V^\perp$.

For any nonzero $v_0 \in V^\perp$, the map $v \mapsto (q(v)/q(v_0))^{1/2}$ is a k-linear injection of $V^\perp$ into $k^{1/2}$ with image $\mathscr{V}$ containing 1 and generating K as a k-algebra. If we replace v_0 with a nonzero $v_1 \in V^\perp$ then the associated k-subspace of K is $\lambda\mathscr{V}$ where $\lambda = (q(v_0)/q(v_1))^{1/2} \in K^\times$. In particular, the case $K \neq k$ occurs if and only if $\dim V^\perp \geqslant 2$, which is precisely when the regular q is *degenerate*, and always $[k : k^2] = [k^{1/2} : k] \geqslant \dim V^\perp$. If $V^\perp = K$, as happens whenever $[k : k^2] = 2$, then $\mathrm{SO}(q)$ is the quotient of a "basic exotic" k-group [CGP, §7.2] modulo its center. The $\mathrm{SO}(q)$'s with $V^\perp \neq K$ (so $[k : k^2] > 2$) are a new class of absolutely pseudo-simple k-groups of type B_n (with trivial center); for $n = 1$ and isotropic q these are the type-A_1 groups $PH_{V^\perp, K/k}$ built in §3.1.

In §7.2–§7.3 we show that every k-isomorphism $\mathrm{SO}(q') \simeq \mathrm{SO}(q)$ arises from a conformal isometry $q' \simeq q$ and use this to construct more absolutely pseudo-simple k-groups of type B with trivial center via geometrically integral non-smooth quadrics in Severi–Brauer varieties associated to certain elements of order 2 in the Brauer group $\mathrm{Br}(k)$. Remarkably, this accounts for *all* non-reductive pseudo-reductive groups whose Cartan subgroups are tori (see Proposition 7.3.7), and when combined with the exceptional isogeny $\mathrm{Sp}_{2n} \to \mathrm{SO}_{2n+1}$ in characteristic 2 via a fiber product construction it yields (in §8.2) new absolutely pseudo-simple groups of type C_n when $n \geqslant 2$ and $[k : k^2] > 2$ (with short root groups over k_s of dimension $[K : k]$ and long root groups over k_s of dimension $\dim V^{\perp}$). A generalization in §8.3 gives *even more* such k-groups for $n = 2$ if $[k : k^2] > 8$ (using that $\mathrm{B}_2 = \mathrm{C}_2$). In §1.5–§1.6 we provide a context for this zoo of constructions.

1.4 Tame central extensions

A new ingredient in this monograph is a generalization of the "standard construction" (from §2.1) that is better-suited to the peculiar demands of characteristic 2. Before we address that, it is instructive to recall the principle underlying the ubiquity of standardness *away from* the case $\mathrm{char}(k) = 2$ with $[k : k^2] > 2$, via splitting results for certain classes of central extensions. We now review the most basic instance of such splitting, to see why it breaks down completely (and hence new methods are required) when $\mathrm{char}(k) = 2$ with $[k : k^2] > 2$ (see 1.4.2).

1.4.1. Let G be an absolutely pseudo-simple k-group with minimal field of definition K/k for $\mathscr{R}_u(G_{\overline{k}}) \subset G_{\overline{k}}$, and let $G' := G_K^{\mathrm{ss}}$ be the maximal semisimple quotient of G_K. For the simply connected central cover $q : \widetilde{G}' \to G'$ and $\mu := \ker q \subset Z_{\widetilde{G}'}$, there is (as in [CGP, Def. 5.3.5]) a canonical k-homomorphism

$$\xi_G : G \to \mathscr{D}(\mathrm{R}_{K/k}(G')) = \mathrm{R}_{K/k}(\widetilde{G}')/\mathrm{R}_{K/k}(\mu) \qquad (1.4.1.1)$$

induced by the natural map $i_G : G \to \mathrm{R}_{K/k}(G')$. The map ξ_G makes sense for any pseudo-reductive G but (as in [CGP]) it is of interest only for absolutely pseudo-simple G. By Proposition 2.3.4, $\ker \xi_G$ is central if $\mathrm{char}(k) \neq 2$.

The key to the proof that G is standard if $\mathrm{char}(k) \neq 2, 3$ is the surjectivity of ξ_G for such k, as then (1.4.1.1) pulls back to a central extension E of $\mathrm{R}_{K/k}(\widetilde{G}')$ by $\ker \xi_G$. This central extension is *split* due to a general fact: if k'/k is an arbitrary finite extension of fields and $\mathscr{G}'$ a connected semisimple k'-group that is *simply connected* then for any commutative affine k-group scheme Z of finite type with no nontrivial smooth connected k-subgroup (e.g., $Z = \ker \xi_G$ as above), every central extension of k-group schemes

$$1 \to Z \to E \to \mathrm{R}_{k'/k}(\mathscr{G}') \to 1 \tag{1.4.1.2}$$

is (uniquely) split over k (see [CGP, Ex. 5.1.4]). In contrast, for many imperfect k and k-finite $k' \subset k^{1/p}$ for $p = \mathrm{char}(k)$, the k-group $\mathrm{R}_{k'/k}(\mathrm{SL}_n)$ seems to admit non-split central extensions by $\mathbf{G}_{\mathrm{a}}$ when $n > 2$ [CGP, Rem. 5.1.5].

1.4.2. For an absolutely pseudo-simple k-group G, two substantial difficulties arise if ξ_G is not surjective (so $\mathrm{char}(k) = 2, 3$) or if G_{k_s} has a non-reduced root system (which can occur only if the field k is imperfect and of characteristic 2):

(i) Assume G_{k_s} has a *reduced* root system (so $\ker \xi_G$ is central in G, by Proposition 2.3.4) but that ξ_G is not surjective. The possibilities for $\xi_G(G)$ force us to go beyond the simply connected semisimple central cover $\widetilde{G}'$ of G_K^{ss} and consider a wider class of absolutely pseudo-simple groups over finite extensions of k, called *generalized basic exotic* and *basic exceptional*, building on §1.3; see Chapter 8. (The maximal geometric semisimple quotient of these new groups is simply connected, and the basic exceptional case – which occurs over k if and only if $\mathrm{char}(k) = 2$ with $[k : k^2] > 8$ – rests on the equality $\mathrm{B}_2 = \mathrm{C}_2$.)

(ii) Assume k is imperfect with $\mathrm{char}(k) = 2$. If $[k : k^2] = 2$ then every pseudo-reductive k-group uniquely has the form $H \times \prod H_i$ where H_{k_s} has a reduced root system and each H_i is absolutely pseudo-simple over k with a non-reduced root system over k_s [CGP, Prop. 10.1.4, Prop. 10.1.6] (and each H_i is pseudo-split, has trivial center, and $\mathrm{Aut}_k(H_i) = H_i(k)$ [CGP, Thm. 9.9.3]).

In contrast, when $[k : k^2] > 2$ it is generally impossible to split off (as a direct factor) the contribution from non-reduced irreducible components of the root system over k_s; see Example 6.1.5. Moreover, as is explained in [CGP, §9.8–§9.9], the classification of pseudo-split absolutely pseudo-simple k-groups with a non-reduced root system over k_s rests on invariants of linear algebraic nature that do not arise (in nontrivial ways) when $[k : k^2] = 2$.

1.4.3. To classify absolutely pseudo-semisimple G over any k whatsoever, we shall use the following new construction. A commutative affine k-group scheme of finite type is *k-tame* if it does not contain a nontrivial unipotent k-subgroup scheme. For example, if k'/k is an extension of finite degree and μ' is a k'-group scheme of multiplicative type then $\mathrm{R}_{k'/k}(\mu')$ is k-tame. If G is a perfect smooth connected affine group over a field k then a central extension

$$1 \to Z \to E \to G \to 1$$

with affine E of finite type is called *k-tame* if Z is k-tame. In Theorem 5.1.3 we show that for any such G, if K/k is the minimal field of definition for $\mathscr{R}_u(G_{\overline{k}}) \subset G_{\overline{k}}$ then the category of k-tame central extensions E of G that are smooth, connected, and perfect is equivalent to the category of connected semisimple central extensions of $G' := G_K^{\rm ss}$ over K via $E \rightsquigarrow E' := E_K/\mathscr{R}_{u,K}(E_K)$.

The perfect smooth connected k-tame central extension $\widetilde{G}$ of G for which the associated connected semisimple central extension of G' is simply connected is called the *universal smooth k-tame central extension* of G (it is initial among smooth k-tame central extensions of G). It is elementary that if G is pseudo-reductive then so is $\widetilde{G}$ and that if $G := \mathscr{D}(\mathrm{R}_{k'/k}(G'))$ for a finite extension k'/k and connected semisimple k'-group G' then $\widetilde{G} = \mathrm{R}_{k'/k}(\widetilde{G}')$ for the simply connected central cover $\widetilde{G}' \to G'$. In proofs of general theorems it is often possible to pass from G to $\widetilde{G}$ (which has better properties), and by Theorem 9.2.1 (and Proposition 5.3.3) the k-group $\widetilde{G}$ is described in terms of a known list of constructions when $\widetilde{G}$ is of "minimal type" in a sense discussed in §1.6.

1.5 Generalized standard groups

In [CGP, Ch. 7–8], we constructed a class of pseudo-semisimple groups over any imperfect field k of characteristic $p \in \{2,3\}$ by using certain non-standard absolutely pseudo-simple groups G' – called "basic exotic" – over finite extensions k'/k such that $(G'_{\overline{k'}})^{\rm ss}$ is simply connected and the irreducible root system Φ of $G'_{k'_s}$ is reduced with an edge of multiplicity p: such Φ can be F_4, B_n, or C_n in characteristic 2 (with any $n \geqslant 2$) and G_2 in characteristic 3. Letting K'/k' be the minimal field of definition for $\mathscr{R}_u(G'_{\overline{k'}}) \subset G'_{\overline{k'}}$, we have $k' \subsetneq K' \subset k'^{1/p}$ and over k'_s the long root groups are 1-dimensional whereas short root groups have dimension $[K' : k'] > 1$ (short root groups are isomorphic to $\mathrm{R}_{K'_s/k'_s}(\mathbf{G}_{\rm a})$).

Going beyond these constructions, in [CGP, Ch. 9] pseudo-split absolutely pseudo-simple groups G' with root system BC_n (for any $n \geqslant 1$) are constructed over any imperfect field k' with characteristic 2. If $[k' : k'^2] = 2$ then by [CGP, Thm. 9.9.3(1)] the k'-group G' is classified up to k'-isomorphism by the rank $n \geqslant 1$ and the minimal field of definition K'/k' for $\mathscr{R}_u(G'_{\overline{k'}}) \subset G'_{\overline{k'}}$; here, n can be arbitrary and K'/k' can be any nontrivial purely inseparable finite extension.

For imperfect k of characteristic 2 or 3, with $[k : k^2] = 2$ in the BC_n-cases, Weil restrictions to k of groups G' as above over finite extensions k'/k are perfect and satisfy the splitting result for central extensions as in (1.4.1.2); see [CGP, Prop. 8.1.2, Thm. 9.9.3(3)]. (That splitting result fails in some BC_n-cases with $k' = k$ and $n \geqslant 1$ whenever $[k : k^2] > 2$; see Examples B.4.1 and B.4.3.)

If k is imperfect and either $\mathrm{char}(k) = 3$ or $\mathrm{char}(k) = 2$ with $[k : k^2] = 2$ then the preceding constructions capture *all* deviations from standardness over k (see [CGP, Thm. 10.2.1]). However, over any field k of characteristic 2 with $[k : k^2] > 2$ there exist many other pseudo-reductive groups, starting with:

Example 1.5.1. Consider imperfect k of characteristic 2 and a pseudo-split absolutely pseudo-simple k-group G with a *reduced* root system such that $G^{\mathrm{ss}}_{\overline{k}} \simeq \mathrm{SL}_2$. If $[k : k^2] = 2$ then $G \simeq \mathrm{R}_{K/k}(\mathrm{SL}_2)$ for a purely inseparable finite extension K/k (see [CGP, Prop. 9.2.4]). In contrast, as we review in §3.1, if $[k : k^2] > 2$ then many more possibilities for G occur: in addition to the field invariant K/k, there are linear algebra invariants (such as certain $K^\times$-homothety classes of nonzero kK^2-subspaces V of K, with the case $V \neq K$ occurring if $[k : k^2] > 2$).

The groups in Example 1.5.1 can be used to "shrink" short root groups (for type B) or "fatten" long root groups (for type C) in pseudo-split basic exotic k-groups with rank $n \geqslant 2$. When $[k : k^2] > 2$, this relates the new classes of absolutely pseudo-simple groups G mentioned in 1.4.2(i) to the basic exotic cases. For these additional constructions (and the derived groups of their Weil restrictions through finite extensions of the ground field) we prove a splitting result for central extensions as in (1.4.1.2) when $[k : k^2] \leqslant 8$ (see Proposition B.3.4), but this splitting result *fails* whenever $[k : k^2] > 8$ (see §B.1–§B.2).

For any imperfect field k of characteristic 2, the data classifying pseudo-split absolutely pseudo-simple k-groups G with root system BC_n ($n \geqslant 1$) is much more intricate when $[k : k^2] > 2$ than when $[k : k^2] = 2$, and one encounters new behavior when $[k : k^2] > 2$ that never occurs when $[k : k^2] = 2$. For instance, if K/k is the minimal field of definition for $\mathscr{R}_u(G_{\overline{k}}) \subset G_{\overline{k}}$ and $[k : k^2] > 2$ then there can be proper subfields $F \subset K$ over k such that the non-reductive maximal pseudo-reductive quotient $G_F/\mathscr{R}_{u,F}(G_F)$ of G_F has a *reduced* root system (see [CGP, Ex. 9.1.8]); this never happens if $[k : k^2] = 2$.

Weil restrictions to k of generalized basic exotic groups, basic exceptional groups, and the constructions in [CGP, Ch. 9] with non-reduced root systems are used to define a *generalized standard* construction over any field k in Definitions 9.1.5 and 9.1.7. This construction satisfies many nice properties (see §9.1). The information required to make a "generalized standard presentation" of a pseudo-reductive k-group G (if it admits such a presentation at all!) consists of data that is uniquely functorial with respect to isomorphisms in the pair (G, T) where T is a maximal k-torus of G (see Proposition 9.1.12); any T may be used.

1.6 Minimal type and general structure theorem

For a pseudo-reductive group G over any field k, the important notion of G being of *minimal type* was introduced in [CGP, Def. 9.4.4] and is reviewed in §2.3. Every pseudo-reductive k-group G admits a canonical pseudo-reductive central quotient $\overline{G}$ of minimal type with the *same* root datum as G (over k_s) [CGP, Prop. 9.4.2(iii)], and the central quotient G/Z_G is always pseudo-reductive and of minimal type (see Proposition 4.1.3).

Many of our results for general G rest on a classification and structure theorem for pseudo-split absolutely pseudo-simple groups of minimal type (over any field) given in Theorem 3.4.1 in the spirit of the Existence and Isomorphism Theorems for split connected semisimple groups. This classification in the pseudo-split minimal-type case supplements the root datum with additional field-theoretic data, as well as linear-algebraic data in characteristic 2. To prove theorems about general pseudo-reductive groups, it is often harmless and genuinely useful to pass to the minimal type case (e.g., see the proofs of Proposition 3.4.4, Proposition 6.1.4, Proposition B.3.1, and Theorem C.2.10).

Generalized basic exotic and basic exceptional k-groups from 1.4.2(i) are of minimal type and admit an intrinsic characterization via this condition; see Theorem 8.4.5 (and Definition 8.4.1). However, a pseudo-reductive central quotient of a pseudo-reductive group of minimal type is generally *not* of minimal type; absolutely pseudo-simple counterexamples that are standard exist over *every* imperfect field (see Example 2.3.5). Hence, a general structure theorem for pseudo-reductive groups must go beyond the minimal type case.

There is a weaker condition on a pseudo-reductive k-group G that we call *locally of minimal type*: for a maximal k-torus $T \subset G$, this is the property that for all non-divisible roots a of (G_{k_s}, T_{k_s}), the pseudo-simple k_s-group $(G_{k_s})_a$ of rank 1 generated by the $\pm a$-root groups admits a pseudo-simple central extension of minimal type. This notion might appear to be ad hoc, but it is not because it admits an elegant global characterization in the pseudo-semisimple case: such a k-group G is locally of minimal type if and only if its universal smooth k-tame central extension $\widetilde{G}$ is of minimal type (Proposition 5.3.3). In particular, for *every* pseudo-semisimple G, the universal smooth k-tame central extension of G/Z_G is always of minimal type; this is convenient in general proofs. By design, if G is pseudo-reductive and locally of minimal type then so is any pseudo-reductive central quotient of G. It is also easy to check that all generalized standard pseudo-reductive groups are locally of minimal type.

Example 1.6.1. Rank-1 absolutely pseudo-simple k_s-groups are classified in

Proposition 3.1.9 if $\mathrm{char}(k) \neq 2$ and in [CGP, Prop. 9.2.4, Thm. 9.9.3(1)] if $\mathrm{char}(k) = 2$ with $[k : k^2] \leqslant 2$. This classification implies that over such k *every* pseudo-reductive k-group is locally of minimal type.

Example 1.6.1 is optimal in the k-aspect: if $\mathrm{char}(k) = 2$ with $[k : k^2] > 2$ then for any $n \geqslant 1$ there are pseudo-split pseudo-simple k-groups with root system BC_n that are *not* locally of minimal type (see §B.4), and likewise (see 4.2.2 and §B.1–§B.2) for pseudo-split pseudo-simple k-groups with root systems B_n and C_n for any $n \geqslant 1$ when $[k : k^2] > 8$ (optimal by Proposition B.3.1); here B_1 and C_1 mean A_1. These examples suggest that there is no analogue of the "standard construction" beyond the locally minimal type class.

Since "locally of minimal type" is more robust than "minimal type", we seek to describe all pseudo-reductive groups locally of minimal type. One of our main results (Theorem 9.2.1) is a converse to the elementary fact that generalized standard pseudo-reductive groups (see §1.5) are locally of minimal type:

Structure Theorem. *A pseudo-reductive group locally of minimal type is generalized standard. In particular, if G is an arbitrary pseudo-reductive group then G/Z_G is generalized standard.*

The novelty is that when $\mathrm{char}(k) = 2$ there is no restriction on $[k : k^2]$. The cases $\mathrm{char}(k) \neq 2$ or $\mathrm{char}(k) = 2$ with $[k : k^2] \leqslant 2$ are part of [CGP, Thm. 10.2.1]; that earlier result is reproved in a new way in this monograph (using certain inputs from [CGP]) in the course of proving the above more general theorem.

1.7 Galois-twisted forms and Tits classification

A *k_s/k-form* of a pseudo-reductive group G over a field k is a pseudo-reductive k-group H such that $H_{k_s} \simeq G_{k_s}$. The theory of Chevalley groups ensures that if G is reductive then it admits a unique split k_s/k-form. Uniqueness of pseudo-split k_s/k-forms holds in the pseudo-reductive case (Proposition C.1.1). Existence of a pseudo-split k_s/k-form seems to be intractable for commutative pseudo-reductive G, so now consider pseudo-semisimple G.

Over many imperfect k (with arbitrary characteristic $p > 0$) there are pseudo-semisimple G *without* a pseudo-split k_s/k-form, due to a field-theoretic obstruction that cannot arise if G is absolutely pseudo-simple or if $[k : k^p] = p$; see Example C.1.2. Additional examples allowing $[k : k^p] = p$ are given in Example C.1.6, but those are also not absolutely pseudo-simple.

Existence result: For any *absolutely* pseudo-simple G, a pseudo-split k_s/k-form exists if $\mathrm{char}(k) \neq 2$ and also if $\mathrm{char}(k) = 2$ with $[k : k^2] \leqslant 4$ except possibly (for the latter case) if G is *standard* of type D_{2n} with $n \geqslant 2$ and k admits a quadratic Galois extension (or a cubic Galois extension when $n = 2$); see Proposition C.1.3 and Corollary C.2.12. The same conclusion holds in the *standard* absolutely pseudo-simple case when $\mathrm{char}(k) = 2$ without restriction on $[k : k^2]$ (subject to the same exceptions for type D_{2n}).

Avoidance of type D_{2n} ($n \geqslant 2$) is necessary because for all $n \geqslant 2$ and imperfect k of characteristic 2 that admits a quadratic (or cubic when $n = 2$) Galois extension there exists a standard absolutely pseudo-simple k-group G of type D_{2n} with no pseudo-split k_s/k-form; see Proposition C.1.4 and Remark C.1.5.

More counterexamples: What about the *non-standard* case if $\mathrm{char}(k) = 2$ and $[k : k^2] > 4$? If $[k : k^2] > 4$ and k has sufficiently rich Galois theory then in Example C.3.1 we make (non-standard) absolutely pseudo-simple groups of type A_1 over k *without* a pseudo-split k_s/k-form. These are used in §C.4 to make many more non-standard absolutely pseudo-simple k-groups without a pseudo-split k_s/k-form: generalized basic exotic k-groups whose root system over k_s is B_n or C_n for any $n \geqslant 2$, and absolutely pseudo-simple k-groups of minimal type whose root system over k_s is BC_n for any $n \geqslant 1$.

Going beyond the study of pseudo-split k_s/k-forms, it is natural to seek a pseudo-reductive analogue of the existence and uniqueness of quasi-split *inner* forms for connected reductive groups. Recall that for connected reductive G the notion of inner form involves Galois-twisting against the action of the identity component G/Z_G of the automorphism scheme of G. Due to the mysterious nature of commutative pseudo-reductive groups, for an analogous result in the pseudo-reductive case we shall restrict attention to pseudo-semisimple G.

The analogue of "inner form" for pseudo-semisimple groups G is *not* defined via the action of G/Z_G, but rather via the identity component of the maximal smooth closed k-subgroup of the automorphism scheme of G. To be precise, in §6.2 we prove for pseudo-semisimple G that the automorphism functor of G on the category of k-algebras is represented by an affine k-group scheme $\mathrm{Aut}_{G/k}$ of finite type (this functor is often *not* representable for commutative G; see Example 6.2.1). In general $\mathrm{Aut}_{G/k}$ is *not* k-smooth (Example 6.2.3), but its maximal smooth closed k-subgroup $\mathrm{Aut}^{\mathrm{sm}}_{G/k}$ has structure analogous to the semisimple case (see Propositions 6.2.4 and 6.3.10): $(\mathrm{Aut}^{\mathrm{sm}}_{G/k})^0$ is pseudo-reductive and its *derived group* is G/Z_G. (Absolutely pseudo-simple G with $G/Z_G \neq (\mathrm{Aut}^{\mathrm{sm}}_{G/k})^0$ arise over every imperfect field; see Remark 6.2.5.)

Inspired by the semisimple case, for pseudo-semisimple G we prove that $\pi_0(\mathrm{Aut}^{\mathrm{sm}}_{G/k})(k_s)$ is a subgroup of the automorphism group of the based root datum over k_s (Remark 6.3.6). For absolutely pseudo-simple G we show that this subgroup inclusion is often an equality. Counterexamples to equality in the absolutely pseudo-simple case exist *precisely* for type D_{2n} ($n \geqslant 2$) with k imperfect of characteristic 2 (for the same reason that such cases may not have pseudo-split k_s/k-form); see Proposition 6.3.10.

In §6.3 we use our study of the structure of $\mathrm{Aut}^{\mathrm{sm}}_{G/k}$ (including its behavior under passage to pseudo-reductive central quotients of G) to prove a Tits-style classification theorem in the general pseudo-semisimple case (no minimal-type hypothesis!), recovering the well-known result due to Tits in the semisimple case. As an illustration of the method, in Appendix D we show that if k is imperfect of characteristic 2 then absolutely pseudo-simple k-groups of type F_4 that are not pseudo-split cannot have k-rank 3 whereas they *can* have k-rank 2 (in contrast with the semisimple case!); all instances of the latter are described via anisotropic quadratic forms over k (with examples given over specific k).

In §C.2 the notion of *pseudo-inner* form of a pseudo-reductive k-group G is defined in terms of $(\mathrm{Aut}^{\mathrm{sm}}_{\mathscr{D}(G)/k})^0$. We use the structure of $(\mathrm{Aut}^{\mathrm{sm}}_{\mathscr{D}(G)/k})^0$ to prove uniqueness of pseudo-inner k_s/k-forms that are *quasi-split* (i.e., admit a solvable pseudo-parabolic k-subgroup). The existence of such k_s/k-forms is proved assuming when $\mathrm{char}(k) = 2$ that $[k : k^2] \leqslant 4$ or G is standard (Theorem C.2.10). This is optimal because if $\mathrm{char}(k) = 2$, $[k : k^2] > 4$, and k has sufficiently rich Galois theory (more precisely, k admits a quadratic Galois extension k' such that $\ker(\mathrm{Br}(k) \to \mathrm{Br}(k')) \neq 1$) then for every $n \geqslant 1$ there exist non-standard absolutely pseudo-simple k-groups of types B_n, C_n, and BC_n over k_s *without* a quasi-split k_s/k-form: examples without a pseudo-split k_s/k-form (see §C.3–§C.4) do the job, by Lemma C.2.2.

1.8 Background, notation, and acknowledgments

In this monograph we use many constructions and results from [CGP]. Familiarity with Chapters 1–5, §7.1–§7.2, §8.1, Chapter 9, parts of Appendix A (A.5, A.7, A.8), Theorem B.3.4, and Appendix C.2 (especially Theorem C.2.29) of [CGP] is sufficient for understanding our main techniques. Chapter 9 in the first edition of [CGP] has been completely rewritten in the second edition, incorporating significant improvements that are used throughout this monograph, and some results outside Chapter 9 of [CGP] are improved in the second edition and used here. We provide many cross-references to aid the reader. (All numerical labeling in the first edition of [CGP] is unchanged in the second edition except in Chapter 9, apart from an equation label in [CGP, Ex. 1.6.4].)

For a scheme X of finite type over a field k and a closed subscheme Z of

X_K for an extension field K/k, the intersection of all subfields $k' \subset K$ over k such that Z descends (necessarily uniquely) to a closed subscheme of $X_{k'}$ is also such a subfield, called the *minimal field of definition* for $Z \subset X_K$ relative to k; see [EGA, IV$_2$, §4.8] for a detailed discussion of the existence of such a field. The behavior of K/k with respect to extension of k is addressed in [CGP, Lemma 1.1.8]. (In [CGP] the phrase "field of definition" is understood to require minimality, but in this monograph we keep "minimality" in the terminology.)

For an automorphism σ of a field k and a k-scheme X, ${}^{\sigma}X$ denotes the k-scheme $k \otimes_{\sigma,k} X$. For a map $f : X \to Y$ of k-schemes, ${}^{\sigma}f$ denotes the induced map ${}^{\sigma}X \to {}^{\sigma}Y$ over k.

The maximal smooth closed k-subgroup of a k-group scheme H of finite type is denoted H^{sm} (see [CGP, Rem. C.4.2]). For a smooth connected affine k-group G and closed k-subgroup scheme $H \subset G$ that is either smooth or of multiplicative type, the *scheme-theoretic centralizer* $Z_G(H)$ for the H-action on G via conjugation is defined as in [CGP, A.1.9ff., A.8.10]; the *scheme-theoretic center* is $Z_G := Z_G(G)$ [CGP, A.1.10]. The k*-unipotent radical* $\mathscr{R}_{u,k}(G)$ is the maximal unipotent smooth connected normal k-subgroup of G. The k*-radical* $\mathscr{R}_k(G)$ is the maximal solvable smooth connected normal k-subgroup of G.

If K/k denotes the minimal field of definition for $\mathscr{R}_u(G_{\overline{k}}) \subset G_{\overline{k}}$ then we define G_K^{red} to be the quotient $G_K/\mathscr{R}_{u,K}(G_K)$ that is a K-descent of the maximal reductive quotient $G_{\overline{k}}^{\mathrm{red}}$ of $G_{\overline{k}}$. Taking K/k instead to be the minimal field of definition for $\mathscr{R}(G_{\overline{k}}) \subset G_{\overline{k}}$ yields the quotient $G_K^{\mathrm{ss}} := G_K/\mathscr{R}_K(G_K)$ of G_K as a K-descent of the maximal semisimple quotient of $G_{\overline{k}}$. A *Levi* k*-subgroup* of G is a smooth closed k-subgroup $L \subset G$ such that $L_{\overline{k}} \to G_{\overline{k}}^{\mathrm{red}}$ is an isomorphism.

The *Weyl group* of a root system Φ is is denoted $W(\Phi)$. If Φ is irreducible and not simply laced then for any basis Δ of Φ we denote by $\Delta_{>}$ and $\Delta_{<}$ the respective subsets of longer and shorter roots in Δ.

We thank Ofer Gabber for his very illuminating advice and suggestions, Kęstutis Česnavičius and the referees for helpful comments, and Stella Gastineau for typesetting assistance. We thank Indu Prasad for her encouragement, hospitality, and support over the years. G.P. thanks the Institute for Advanced Study (Princeton), and his host Peter Sarnak, for hospitality and support during 2012–13. He also thanks the Mathematics Research Center at Stanford University for support during a visit in the summer of 2013 and RIMS (Kyoto) for its hospitality during July 2014. B.C. is grateful to IAS for its hospitality during several visits. We thank Vickie Kearn, Betsy Blumenthal, Nathan Carr, and Glenda Krupa for their editorial work. B.C. was supported by NSF grant DMS-1100784 and G.P. was supported by NSF grants DMS-1001748 and DMS-1401380.

2

Preliminary notions

2.1 Standard groups, Levi subgroups, and root systems

2.1.1. Let k be a field. The "standard construction" of pseudo-reductive k-groups, as reviewed below, provides both a general structure theorem for pseudo-reductive k-groups when $\mathrm{char}(k) \neq 2$ as well as a guide for the main results we shall prove in this monograph.

As motivation, consider a smooth connected affine k-group G with minimal field of definition K/k for its geometric unipotent radical. The extension K/k is purely inseparable of finite degree and $\mathscr{R}_u(G_{\overline{k}})$ descends to the maximal smooth connected unipotent normal K-subgroup $\mathscr{R}_{u,K}(G_K) \subset G_K$. Consider the maximal reductive quotient $G' := G_K/\mathscr{R}_{u,K}(G_K)$ over K. The natural map

$$i_G : G \to \mathrm{R}_{K/k}(G') \tag{2.1.1}$$

is a first attempt to relate G to a canonically associated Weil restriction of a connected reductive group. The image has good properties, given in (ii) below:

Proposition 2.1.2. *Let G, K/k, and G' be as above with G pseudo-reductive.*

(i) *Let $L \subset G$ be a Levi k-subgroup. Any smooth connected k-subgroup $H \subset G$ containing L is pseudo-reductive, L is a Levi k-subgroup of H, and $H_{\overline{k}} \cap \mathscr{R}_u(G_{\overline{k}}) = \mathscr{R}_u(H_{\overline{k}})$.*

(ii) *The image $\overline{G} := i_G(G) \subset \mathrm{R}_{K/k}(G')$ is pseudo-reductive, K/k is the minimal field of definition for its geometric unipotent radical, and the inclusion of $\overline{G}$ into $\mathrm{R}_{K/k}(G')$ is $i_{\overline{G}}$.*

Part (i) is [CGP, Lemma 7.2.4], and we give its proof for the convenience of the reader. If G is pseudo-split then L as in (i) exists by [CGP, Thm. 3.4.6].

Proof. We first prove (i). Since $L_{\overline{k}} \to G_{\overline{k}}/\mathscr{R}_u(G_{\overline{k}})$ is an isomorphism,

$$H_{\overline{k}} = L_{\overline{k}} \ltimes (H_{\overline{k}} \cap \mathscr{R}_u(G_{\overline{k}}))$$

as schemes. Thus, the smoothness and connectedness of $H_{\overline{k}}$ imply that the unipotent normal subgroup scheme $H_{\overline{k}} \cap \mathscr{R}_u(G_{\overline{k}})$ in $H_{\overline{k}}$ is smooth and connected, with quotient by this isomorphic to $L_{\overline{k}}$. Hence, $\mathscr{R}_u(H_{\overline{k}}) = H_{\overline{k}} \cap \mathscr{R}_u(G_{\overline{k}})$, so $\mathscr{R}_{u,k}(H) \subseteq \mathscr{R}_{u,k}(G) = 1$ and hence H is pseudo-reductive. This proves (i).

Since an equality among purely inseparable extensions of k can be checked after scalar extension to k_s, to prove (ii) we may and do assume $k = k_s$. Now G is pseudo-split, so by [CGP, Thm. 3.4.6] it contains a Levi k-subgroup L. By definition of a Levi subgroup, $L_K \to G'$ is an isomorphism. The natural map $\mathrm{R}_{K/k}(G')_K \to G'$ is a K-descent of the maximal geometric reductive quotient (by [CGP, Prop. A.5.11(1),(2)]), so the homomorphism $L \to \mathrm{R}_{K/k}(G')$ induced by i_G is a Levi k-subgroup inclusion. Thus, we may apply (i) to the ambient group $\mathrm{R}_{K/k}(G')$ and its k-subgroup $\overline{G} = i_G(G)$ to deduce that $\overline{G}$ is pseudo-reductive with geometric unipotent radical defined over K and with L as a Levi k-subgroup, so the evident map $\overline{G}_K \to G'$ is a K-descent of the maximal geometric reductive quotient. Hence, the *minimal* field of definition F/k for the geometric unipotent radical of $\overline{G}$ is a subextension of K/k and it remains to show that the inclusion $F \subset K$ over k is an equality.

By the definition of K/k, it is sufficient to prove that the geometric unipotent radical of G is defined over F. That is, it suffices to construct a smooth connected unipotent normal F-subgroup U of G_F such that G_F/U is reductive. Consider the composite F-homomorphism

$$q : G_F \twoheadrightarrow \overline{G}_F \twoheadrightarrow \overline{G}_F^{\mathrm{red}}.$$

It is sufficient to prove that the kernel of this map is smooth, connected, and unipotent. For this purpose we may extend scalars to K, but the K-descent of the maximal geometric reductive quotient of $\overline{G}$ has been identified with the map $\overline{G}_K \to G'$ corresponding to the given inclusion $j : \overline{G} \hookrightarrow \mathrm{R}_{K/k}(G')$. Hence, q_K is the map $G_K \to G'$ corresponding to the composition of $G \twoheadrightarrow \overline{G}$ with j, which is to say that q_K corresponds to $i_G : G \to \mathrm{R}_{K/k}(G')$. By the definition of i_G in terms of the universal property of Weil restriction, this implies that q_K is the canonical quotient map $G_K \to G'$ whose kernel is a K-descent of $\mathscr{R}_u(G_{\overline{k}})$. □

For pseudo-reductive G, the map i_G is problematic for two reasons:

(i) if G is non-commutative (hence non-solvable) then the field K is a coarser invariant than the collection of minimal fields of definition K_j/k for the kernels of projections of $G_{\overline{k}}$ onto the simple factors H_j of the adjoint central quotient $G_{\overline{k}}^{\rm ad} \neq 1$ of the maximal reductive quotient $G_{\overline{k}}^{\rm red}$,

(ii) i_G might have nontrivial kernel.

The following construction bypasses both of these problems.

Let k' be a nonzero finite reduced k-algebra. Let $G' \to \mathrm{Spec}(k')$ be a smooth affine group scheme with connected reductive fibers. Write $k' = \prod k_i'$ for fields k_i' and let G_i' denote the k_i'-fiber of G', so $\mathrm{R}_{k'/k}(G') = \prod \mathrm{R}_{k_i'/k}(G_i')$. Let T' be a maximal k'-torus in G', and define $\overline{T}' := T'/Z_{G'}$ to be the associated maximal k'-torus in $\overline{G}' := G'/Z_{G'}$, where $Z_{G'}$ denotes the scheme-theoretic center of G'. Suppose there is given a commutative pseudo-reductive k-group C and a factorization in k-homomorphisms

$$\mathrm{R}_{k'/k}(T') \stackrel{\phi}{\to} C \to \mathrm{R}_{k'/k}(\overline{T}') \qquad (2.1.2.1)$$

of the Weil restriction to k of the canonical quotient map $q : T' \to \overline{T}'$ over k'. (Beware that $\mathrm{R}_{k'/k}(q)$ may not be surjective when k' is not k-étale, and we do not require ϕ to be surjective.)

The natural $\overline{G}'$-action on G' over k' defines a natural $\mathrm{R}_{k'/k}(\overline{G}')$-action on $\mathrm{R}_{k'/k}(G')$ over k, and hence a natural action of C on $\mathrm{R}_{k'/k}(G')$ via composition with $C \to \mathrm{R}_{k'/k}(\overline{T}') \to \mathrm{R}_{k'/k}(\overline{G}')$. The anti-diagonal map

$$\mathrm{R}_{k'/k}(T') \to \mathrm{R}_{k'/k}(G') \rtimes C$$

is an inclusion with central image, and the associated central quotient

$$G = (\mathrm{R}_{k'/k}(G') \rtimes C)/\mathrm{R}_{k'/k}(T') \qquad (2.1.2.2)$$

is always pseudo-reductive [CGP, Prop. 1.4.3]. Informally, G is obtained from $\mathrm{R}_{k'/k}(G')$ by replacing the Cartan k-subgroup $\mathrm{R}_{k'/k}(T')$ with the commutative pseudo-reductive k-group C (whose structure we treat as a black box); G is a pushout of $\mathrm{R}_{k'/k}(G')$ along ϕ. Note that C is a Cartan k-subgroup of G, since $\mathrm{R}_{k'/k}(T') \times C$ is a Cartan k-subgroup of $\mathrm{R}_{k'/k}(G') \rtimes C$.

Definition 2.1.3. A pseudo-reductive k-group is *standard* if it is k-isomorphic to a k-group as on the right side of (2.1.2.2).

Every commutative pseudo-reductive k-group G is standard, by using $k' = k$, $T' = G' = 1$, and $C = G$. By [CGP, Thm. 4.1.1], any non-commutative standard pseudo-reductive k-group G admits a *standard presentation*: a description as in (2.1.2.2) using a 4-tuple $(G', k'/k, T', C)$ as above (including a specified factorization (2.1.2.1) that we suppress from the notation) for which the fibers of $G' \to \mathrm{Spec}(k')$ are semisimple, absolutely simple, and simply connected.

If G is a non-commutative standard pseudo-reductive group and the map $j : \mathrm{R}_{k'/k}(G') \to G$ arises from a standard presentation of G then the triple $(G', k'/k, j)$ is uniquely determined by G up to unique isomorphism [CGP, Prop. 4.2.4] and there is a natural bijection between the sets of maximal k'-tori T' of G' and Cartan k-subgroups C of G via the relation $j(\mathrm{R}_{k'/k}(T')) \subset C$ [CGP, Prop. 4.1.4(2),(3)]. In this sense, a standard presentation of a non-commutative standard pseudo-reductive k-group G amounts to a choice of a Cartan k-subgroup of G (or equivalently, a choice of maximal k-torus of G).

Let G be a pseudo-split pseudo-reductive k-group and T a split maximal k-torus in G, so $S := T \cap \mathscr{D}(G)$ is a split maximal k-torus of $\mathscr{D}(G)$ and T is an almost direct product of S and the maximal central k-torus Z of G [CGP, Lemma 1.2.5]. Consider the T-action and S-action on $\mathrm{Lie}(G)$. As is explained in the proof of [CGP, Thm. 2.3.10], the weight spaces in $\mathrm{Lie}(G)$ for nontrivial T-weights are supported inside $\mathrm{Lie}(\mathscr{D}(G))$ and coincide with the weight spaces for nontrivial S-weights. These weight spaces can have very high dimension, but nonetheless the pair $(\mathrm{X}(T), \Phi(G,T))$ can be naturally enhanced to a root datum $R(G,T)$ such that the $\mathbf{Q}$-span of $\Phi(G,T)$ maps isomorphically onto the quotient $\mathrm{X}(S)_{\mathbf{Q}}$ of $\mathrm{X}(T)_{\mathbf{Q}}$ [CGP, §3.2].

The root system $\Phi(G,T)$ is reduced whenever $\mathrm{char}(k) \neq 2$ and also when $\mathrm{char}(k) = 2$ provided that $G_{\overline{k}}^{\mathrm{red}}$ has no connected semisimple normal subgroup that is simple and simply connected of type C (where $\mathrm{C}_1 = \mathrm{A}_1$). In general the set of non-multipliable roots in $\Phi(G,T)$ is equal to $\Phi(G_{\overline{k}}^{\mathrm{red}}, T_{\overline{k}})$ and this maps isomorphically onto a reduced root system in $\mathrm{X}(S)_{\mathbf{Q}}$ [CGP, Thm. 2.3.10].

Lemma 2.1.4. *Let $H \subset H'$ be an inclusion of pseudo-reductive groups over a field k, and assume that their maximal tori have the same dimension. Let T be a maximal k-torus of H, and assume that $\Phi(H_{k_s}, T_{k_s}) = \Phi(H'_{k_s}, T_{k_s})$. A connected reductive k-subgroup $L \subset H$ is a Levi k-subgroup of H if and only if it is a Levi k-subgroup of H'.*

Proof. We may and do assume $k = k_s$. By Proposition 2.1.2(i), if L is a Levi k-subgroup of H' then it is a Levi k-subgroup of H (without needing to assume the equality of root systems, which could fail when $\Phi(H',T)$ is non-reduced).

Assume instead that L is a Levi k-subgroup of H, and let Φ denote the common root system for (H,T) and (H',T). Fix a positive system of roots Φ^+ and let Φ' denote the set of non-multipliable roots in Φ, so $\Phi' = \Phi(L,T)$. Then $\Phi'^+ := \Phi^+ \cap \Phi'$ is a positive system of roots in Φ'; let Δ' be the basis of simple roots in Φ'^{+}. By [CGP, Thm. 3.4.6], Levi k-subgroups of H' containing T are uniquely determined by their root groups E'_a for $a \in \Delta'$, and each E'_a may be chosen arbitrarily among the 1-dimensional T-stable smooth connected k-subgroups of the a-root group of (H', T) (defined as in [CGP, Def. 2.3.13]: its Lie algebra is the a-weight space since a is not multipliable). Hence, there is a unique Levi k-subgroup L' of H' containing T such that its a-root group E'_a coincides with the a-root group E_a of (L, T) for all $a \in \Delta'$. Our task is to prove $L' = L$ as k-subgroups of H'.

It suffices to prove $E'_{-a} = E_{-a}$ for all $a \in \Delta'$ because any connected reductive group equipped with a chosen split maximal torus is generated by that maximal torus and its root groups for the simple positive and negative roots relative to a choice of positive system of roots in the root system. Let U'_{-a} be the $-a$-root group for (H', T), and let U_{-a} be the $-a$-root group for (H, T). By the dynamic construction of such root groups, $U'_{-a} \cap H = U_{-a}$. Choose a nontrivial element $x \in E_a(k) = E'_a(k)$. By [CGP, Prop. 3.4.2], there are unique elements $u', v' \in U'_{-a}(k) - \{1\}$ such that $u'xv' \in N_{H'}(T)(k)$ and unique $u, v \in U_{-a}(k) - \{1\}$ such that $uxv \in N_H(T)(k)$. From the uniqueness we conclude that $u' = u$ and $v' = v$, so $u', v' \in U_{-a}(k) - \{1\}$. The same reasoning applies to (L, T) and (L', T), so in fact $u', v' \in E'_{-a}(k) \cap E_{-a}(k)$. The T-orbits of these nontrivial elements under conjugation exhaust $E'_{-a} - \{1\}$ and $E_{-a} - \{1\}$, so $E'_{-a} = E_{-a}$ as desired. □

2.2 The basic exotic construction

The only nontrivial multiplicities that occur for the edges of Dynkin diagrams of reduced irreducible root systems are 2 and 3. This underlies the existence of exceptional isogenies that only arise in characteristics 2 and 3, and such isogenies are used to build non-standard absolutely pseudo-simple groups called "basic exotic" (see Definition 2.2.2). One of our main tasks (see Chapters 7–8) is to generalize the "basic exotic" construction so that we can prove a structure theorem over any imperfect k of characteristic 2, without restriction on $[k : k^2]$.

2.2.1. Let k be an imperfect field with $p := \mathrm{char}(k) \in \{2, 3\}$, and let k'/k be a nontrivial finite extension satisfying $k'^p \subset k$. Let G' be a connected semisimple k'-group that is absolutely simple and simply connected with an edge of multiplicity p in its Dynkin diagram (so G' has type G_2 if $p = 3$ and type F_4 or B_n or C_n with $n \geqslant 2$ if $p = 2$). By [CGP, Lemma 7.1.2], there is a unique minimal non-central normal k'-subgroup scheme $N' \subset G'$ whose relative Frobenius morphism is trivial; we call $\pi' : G' \to \overline{G}' := G'/N'$ the *very special isogeny* of G'.

Let $T' \subset G'$ be a maximal k-torus, and $\overline{T}' = \pi(T')$. By [CGP, Prop. 7.1.5], π' carries long root groups for (G', T') isomorphically onto short root groups for $(\overline{G}', \overline{T}')$ and carries short root groups for (G', T') onto long root groups of $(\overline{G}', \overline{T}')$ via a Frobenius isogeny. Moreover, $\overline{G}'$ is simply connected with root system dual to that of G', and the factorization of the Frobenius isogeny $F_{G'/k'} : G' \to G'^{(p)}$ through π' is via an isogeny $\overline{\pi}' : \overline{G}' \to G'^{(p)}$ that is the very special isogeny of $\overline{G}'$. For types F_4 in characteristic 2 and G_2 in characteristic 3, this provides the unique nontrivial factorization of $F_{G'/k'}$ [CGP, Lemma 7.1.2].

Consider the Weil restriction $f = \mathrm{R}_{k'/k}(\pi') : \mathrm{R}_{k'/k}(G') \to \mathrm{R}_{k'/k}(\overline{G}')$; this is *not* surjective. By definition of Levi k-subgroups, for any Levi k-subgroup $\overline{\jmath} : \overline{G} \hookrightarrow \mathrm{R}_{k'/k}(\overline{G}')$ (if one exists) the associated map $\overline{G}_{k'} \to \overline{G}'$ is an isomorphism. (For a link between such Levi k-subgroups and k-descents of $\overline{G}'$, see [CGP, Lemma 7.2.1].) By [CGP, Prop. 7.3.1] the following are equivalent:

(i) $\overline{G}$ lies inside the image of f,
(ii) $f^{-1}(\overline{G})$ is smooth,
(ii) $f^{-1}(\overline{G})_{k_s}$ contains a Levi k_s-subgroup of $\mathrm{R}_{k'/k}(G')_{k_s}$.

When these equivalent conditions hold, by [CGP, 7.2.6–7.2.7] the fiber product $\mathscr{G} := f^{-1}(\overline{G})$ in the diagram

$$\begin{array}{ccc} \mathscr{G} & \xrightarrow{\ j\ } & \mathrm{R}_{k'/k}(G') \\ \downarrow & & \downarrow{\scriptstyle f} \\ \overline{G} & \xrightarrow[\ \overline{\jmath}\ ]{} & \mathrm{R}_{k'/k}(\overline{G}') \end{array} \tag{2.2.1}$$

is absolutely pseudo-simple, k'/k is the minimal field of definition for its geometric radical, and the given inclusion $\mathscr{G} \hookrightarrow \mathrm{R}_{k'/k}(G')$ corresponds to the maximal reductive quotient $\mathscr{G}_{k'} \twoheadrightarrow G'$. Also, $\mathscr{G}$ is not standard [CGP, Prop. 8.1.1].

Definition 2.2.2. A pseudo-reductive k-group $\mathscr{G}$ as in (2.2.1) is *basic exotic.*

By [CGP, Prop. 7.2.7(3)], if E/k is a separable extension of fields then a pseudo-reductive k-group H is basic exotic if and only if H_E is basic exotic.

If K is a nonzero finite reduced k-algebra, so $K = \prod K_i$ for finite extension fields K_i/k, and if $\mathscr{G}_i$ is a basic exotic K_i-group for each i, then for $\mathscr{G} := \coprod \mathscr{G}_i$ over $\mathrm{Spec}(K)$ the pseudo-reductive k-group $\mathrm{R}_{K/k}(\mathscr{G}) = \prod \mathrm{R}_{K_i/k}(\mathscr{G}_i)$ is perfect by [CGP, Prop. 8.1.2]. This k-group uniquely determines the pair $(\mathscr{G}, K/k)$ up to unique k-isomorphism in the following sense: if $(\mathscr{H}, L/k)$ is another such pair then any k-isomorphism $\mathrm{R}_{K/k}(\mathscr{G}) \simeq \mathrm{R}_{L/k}(\mathscr{H})$ arises from a uniquely determined pair (σ, φ) consisting of a k-algebra isomorphism $\sigma : K \simeq L$ and group isomorphism $\varphi : \mathscr{G} \simeq \mathscr{H}$ over σ. The existence and uniqueness of (σ, φ) rests on intrinsically characterizing $(\mathscr{G}, K/k)$ in terms of the k-group $\mathrm{R}_{K/k}(\mathscr{G})$; this is carried out as part of [CGP, Prop. 8.2.4], but an alternative approach is given in Corollary 3.3.9 as an application of ideas developed in §2.3 and §3.3.

Definition 2.2.3. A k-group of the form $\mathrm{R}_{K/k}(\mathscr{G})$ for a pair $(\mathscr{G}, K/k)$ as above is called an *exotic* pseudo-reductive k-group.

2.3 Minimal type

2.3.1. Let G be a pseudo-reductive group over a field k, and recall the map i_G as in (2.1.1). It is not clear how to describe the possibilities for $\ker i_G$ in general; the most favorable situation is when $(\ker i_G)^T = 1$ for a maximal k-torus T of G, as then we see using [CGP, Prop. 2.1.12(2)] that $\ker i_G$ is a connected group scheme and we can try to understand its structure in terms of the intersection of $(\ker i_G)_{k_s}$ with root groups of (G_{k_s}, T_{k_s}). Since $(\ker i_G)^T = \ker i_G \cap Z_G(T)$, we are led to consider the intersection of the unipotent k-group scheme $\ker i_G$ with a Cartan k-subgroup $C = Z_G(T)$ of G.

The intersection $(\ker i_G) \cap C$ has remarkable properties (proved in [CGP, Prop. 9.4.2, Cor. 9.4.3]): it is independent of C (so we denote it as $\mathscr{C}_G$), it is central in G, the central quotient $G/\mathscr{C}_G$ is pseudo-reductive with the *same* root datum as G over k_s and the *same* minimal field of definition over k for its geometric unipotent radical, and i_G is the composition of the central quotient map $G \twoheadrightarrow G/\mathscr{C}_G$ and $i_{G/\mathscr{C}_G}$. Thus, $(\ker i_G)/\mathscr{C}_G = \ker i_{G/\mathscr{C}_G}$, so $\mathscr{C}_{G/\mathscr{C}_G} = 1$. The k-group $\mathscr{C}_G$ has no nontrivial smooth connected k-subgroup since $\ker i_G$ contains no such k-subgroup (as $(\ker i_G)_{\overline{k}} \subset \mathscr{R}_u(G_{\overline{k}})$ and G is pseudo-reductive).

Definition 2.3.2. A pseudo-reductive k-group G is of *minimal type* if $\mathscr{C}_G$ is trivial.

Example 2.3.3. For any basic exotic pseudo-simple group G over an imperfect field of characteristic 2 or 3, $\ker i_G$ is trivial by [CGP, Prop. 7.2.7(1),(2)] and so such G are of minimal type.

Proposition 2.3.4. *Let G be a pseudo-reductive k-group whose root system over k_s is reduced. Let K be the minimal field of definition over k for the geometric unipotent radical $\mathscr{R}_u(G_{\overline{k}})$ of G, and let $G' = G_K/\mathscr{R}_{u,K}(G_K)$. The kernel of the homomorphism $i_G : G \to \mathrm{R}_{K/k}(G')$ is central (so* $\ker i_G$ *is trivial if G is* *of minimal type) and it does not contain any nontrivial smooth connected k-subgroup.*

If G' is semisimple and simply connected and i_G is surjective then i_G is an isomorphism, so in such cases G is of minimal type and standard.

The reducedness hypothesis on the root system cannot be dropped: if k is imperfect of characteristic 2 then there exist absolutely pseudo-simple k-groups G of minimal type whose root system over k_s is non-reduced, and for any such G the kernel of i_G is connected and commutative but not central (and such G exist for which i_G is surjective); see [CGP, Thm. 9.4.7, Thm. 9.8.1(4)].

Proof. To prove the proposition we may and do replace k with k_s so that k is separably closed. Let T be a maximal k-torus in G and let $\Phi := \Phi(G, T)$ be the root system of G with respect to T. We view $T' := T_K$ as a maximal K-torus of G'. As Φ has been assumed to be reduced, $\Phi(G', T') = \Phi$.

Let $\lambda : \mathrm{GL}_1 \to T$ be a 1-parameter k-subgroup such that $\langle a, \lambda \rangle \neq 0$ for every $a \in \Phi$, so $Z_G(\lambda) = Z_G(T)$. For $a \in \Phi$, let U_a be the corresponding root group in G. According to [CGP, Prop. 2.3.11], U_a is a k-vector group admitting a unique linear structure with respect to which T acts linearly, and $\mathrm{Lie}(U_a)$ is the a-root space for T in $\mathrm{Lie}(G)$. Since $a \in \Phi(G', T')$, the restriction of i_G to U_a is a T-equivariant homomorphism between a-root groups. The 1-dimensional a_K-root group of G' has a unique K-linear structure with respect to which T' acts linearly, and the map induced by i_G between the a-root groups respects the linear structures since GL_1-compatibility follows from the nontriviality of a. Thus, $\ker(i_G|_{U_a})$ is also a vector group, so it is a smooth connected k-subgroup of G whose geometric fiber lies in the unipotent radical. By pseudo-reductivity of G it follows that $\ker(i_G|_{U_a}) = 1$, so the a-weight space in $\mathrm{Lie}(\ker i_G)$ vanishes.

We conclude that the k-subgroups $U_{\ker i_G}(\pm\lambda)$ $(= U_G(\pm\lambda) \cap \ker i_G)$ are trivial. Now using [CGP, Prop. 2.1.12 (2)] for $H := (\ker i_G)_{\overline{k}}$ contained in the connected unipotent group $\mathscr{R}_u(G_{\overline{k}})$, it follows that $\ker i_G = Z_{\ker i_G}(\lambda) = \ker i_G \cap Z_G(\lambda) \subset Z_G(T)$. As T was an arbitrary maximal k-torus of G and the Cartan subgroup $C = Z_G(T)$ is commutative, we infer that $\ker i_G$ commutes with C and with the k-subgroup G_t of G generated by the k-tori of G. Since $\mathscr{D}(G)$ is perfect, it is contained in G_t and thus $G = C \cdot G_t$. Hence, $\ker i_G$ is central.

Since $(\ker i_G)_{\overline{k}} \subset \mathscr{R}_u(G_{\overline{k}})$, by applying [CGP, Lemma 1.2.1] we see that $\ker i_G$ does not contain any nontrivial smooth connected k-subgroup. Now if G' is semisimple and simply connected and $i_G : G \to \mathrm{R}_{K/k}(G')$ is surjective then

$$1 \to \ker i_G \to G \to \mathrm{R}_{K/k}(G') \to 1$$

is a central extension of $\mathrm{R}_{K/k}(G')$ by the k-group scheme $\ker i_G$ which does not contain a nontrivial smooth connected k-subgroup. By [CGP, Prop. 5.1.3, Ex. 5.1.4] such a central extension must be trivial. But G is smooth and connected, so it follows that $\ker i_G = 1$ and hence $i_G : G \to \mathrm{R}_{K/k}(G')$ is an isomorphism. Thus, G is of minimal type. □

As noted above, if G is a pseudo-reductive k-group then $\mathscr{C}_{G/\mathscr{C}_G} = 1$. Thus, the pseudo-reductive k-group $G/\mathscr{C}_G$ is of minimal type. If the common root system of G_{k_s} and $(G/\mathscr{C}_G)_{k_s}$ is *reduced* then $\ker i_G = \mathscr{C}_G$ and $\ker i_{G/\mathscr{C}_G} = 1$ by Proposition 2.3.4 (applied to $G/\mathscr{C}_G$), so in such cases the image $i_G(G) = G/\mathscr{C}_G$ (which is pseudo-reductive with the same minimal field of definition over k for its geometric unipotent radical as G, by Proposition 2.1.2(ii)) is of minimal type.

Over any imperfect field k of characteristic $p > 0$ there exist standard absolutely pseudo-simple groups not of minimal type:

Example 2.3.5. Let k be imperfect of characteristic $p > 0$, and let $k' = k(a^{1/p^2})$ and $\mathrm{k} = k(a^{1/p}) = k' \cap k^{1/p}$ with $a \in k - k^p$. Consider the k-group

$$G = \mathrm{R}_{k'/k}(\mathrm{SL}_p)/\mathrm{R}_{\mathrm{k}/k}(\mu_p) \tag{2.3.5}$$

of rank $p-1$ (so rank 1 when $p = 2$). By [CGP, Ex. 5.3.7] G is a standard pseudo-simple k-group and $\dim \ker i_G = (p-1)^2 > 0$. Nontriviality of $\ker i_G$ implies that G is not of minimal type since its root system is reduced.

For pseudo-reductive k-groups G_1 and G_2 it is clear that $\mathscr{C}_{G_1} \times \mathscr{C}_{G_2} = \mathscr{C}_{G_1 \times G_2}$ inside $G_1 \times G_2$, so $G_1 \times G_2$ is of minimal type if and only if G_1 and G_2 are of minimal type. As the following lemma shows, the formation of $\mathscr{C}_G$ (and hence of the minimal type property) behaves well with respect to separable extension of the ground field (such as scalar extension to k_s, or from a global function field to its completion at some place):

Lemma 2.3.6. *Let G be a pseudo-reductive group over a field k. For any separable extension k'/k, $(\mathscr{C}_G)_{k'} = \mathscr{C}_{G_{k'}}$ inside $G_{k'}$. In particular, G is of minimal type if and only if $G_{k'}$ is of minimal type.*

Proof. Let K/k be the minimal field of definition for the geometric unipotent radical of G. Clearly K is purely inseparable of finite degree over k and $K' := k' \otimes_k K$ is a field separable over K. By [CGP, Prop. 1.1.9(2)], K'/k' is the minimal field of definition for the geometric unipotent radical of $G_{k'}$. Since $(G_K^{\mathrm{red}}) \otimes_K K' = (G_{k'})_{K'}^{\mathrm{red}}$, it follows that $(i_G)_{k'} = i_{G_{k'}}$. For any Cartan k-subgroup C of G, $C_{k'}$ is a Cartan k'-subgroup of $G_{k'}$ and $(\mathscr{C}_G)_{k'} = (\ker i_G \cap C)_{k'} = \ker i_{G_{k'}} \cap C_{k'} = \mathscr{C}_{G_{k'}}$. □

The following proposition provides an alternative description of the central unipotent k-subgroup scheme $\mathscr{C}_G$.

Proposition 2.3.7. *Let G be a pseudo-reductive k-group. The central unipotent k-subgroup scheme $\mathscr{C}_G$ of G contains every central unipotent k-subgroup scheme of G.*

Proof. Let $\mathscr{G} = G/\mathscr{C}_G$. Recall that $\mathscr{G}$ is a pseudo-reductive k-group of minimal type. It suffices to show that any central unipotent k-subgroup scheme of $\mathscr{G}$ (such as the image of a central unipotent k-subgroup scheme of G) is trivial.

Let $\mathscr{T}$ be a maximal k-torus of $\mathscr{G}$ and let $\mathscr{U}$ be a central unipotent k-subgroup scheme of $\mathscr{G}$, so $\mathscr{U}$ is contained in the Cartan subgroup $Z_{\mathscr{G}}(\mathscr{T})$ of $\mathscr{G}$. Let K/k be the minimal field of definition for the geometric unipotent radical of $\mathscr{G}$, and define $\mathscr{G}' = \mathscr{G}_K^{\mathrm{red}}$. As $\mathscr{G}$ is of minimal type, the restriction of $i_{\mathscr{G}} : \mathscr{G} \to \mathrm{R}_{K/k}(\mathscr{G}')$ to the Cartan subgroup $Z_{\mathscr{G}}(\mathscr{T})$ has trivial kernel. If $\mathscr{T}'$ is the image of $\mathscr{T}_K$ in $\mathscr{G}'$ then $i_{\mathscr{G}}$ carries the Cartan k-subgroup $Z_{\mathscr{G}}(\mathscr{T})$, and hence also $\mathscr{U}$, isomorphically onto its image in the Cartan k-subgroup $\mathrm{R}_{K/k}(\mathscr{T}')$ of $\mathrm{R}_{K/k}(\mathscr{G}')$. Thus, it suffices to show that every unipotent k-subgroup scheme U of $\mathrm{R}_{K/k}(\mathscr{T}')$ is trivial. The inclusion $j : U \hookrightarrow \mathrm{R}_{K/k}(\mathscr{T}')$ corresponds to a K-homomorphism $\mathscr{U}_K \to \mathscr{T}'$, and this latter homomorphism is trivial since $\mathscr{U}_K$ is unipotent and $\mathscr{T}'$ is a K-torus. Hence, j is trivial, so U is trivial as desired. □

Remark 2.3.8. Let $f : G \to \overline{G}$ be a quotient homomorphism between pseudo-reductive groups. The preceding proposition implies that $f(\mathscr{C}_G) \subset \mathscr{C}_{\overline{G}}$. Hence, $G/\mathscr{C}_G$ uniquely dominates every minimal type pseudo-reductive quotient of G (i.e., it is the unique maximal such quotient) and f induces a homomorphism between the maximal pseudo-reductive minimal type quotients $G/\mathscr{C}_G$ and $\overline{G}/\mathscr{C}_{\overline{G}}$ of G and $\overline{G}$ respectively.

Example 2.3.9. If F is a nonzero finite reduced k-algebra and $\mathscr{G}$ is a smooth affine F-group with pseudo-reductive fibers of minimal type over the factor fields of F then the pseudo-reductive k-group $G := \mathrm{R}_{F/k}(\mathscr{G})$ is also of minimal type. To see this, we note that if $\mathscr{Z}$ is the center of $\mathscr{G}$, then by [CGP, Prop. A.5.15(1)] the center of G is $\mathrm{R}_{F/k}(\mathscr{Z})$. Now if $\mathscr{G}$ is of minimal type then its center $\mathscr{Z}$ does not contain a nontrivial unipotent F-subgroup scheme. This implies that the center $\mathrm{R}_{F/k}(\mathscr{Z})$ of G does not contain any nontrivial unipotent k-subgroup scheme (see the argument used towards the end of the proof of Proposition 2.3.7), so G is of minimal type as claimed. Thus, by Example 2.3.3, if k is imperfect of characteristic 2 or 3 then every exotic pseudo-semisimple k-group is of minimal type.

Later arguments with torus centralizers and their derived groups will require good interaction of such constructions with the minimal-type property:

Lemma 2.3.10. *Let G be a pseudo-reductive k-group of minimal type. Any smooth connected normal k-subgroup N of G is of minimal type, as is $Z_G(\mu)^0$ for any closed k-subgroup scheme μ of a k-torus in G. In particular, the pseudo-semisimple k-group $\mathscr{D}(Z_G(\mu)^0)$ is of minimal type.*

Pseudo-reductivity of $Z_G(\mu)^0$ is a special case of [CGP, Prop. A.8.14(2)].

Proof. We may assume $k = k_s$. Now all k-tori are split, so by [CGP, Prop. 9.4.5] both N and $Z_G(\mu)^0$ are of minimal type. Since $\mathscr{D}(Z_G(\mu)^0)$ is normal in $Z_G(\mu)^0$, it is of minimal type since the same holds for $Z_G(\mu)^0$. □

2.3.11. We finish §2.3 by analyzing maximal pseudo-reductive quotients of certain k-groups. As motivation, for a finite purely inseparable extension K/k and pseudo-reductive K-group G', note that the natural map $\mathrm{R}_{K/k}(G')_K \to G'$ has smooth connected unipotent kernel [CGP, Prop. A.5.11(1),(2)], so it is the maximal pseudo-reductive quotient over K. Thus, given the k-group $\mathrm{R}_{K/k}(G')$ and the extension K of k we can canonically reconstruct the K-group G'.

We seek a refinement in which G' is pseudo-semisimple and the Weil restriction $\mathrm{R}_{K/k}(G')$ (which is generally not perfect) is replaced with its derived group. That is, is the pseudo-reductive quotient $\mathscr{D}(\mathrm{R}_{K/k}(G'))_K = \mathscr{D}(\mathrm{R}_{K/k}(G')_K) \twoheadrightarrow G'$ also maximal? If k is imperfect of characteristic 2 and either $[k;k^2] \geqslant 16$ or G'_{K_s} has a non-reduced root system then the answer may in general be "no" (in all other cases the answer is "yes", by Proposition B.3.5), so we shall consider a finer condition than pseudo-semisimplicity in order to recover G' from $\mathscr{D}(\mathrm{R}_{K/k}(G'))$ and K/k in general. A suitable refinement involves incorporating the "minimal type" property. We shall present this in a wider setting, as follows.

Let H be a smooth connected affine group over a field k (such as $H = \mathscr{D}(\mathrm{R}_{K/k}(G'))$ for $(K/k, G')$ as above), and let $H^{\mathrm{pred}} := H/\mathscr{R}_{u,k}(H)$ denote its maximal pseudo-reductive quotient. The further quotient

$$H^{\mathrm{prmt}} := H^{\mathrm{pred}}/\mathscr{C}_{H^{\mathrm{pred}}} \tag{2.3.11}$$

is pseudo-reductive of minimal type and is maximal among such quotients of H; its formation commutes with separable extension on k by Lemma 2.3.6.

Remark 2.3.12. If $H^{\mathrm{pred}}_{k_s}$ has a reduced root system then there is a useful alternative description of H^{prmt}: we claim that it is the image of the natural map

$$f_H : H \to \mathrm{R}_{E/k}(H_E/\mathscr{R}_{u,E}(H_E))$$

for any finite extension E/k such that $\mathscr{R}_u(H_{\overline{k}})$ descends to an E-subgroup of H_E. Since $\mathscr{R}_{u,k}(H)_E \subset \mathscr{R}_{u,E}(H_E)$, for $H' := H_E/\mathscr{R}_{u,E}(H_E)$ we have naturally $H' = (H^{\rm pred})'$ as quotients of H_E and $f_H = f_{H^{\rm pred}} \circ q_H$ for the quotient map $q_H : H \twoheadrightarrow H^{\rm pred}$. Hence, we may replace H with $H^{\rm pred}$ so that now H is pseudo-reductive with a reduced root system over k_s.

If K/k is the minimal field of definition for the geometric unipotent radical of H then $K \subset E$ and $\mathscr{R}_{u,E}(H_E)$ descends to the K-subgroup $\mathscr{R}_{u,K}(H_K)$ of H_K. Hence, f_H is the composition of i_H and the inclusion

$$\mathrm{R}_{K/k}(H_K^{\rm red}) \hookrightarrow \mathrm{R}_{K/k}(\mathrm{R}_{E/K}(H_E^{\rm red})) = \mathrm{R}_{E/k}(H_E^{\rm red}),$$

so $\ker f_H = \ker i_H$ and therefore $\mathscr{C}_H = (\ker f_H) \cap Z_H(T)$ for a maximal k-torus T in H. Via the identification of H' and $(H/\mathscr{C}_H)'$ we may thereby replace H with $H/\mathscr{C}_H = H^{\rm prmt}$ (which has the same root system over k_s as H) to reduce to the case where H is also of minimal type. But then $\ker f_H$ $(= \ker i_H)$ is trivial since H_{k_s} has a reduced root system (see Proposition 2.3.4).

Since the quotient map $H_{\overline{k}} \twoheadrightarrow H_{\overline{k}}^{\rm red}$ kills any smooth connected unipotent normal subgroup as well as any central unipotent subgroup scheme, $(H^{\rm prmt})_{\overline{k}}$ dominates $H_{\overline{k}}^{\rm red}$ as quotients of $H_{\overline{k}}$. In particular, $H \to H^{\rm prmt}$ induces an isomorphism between maximal geometric reductive quotients.

If N is a smooth connected normal k-subgroup of H then its image in $H^{\rm prmt}$ is pseudo-reductive of minimal type by Lemma 2.3.10, so the map $N \to H^{\rm prmt}$ factors through $N \twoheadrightarrow N^{\rm prmt}$, yielding a canonical k-homomorphism $N^{\rm prmt} \to H^{\rm prmt}$. It is very important that $N^{\rm prmt} \to H^{\rm prmt}$ has trivial kernel:

Proposition 2.3.13. *Let H be a smooth connected affine group over a field k, and N a smooth connected normal k-subgroup. The canonical map $N^{\rm prmt} \to H^{\rm prmt}$ has trivial kernel.*

In particular, if K/k is a purely inseparable finite extension of fields and G' is a pseudo-semisimple K-group of minimal type then the natural surjective homomorphism

$$\mathscr{D}(\mathrm{R}_{K/k}(G'))_K = \mathscr{D}(\mathrm{R}_{K/k}(G')_K) \twoheadrightarrow \mathscr{D}(G') = G' \qquad (2.3.13)$$

is the maximal pseudo-reductive quotient over K of minimal type, so naturally $\mathscr{D}(\mathrm{R}_{K/k}(G'))_{\overline{k}}^{\rm ss} \simeq (G')_{\overline{k}}^{\rm ss}$ and we can canonically reconstruct G' from $\mathscr{D}(\mathrm{R}_{K/k}(G'))$ and the extension K/k.

If $H_{k_s}^{\rm pred}$ has a reduced root system then this result admits an elementary proof via Remark 2.3.12 (using [CGP, Prop. A.4.8]). However, we will later need the case of non-reduced root systems. See Proposition B.3.5 for a stronger maximality property of (2.3.13) *without* assuming G' is of minimal type, provided that if char$(k) = 2$ then $[k : k^2] \leqslant 8$ and G'_{K_s} has a reduced root system.

Proof. We may assume $k = k_s$, so $H(k)$ is schematically dense in H (in the sense of [EGA, IV$_3$, 11.10.2]). Since the (possibly non-smooth) schematic kernel $U = \ker(N \twoheadrightarrow N^{\mathrm{prmt}})$ is normalized by $H(k)$, it follows that U is a normal k-subgroup scheme of H (normality says that for every k-algebra A and $u \in U(A)$, the A-morphism $H_A \to (H/U)_A$ defined by $h \mapsto huh^{-1} \bmod U$ is trivial, which can be checked on $H(k)$; see [EGA, IV$_3$, 11.10.9]). We have seen that $U \subset \ker(H \twoheadrightarrow H^{\mathrm{prmt}})$, so by passing to the quotient k-groups N/U and H/U we may assume N is pseudo-reductive of minimal type.

The commutator subgroup $(N, \mathscr{R}_{u,k}(H))$ is a smooth connected normal k-subgroup of N contained in $\mathscr{R}_{u,k}(H)$, so $(N, \mathscr{R}_{u,k}(H)) \subset \mathscr{R}_{u,k}(N)$. But as N is pseudo-reductive, its k-unipotent radical is trivial. Thus, $\mathscr{R}_{u,k}(H)$ commutes with N, so $\mathscr{R}_{u,k}(H) \cap N$ is a central unipotent k-subgroup scheme of N. But as N is of minimal type, such a subgroup scheme is trivial (Proposition 2.3.7). Hence, by passing to the quotient $H/\mathscr{R}_{u,k}(H)$ we can assume that H is pseudo-reductive. Now both H and N are pseudo-reductive and N is of minimal type. The intersection $\mathscr{C}_H \cap N$ is a central unipotent k-subgroup scheme of N, so (as N is of minimal type) this subgroup scheme is trivial. Therefore, the homomorphism $N \to H/\mathscr{C}_H = H^{\mathrm{prmt}}$ has trivial kernel. $\square$

Remark 2.3.14. The minimal-type condition on the quotients of N and H in Proposition 2.3.13 cannot be dropped: over any imperfect field k of characteristic $p > 0$ it can happen that the k-homomorphism $N^{\mathrm{pred}} \to H^{\mathrm{pred}}$ has nontrivial kernel. To build such examples with N even pseudo-reductive (hence necessarily not of minimal type, due to Proposition 2.3.13), let $k'/\mathrm{k}/k$ be the tower of purely inseparable extensions of degree p as in Example 2.3.5 and define N to be the absolutely pseudo-simple k-group G as defined there. Note that N contains the central k-subgroup scheme $Q := \mathrm{R}_{k'/k}(\mu_{p^2})/\mathrm{R}_{\mathrm{k}/k}(\mu_{p^2})$. (By [CGP, Ex. 5.3.7] we have $\ker i_N = Q$ and $\dim Q = p-1 > 0$, so indeed N is not of minimal type.)

The k-group $\mathscr{N} := \mathrm{R}_{\mathrm{k}/k}(\mathrm{R}_{k'/\mathrm{k}}(\mathrm{SL}_{p^2})/\mu_{p^2})$ has the smooth connected central k-subgroup $U := \mathrm{R}_{\mathrm{k}/k}(\mathrm{R}_{k'/\mathrm{k}}(\mu_p)/\mu_p) = \mathrm{R}_{\mathrm{k}/k}(\mathrm{R}_{k'/\mathrm{k}}(\mathrm{GL}_1)/\mathrm{GL}_1)$ that is unipotent of dimension $p(p-1)$. Left-exactness of $\mathrm{R}_{\mathrm{k}/k}$ provides a central inclusion $j : Q \hookrightarrow \mathscr{N}$, and $j(Q)$ contains $Z := \mathrm{R}_{k'/k}(\mu_p)/\mathrm{R}_{\mathrm{k}/k}(\mu_p) \subset U$, so the smooth connected central pushout $H := N \times^Q \mathscr{N}$ contains N and $\mathscr{N}$ as normal k-subgroups with $N \cap \mathscr{N} = Q$ and $N \cap \mathscr{R}_{u,k}(H) \supset N \cap U \supset Z \neq 1$. Hence, $\ker(N = N^{\mathrm{pred}} \to H^{\mathrm{pred}})$ is nontrivial.

3

Field-theoretic and linear-algebraic invariants

3.1 A non-standard rank-1 construction

Below we give a construction of *non-standard* pseudo-split absolutely pseudo-simple k-groups with root system A_1 over any imperfect field k of characteristic 2 for which $[k:k^2] > 2$.

3.1.1. Let k be an imperfect field of characteristic 2 and let K/k be a nontrivial purely inseparable finite extension (so $K \neq kK^2$). Let V be a nonzero k-subspace of K. Define $k\langle V\rangle \subset K$ (resp. $k\langle V^2\rangle \subset K$) to be the k-subalgebra generated by ratios v/v' for $v, v' \in V - \{0\}$ (resp. ratios v^2/v'^2 for $v, v' \in V - \{0\}$), so $k\langle V\rangle = k[V]$ if $1 \in V$. As K is an algebraic extension of k, both $k\langle V\rangle$ and $k\langle V^2\rangle$ are subfields of K.

Assume V is a $k\langle V^2\rangle$-subspace of K, so for any $v \in V - \{0\}$ the k-subspace $v^{-2}V$ is independent of v; we denote it as V^-. If $k\langle V\rangle = K$ and $[K:kK^2] = 2$ then $V = K$, whereas if $[K:kK^2] > 2$ then there exists a nonzero proper kK^2-subspace $V \subset K$ satisfying $k\langle V\rangle = K$. If $[k:k^2] = 2$ then $[K:kK^2] = 2$ in all cases, so what follows is of most interest when $[k:k^2] > 2$.

Definition 3.1.2. Let $H_{V,K/k}$ be the k-subgroup of $R_{K/k}(SL_2)$ generated by the k-subgroup U_V^+ of the upper triangular unipotent k-subgroup of $R_{K/k}(SL_2)$ corresponding to the k-subspace $V \subset K$ and the k-subgroup $U_{V^-}^-$ of the lower triangular unipotent subgroup of $R_{K/k}(SL_2)$ corresponding to the k-subspace $V^- \subset K$. Let $PH_{V,K/k}$ denote the image of $H_{V,K/k}$ inside $R_{K/k}(PGL_2)$ (equivalently, it is the k-subgroup of $R_{K/k}(PGL_2)$ generated by the k-groups U_V^+ and $U_{V^-}^-$ viewed inside the upper and lower triangular subgroups of $R_{K/k}(PGL_2)$).

Let L denote either the k-group SL_2 or PGL_2 and H denote the k-subgroup $H_{V,K/k}$ of $R_{K/k}(L_K)$ if $L = SL_2$ and the k-subgroup $PH_{V,K/k}$ of $R_{K/k}(L_K)$

if $L = \mathrm{PGL}_2$. Clearly $1 \in V$ if and only if H contains the canonical k-subgroup $L \subset \mathrm{R}_{K/k}(L_K)$. The detailed structure of the k-group H is worked out in [CGP, §9.1] (where a fixed choice of L within $\{\mathrm{SL}_2, \mathrm{PGL}_2\}$ is made, so the notation "$H_{V,K/k}$" is used there for each choice of L without confusion; we shall use the distinct notations $H_{V,K/k}$ and $\mathrm{P}H_{V,K/k}$ below to distinguish the two cases).

The subfield $k\langle V\rangle \subset K$ is unaffected by $K^\times$-scaling of V, and the K-automorphisms of SL_2 and PGL_2 defined by $\mathrm{diag}(\lambda, 1) \in \mathrm{PGL}_2(K)$ carry $H_{V,K/k}$ to $H_{\lambda V,K/k}$ and carry $\mathrm{P}H_{V,K/k}$ to $\mathrm{P}H_{\lambda V,K/k}$. Thus, by scaling V so that $V \subset k\langle V\rangle$ (it is equivalent that $V \cap k\langle V\rangle \neq 0$), the k-groups $H_{V,k\langle V\rangle/k}$ and $\mathrm{P}H_{V,k\langle V\rangle/k}$ make sense and respectively equal $H_{V,K/k}$ and $\mathrm{P}H_{V,K/k}$. Hence, for general V we have $H_{V,K/k} \simeq H_{V,k\langle V\rangle/k}$ and $\mathrm{P}H_{V,K/k} \simeq \mathrm{P}H_{V,k\langle V\rangle/k}$ as k-groups, so the main case of interest is when $k\langle V\rangle = K$.

If $k\langle V\rangle = K$ and $V' \subset K$ is a nonzero kK^2-subspace then $H_{V',K/k} \simeq H_{V,K/k}$ if and only if $V' = \lambda V$ for some $\lambda \in K^\times$ (see [CGP, Prop. 9.1.7], noting that $k\langle V'\rangle = K$ if $V' = \lambda V$ with $\lambda \in K^\times$). This scaling relation likewise characterizes when $\mathrm{P}H_{V',K/k} \simeq \mathrm{P}H_{V,K/k}$.

The split diagonal k-torus $D \subset L$ is a maximal k-torus in H. To describe the Cartan subgroup $Z_H(D)$, we introduce the following notation.

Definition 3.1.3. We define $V^*_{K/k}$ to be the Zariski closure of the subgroup of $\mathrm{R}_{K/k}(\mathrm{GL}_1)(k) = K^\times$ generated by the ratios v/v' for $v, v' \in V - \{0\}$; $V^*_{K/k}$ will be considered as a k-subgroup of $\mathrm{R}_{K/k}(\mathrm{SL}_2)$ via the map $t \mapsto \mathrm{diag}(t, 1/t)$. When working with subgroups of $\mathrm{R}_{K/k}(\mathrm{PGL}_2)$ we use the same notation to denote the image of that Zariski closure under the natural map $\mathrm{R}_{K/k}(\mathrm{SL}_2) \to \mathrm{R}_{K/k}(\mathrm{PGL}_2)$; the context will always make the intended meaning clear.

The following result, which is [CGP, Prop. 9.1.4], gives the main properties of $H_{V,K/k}$ and $\mathrm{P}H_{V,K/k}$; we reproduce it for the convenience of the reader.

Proposition 3.1.4. *Let H be $H_{V,K/k}$ or $\mathrm{P}H_{V,K/k}$ inside $\mathrm{R}_{K/k}(L_K)$ (with L equal to SL_2 or PGL_2 respectively). The k-group H is absolutely pseudo-simple, its root groups with respect to the diagonal torus D ($\subset \mathrm{R}_{K/k}(D_K)$) are U^+_V and $U^-_{V^-}$ as in Definition 3.1.2, and $Z_H(D) = V^*_{K/k}$. In particular, the k-subgroup $H \subset \mathrm{R}_{K/k}(L_K)$ determines the k-subspace $V \subset K$.*

Moreover, the natural map $\pi : H_K \to L_K$ is a K-descent of the maximal reductive quotient of $H_{\overline{k}}$, and the minimal field of definition over k for the geometric unipotent radical $\mathscr{R}_u(H_{\overline{k}})$ of H is $k\langle V\rangle$. (In particular, for a nonzero k-subspace $V' \subset K$ that is a $k\langle V'^2\rangle$-subspace of K and the corresponding k-subgroup $H' \subset \mathrm{R}_{K/k}(L_K)$, if $H \simeq H'$ as k-groups then $k\langle V\rangle = k\langle V'\rangle$ as purely inseparable extensions of k.)

The structure of the Cartan k-subgroup $V^*_{K/k}$ (even its dimension) is rather mysterious; see [CGP, 9.1.8–9.1.10].

Proof. Conjugation by $\mathrm{diag}(\lambda, 1)$ for $\lambda \in K^\times$ shows that the problems for V and λV are equivalent. (Note that $(\lambda V)^*_{K/k} = V^*_{K/k}$ inside $\mathrm{R}_{K/k}(\mathrm{GL}_1)$.) Thus, we may and do assume $1 \in V$. In particular, the following properties hold: $k\langle V\rangle = k[V]$, $k\langle V^2\rangle = k[V^2]$ (so V is a $k[V^2]$-subspace of K), $V^- = V$, $L \subset H$, $V^2 \subset V$, and $V^*_{K/k}$ contains a Zariski-dense subset generated by the k-points $\mathrm{diag}(v, 1/v)$ for $v \in V - \{0\}$.

By [CGP, Cor. A.5.16], L is a Levi k-subgroup of $\mathrm{R}_{K/k}(L_K)$ via the natural inclusion. As $L \subset H$, Proposition 2.1.2(i) implies that H is pseudo-reductive and L is a Levi k-subgroup of H such that π is a K-descent of the maximal reductive quotient of $H_{\overline{k}}$. Hence, the minimal field of definition over k for $\mathscr{R}_u(H_{\overline{k}})$ is contained in K and $\mathscr{D}(H)$ is absolutely pseudo-simple of type A_1.

For the D-root groups $U^\pm$ of L, the D-root groups of H are given by $U^\pm_H := H \cap \mathrm{R}_{K/k}(U^\pm_K)$. We will show that $U^\pm_H = U^\pm_V$ and $Z_H(D) = V^*_{K/k}$. We first check that $V^*_{K/k} \subset Z_H(D)$, or equivalently that $V^*_{K/k} \subset H$. The subset $V - \{0\} \subset K^\times$ is stable under inversion (since $1 \in V$ and V is a subspace of K over $k\langle V^2\rangle = k[V^2]$), so it suffices to apply the identity

$$\begin{pmatrix} t & 0 \\ 0 & t^{-1} \end{pmatrix} = \begin{pmatrix} 1 & 1 \\ 0 & 1 \end{pmatrix} \begin{pmatrix} 1 & 0 \\ t-1 & 1 \end{pmatrix} \begin{pmatrix} 1 & -1/t \\ 0 & 1 \end{pmatrix} \begin{pmatrix} 1 & 0 \\ t(1-t) & 1 \end{pmatrix}$$

that expresses a diagonal matrix in SL_2 as a product of upper and lower triangular unipotent matrices. (Note that for $t \in V$, $t(1-t) = t - t^2 \in V$ because V contains V^2.) Since $V^2 \cdot V \subset V$, it also follows from the description of $V^*_{K/k}$ as a Zariski closure that it normalizes the smooth connected k-subgroup $U^\pm_V \subset \mathrm{R}_{K/k}(U^\pm_K)$. Hence, it makes sense to define the following subgroups of $H(k)$:

$$\mathscr{Z} = V^*_{K/k}(k),\ \mathscr{U} = U^+_V(k),\ \mathscr{P} = \mathscr{Z} \ltimes \mathscr{U}.$$

Note that $\mathscr{Z}$ is Zariski-dense in $V^*_{K/k}$, due to the definition of $V^*_{K/k}$ as a Zariski closure.

The Bruhat decomposition of the group $\mathrm{R}_{K/k}(L_K)(k) = L(K)$ (with L equal to SL_2 or PGL_2) gives that for

$$n := \begin{pmatrix} 0 & 1 \\ -1 & 0 \end{pmatrix},$$

the map $\mathscr{U} \times \mathscr{P} \to H(k)$ defined by $(u, p) \mapsto unp$ is injective and we have the disjoint union containment

$$\mathscr{U} n \mathscr{P} \bigcup \mathscr{P} \subset H(k).$$

It is easy to check that $\mathscr{H} := \mathscr{U} n \mathscr{P} \bigcup \mathscr{P}$ is stable under inversion and multiplication, so it is a subgroup of $H(k)$. As it contains the subgroups $\mathscr{U}$ and $n\mathscr{U}n^{-1}$ that are Zariski-dense in U_V^+ and U_V^- respectively, and H is generated by U_V^+ and U_V^-, we see that $\mathscr{H}$ is *Zariski-dense* in H.

By [CGP, Prop. 2.1.8(2),(3)] applied to the standard parameterization $\lambda : \mathrm{GL}_1 \simeq D$, the multiplication map

$$n^{-1}U_H^+ n \times Z_H(D) \times U_H^+ \to H$$

is an open immersion, and clearly its left-translate by n meets the Zariski-dense $\mathscr{H}$ in $\mathscr{U} n \mathscr{P}$. Hence, $n^{-1}\mathscr{U} n \mathscr{P}$ is Zariski-dense in H. Therefore, its closure $n^{-1}U_V^+ n \cdot V_{K/k}^* \cdot U_V^+$ inside $n^{-1}U_H^+ n \cdot Z_H(D) \cdot U_H^+$ is full, so $Z_H(D) = V_{K/k}^*$ and $U_H^+ = U_V^+$; hence $U_H^- = nU_H^+ n^{-1} = nU_V^+ n^{-1} = U_V^-$ and H is generated by the vector groups $U_V^{\pm}$ on which D acts linearly through nontrivial characters, so H is perfect. Now it follows from [CGP, Lemma 3.1.2] that H is absolutely pseudo-simple (with root groups $U_V^{\pm} = U_H^{\pm}$ relative to D).

Our remaining task is to show that the minimal field of definition over k for $\ker \pi \subset H_K$ is $k[V]$. For this, we may (and do) assume, after replacing K by $k[V]$, that $K = k[V]$. We will now show that if $F \subset K$ is a subfield containing k such that $\ker \pi$ descends to an F-subgroup scheme $R \subset H_F$ then $F = K$.

Since the Levi k-subgroup L in H yields a Levi F-subgroup L_F in H_F, this F-subgroup maps isomorphically onto the quotient H_F/R. Hence, the K-map $\pi : H_K \to L_K$ descends to an F-map $\pi_0 : H_F \twoheadrightarrow L_F$. (The point is that we have identified the target of the F-descent π_0 of π with the *canonical* F-descent of the target L_K of π.)

Passing to Lie algebras of root groups relative to both the maximal F-torus D_F of H_F and the diagonal F-torus of the target of π_0, we get an abstract F-linear map $F \otimes_k V \to F$ that descends the canonical K-linear map $K \otimes_k V \to K$ induced by $\mathrm{Lie}(\pi)$.

The map π is induced by the canonical quotient map

$$\mathrm{R}_{K/k}(L_K)_K \to L_K$$

that on Lie algebras is the natural multiplication map

$$K \otimes_k \mathfrak{l} \to \mathfrak{l}$$

(with $\mathfrak{l} := \mathrm{Lie}(L_K)$), so the map $K \otimes_k V \to K$ induced by $\mathrm{Lie}(\pi)$ is $c \otimes v \mapsto cv$. Hence, the existence of π_0 implies that $F \cdot V \subset F$ inside of K, and in particular $V \subset F$. But F is a subfield of K containing k, so F contains $k[V] = K$. □

Remark 3.1.5. Since SL_2 is a Levi k-subgroup of $H := H_{V,K/k}$, the schematic center of H is nontrivial. However, the smooth connected Cartan k-subgroup $V^*_{K/k}$ in H may not contain the non-smooth center $\mathrm{R}_{K/k}(\mu_2)$ of $\mathrm{R}_{K/k}(\mathrm{SL}_2)$, due to dimension reasons. For example, suppose $K^2 \subset k$ and let $d = \dim_k V \geqslant 1$, so $\dim V^*_{K/k} \leqslant 1 + d(d-1)/2$ by [CGP, Prop. 9.1.9]. Since $\dim \mathrm{R}_{K/k}(\mu_2) = [K:k] - 1$, it suffices to find V such that $1 + d(d-1)/2 < [K:k] - 1$. If V is the k-span of 1 and a 2-basis of K/k then $[K:k] = 2^{d-1}$, so any such V works when $[K:k] \geqslant 16$.

Example 3.1.6. The behavior of $H_{V,K/k}$ with respect to purely inseparable Weil restriction exhibits some subtleties when $V \neq K$, essentially because the formation of the Cartan k-subgroup $V^*_{K/k}$ can fail to commute with such Weil restriction. To explain this, let $k_0 \subset k$ be a subfield over which k is purely inseparable of finite degree > 1, so we have a k_0-subgroup inclusion

$$\mathrm{R}_{k/k_0}(H_{V,K/k}) \subset \mathrm{R}_{k/k_0}(\mathrm{R}_{K/k}(\mathrm{SL}_2)) = \mathrm{R}_{K/k_0}(\mathrm{SL}_2)$$

that contains the diagonal k_0-torus D_0 and has the same root groups as $H_{V,K/k_0}$. Since the derived group $\mathscr{D}(\mathrm{R}_{k/k_0}(H_{V,K/k}))$ is pseudo-semisimple and hence is generated by its D_0-root groups, which coincide with those of $H_{V,k/k_0}$, we have the general equality

$$\mathscr{D}(\mathrm{R}_{k/k_0}(H_{V,K/k})) = H_{V,K/k_0}. \tag{3.1.6}$$

(Note that the minimal field of definition over k_0 for the geometric unipotent radical of $H_{V,K/k_0}$ is $k_0\langle V\rangle$, and this is equal to $k\langle V\rangle$ inside K since V is a nonzero k-subspace of K.) By the same reasoning, the analogous assertions hold in the PGL_2-case using PH's.

In the special case $V = K$ (which is the only possibility when $[k:k^2] = 2$ and $k\langle V\rangle = K$), which is to say the standard case when $k\langle V\rangle = K$, the inclusions

$$V^*_{K/k} \subset \mathrm{R}_{K/k}(\mathrm{GL}_1), \ \ V^*_{K/k_0} \subset \mathrm{R}_{K/k_0}(\mathrm{GL}_1)$$

are both equalities and so the intervention of a derived group in (3.1.6) is unnecessary; i.e., $\mathrm{R}_{k/k_0}(H_{V,K/k})$ is perfect when $V = K$. But such perfectness can fail in cases with $V \neq K$ (assuming $k\langle V\rangle = K$ and V is a kK^2-subspace of K). More specifically, for any k and $K \subset k^{1/2}$ satisfying $[K:k] = 16$, in

[CGP, 9.1.8–9.1.11] explicit examples are given such that $V^*_{K/k} = \mathrm{R}_{K/k}(\mathrm{GL}_1)$ but $\mathrm{R}_{k/k_0}(H_{V,K/k})$ is *never* perfect for *proper* subfields $k_0 \subset k$ as above (so V^*_{K/k_0} is a proper k_0-subgroup of $\mathrm{R}_{K/k_0}(\mathrm{GL}_1)$ for all $k_0 \neq k$); some such V are recorded in (4.2.2). In these examples, $\dim_k V = 6$ and the maximal k-subalgebra $F \subset K$ over which V is an F-submodule of K coincides with k.

Our interest in the groups $H_{V,K/k}$ and $\mathrm{P}H_{V,K/k}$ is due to [CGP, Prop. 9.1.5], which we reproduce here (allowing any characteristic):

Proposition 3.1.7. *Let k be an infinite field of arbitrary characteristic and let K/k be a purely inseparable finite extension. Let L be either* SL_2 *or* PGL_2*, and let D be the diagonal k-torus of L. Let G be a pseudo-semisimple k-subgroup of* $\mathrm{R}_{K/k}(L_K)$ *that contains the diagonal k-torus D* $(\subset \mathrm{R}_{K/k}(D_K))$.

(i) *If* $\mathrm{char}(k) = 2$ *then there exists a unique nonzero k-subspace $V \subset K$ that is a $k\langle V^2\rangle$-subspace of K and for which $G = H_{V,K/k}$ or $\mathrm{P}H_{V,K/k}$ according as $L = \mathrm{SL}_2$ or PGL_2. In particular, $1 \in V$ if and only if the element $\left(\begin{smallmatrix}1 & 1\\ 0 & 1\end{smallmatrix}\right) \in L(k)$ lies in $G(k)$.*

(ii) *If* $\mathrm{char}(k) \neq 2$ *then there exists a subfield F of K over k and $\lambda \in K^\times$ such that G is the conjugate of the k-subgroup $\mathrm{R}_{F/k}(L_F)$ of $\mathrm{R}_{K/k}(L_K)$ under* $\mathrm{diag}(\lambda, 1)$. *Moreover, G coincides with the k-subgroup $\mathrm{R}_{F/k}(L_F)$ if and only if the element $\left(\begin{smallmatrix}1 & 1\\ 0 & 1\end{smallmatrix}\right) \in L(k)$ lies in $G(k)$.*

Proof. The root groups of (G, D) are the intersections $U^{\pm}_G$ of the root groups $\mathrm{R}_{K/k}(U^{\pm}_K)$ of $\mathrm{R}_{K/k}(L_K)$ with G. These k-subgroups of G are stable under the natural conjugation action of D and are k-subgroups of the respective upper and lower triangular unipotent subgroups of $\mathrm{R}_{K/k}(L_K)$. Thus, by the Zariski-density of $(k^\times)^2$ in GL_1, the k-groups $U^{\pm}_G$ correspond to k-subspaces of K. Hence, $V^{\pm} := U^{\pm}_G(k)$ are k-subspaces of $\mathrm{R}_{K/k}(\mathbf{G}_a)(k) = K$. We have $V^{\pm} \neq 0$ since the split unipotent k-groups $U^{\pm}_G$ are nontrivial, due to the pseudo-simplicity of G. (In assertion (i), if G equals $H_{V,K/k}$ or $\mathrm{P}H_{V,K/k}$ for some V then necessarily $V = V^+$, so uniqueness of V is clear in (i). The problem is to prove existence.)

Applying conjugation by $\mathrm{diag}(\lambda^{-1}, 1)$ for $\lambda \in K^\times$ has no effect on D and when $\mathrm{char}(k) = 2$ it has no effect on whether or not G has the form $H_{V,K/k}$ for a nonzero k-subspace V of K that is a $k\langle V^2\rangle$-subspace, and it replaces $V^{\pm}$ with $\lambda^{\mp 1} V^{\pm}$. Thus, by choosing $\lambda = v$ for $v \in V^+ - \{0\}$ we may reduce to the case $1 \in V^+$; i.e., $\left(\begin{smallmatrix}1 & 1\\ 0 & 1\end{smallmatrix}\right) \in G(k)$.

By [CGP, Prop. 3.4.2], there is a unique element in $N_G(D)(k) - Z_G(D)(k)$ of the form $\nu = u'uu'$, with $u = \left(\begin{smallmatrix}1 & 1\\ 0 & 1\end{smallmatrix}\right)$ and $u' \in U^-_G(k)$. By explicit computation in $\mathrm{SL}_2(K)$ or $\mathrm{PGL}_2(K)$ we see that $u' = \left(\begin{smallmatrix}1 & 0\\ -1 & 1\end{smallmatrix}\right)$ and $\nu = \left(\begin{smallmatrix}0 & 1\\ -1 & 0\end{smallmatrix}\right)$. Thus, the standard Weyl element $n := \left(\begin{smallmatrix}0 & 1\\ -1 & 0\end{smallmatrix}\right)$ lies in $G(k)$. Using conjugation by n we see that $V^+ = V^-$. We will denote this common k-subspace of K by V, so $1 \in V$. Inside the upper and lower triangular unipotent subgroups of $\mathrm{R}_{K/k}(L_K)$, let $U_V^{\pm}$ denote the k-subgroups corresponding to $V \subset K$. Clearly $U_G^{\pm} = U_V^{\pm}$.

Next, we prove that the subset $V - \{0\} \subset K^\times$ is closed under inversion and that $v^2 V \subset V$ for all $v \in V$. Let $Z = Z_G(D)$. The Bruhat decomposition of $G(k)$ relative to the minimal pseudo-parabolic k-subgroup $Z \ltimes U_V^-$ is provided by [CGP, Thm. 3.4.5], and it says

$$G(k) = U_V^-(k) n Z(k) U_V^-(k) \coprod Z(k) U_V^-(k). \tag{3.1.7.1}$$

Since $U_V^+ \cap (Z \ltimes U_V^-) = \{1\}$, we have $U_V^+(k) - \{1\} \subset U_V^-(k) n Z(k) U_V^-(k)$.

Multiplication defines an open immersion $U_V^+ \times Z \times U_V^- \to G$ by [CGP, Prop. 2.1.8(2),(3)], so the equality $U_V^+ = n^{-1} U_V^- n$ and left-translation by n imply that the map $U_V^- \times Z \times U_V^- \to G$ defined by $(u', z, u'') \mapsto u' n z u''$ is an open immersion. In particular, for $u', u'' \in U_V^-(k)$ and $z \in Z(k)$ the product $u' n z u'' \in G(k)$ uniquely determines u', u'', and z. Thus, for each *nonzero* $x \in V$ there exist unique $y_1, y_2 \in V$ and $z \in Z(k)$ such that

$$\begin{pmatrix} 1 & x \\ 0 & 1 \end{pmatrix} = \begin{pmatrix} 1 & 0 \\ y_1 & 1 \end{pmatrix} \cdot \begin{pmatrix} 0 & 1 \\ -1 & 0 \end{pmatrix} \cdot z \cdot \begin{pmatrix} 1 & 0 \\ y_2 & 1 \end{pmatrix} \tag{3.1.7.2}$$

in $\mathrm{SL}_2(K)$ or $\mathrm{PGL}_2(K)$. As all terms in (3.1.7.2) aside from z come from $\mathrm{SL}_2(K)$, when $L = \mathrm{PGL}_2$ we have that z uniquely arises from an element of $\mathrm{SL}_2(K)$ such that (3.1.7.2) holds as an identity in $\mathrm{SL}_2(K)$. Subject to this latter condition in the PGL_2-case, z arises from a unique matrix of the form $\left(\begin{smallmatrix} t^{-1} & 0 \\ 0 & t \end{smallmatrix}\right)$ with $t \in K^\times$. We wish to compute y_1, y_2, and t in terms of x.

Multiplying out the right side of (3.1.7.2) gives

$$\begin{pmatrix} 1 & x \\ 0 & 1 \end{pmatrix} = \begin{pmatrix} t y_2 & t \\ t y_1 y_2 - t^{-1} & t y_1 \end{pmatrix}$$

in $\mathrm{SL}_2(K)$, so $t = x$ and $y_1 = 1/t = y_2$. Hence, $1/x = y_1 \in V$. This shows that the subset $V - \{0\} \subset K^\times$ is stable under inversion and moreover that if $x \in V - \{0\}$ then

$$\begin{pmatrix} x & 0 \\ 0 & x^{-1} \end{pmatrix} \in G(k) \tag{3.1.7.3}$$

inside $\mathrm{SL}_2(K)$ or $\mathrm{PGL}_2(K)$. Thus, for any $x' \in V$ the product

$$\begin{pmatrix} x & 0 \\ 0 & x^{-1} \end{pmatrix} \cdot \begin{pmatrix} 1 & x' \\ 0 & 1 \end{pmatrix} \cdot \begin{pmatrix} x^{-1} & 0 \\ 0 & x \end{pmatrix} = \begin{pmatrix} 1 & x^2 x' \\ 0 & 1 \end{pmatrix}$$

lies in $G(k)$. In other words, $x^2x' \in V$ for all $x, x' \in V$ (as the case $x = 0$ is trivial). This says exactly that $v^2 \cdot V \subset V$ for all $v \in V$.

Assume $\text{char}(k) \neq 2$, so for x, $x' \in V$ both $(1+x)^2x'$ and x^2x' lie in V (recall that $1 \in V$). Thus, $xx' \in V$, so V is a k-subspace of K which contains 1 and is closed under multiplication. Hence, V is a subfield of K containing k; we denote it as F. It is obvious that in this case $G = \mathrm{R}_{F/k}(L_F)$ (as G, being pseudo-semisimple, is generated by its root groups due to [CGP, Lemma 3.1.5]). This proves assertion (ii) of the proposition.

Finally, assume $\text{char}(k) = 2$. Again since G is pseudo-semisimple, it is generated by the root groups $U_G^{\pm} = U_V^{\pm}$, but by definition $H_{V,K/k}$ (resp. $\mathrm{P}H_{V,K/k}$) is the subgroup of $\mathrm{R}_{K/k}(L_K)$ generated by these two subgroups because $V^- = v^{-2}V = V$ for any $v \in V - \{0\}$. Hence, $G = H_{V,K/k}$ (resp. $G = \mathrm{P}H_{V,K/k}$). □

The constructions in Definition 3.1.2 have an intrinsic characterization:

Proposition 3.1.8. *Let G be an absolutely pseudo-simple group over a field k. Assume the rank of G_{k_s} is 1. Let K/k be the minimal field of definition for the geometric unipotent radical of G, and let $G' = G_K/\mathscr{R}_{u,K}(G_K)$. Let $i_G : G \to \mathrm{R}_{K/k}(G')$ be the natural homomorphism.*

(i) *The minimal field of definition for the geometric unipotent radical of $i_G(G)$ is K/k, and if G is of minimal type with root system A_1 over k_s then $i_G : G \to i_G(G)$ is an isomorphism. If* $\text{char}(k) \neq 2$ *then $i_G : G \to \mathrm{R}_{K/k}(G')$ is surjective.*

(ii) *Assume* $\text{char}(k) = 2$ *and G is pseudo-split, and fix a split maximal k-torus T in G as well as an isomorphism of G' onto $L = \mathrm{SL}_2$ or PGL_2 carrying T_K onto the diagonal K-torus. There exists a nonzero kK^2-subspace V of K, unique up to $K^\times$-scaling, such that $i_G(G) = H_{V,K/k}$ in the SL_2-case and $i_G(G) = \mathrm{P}H_{V,K/k}$ in the PGL_2-case, and moreover $k\langle V\rangle = K$.*

Proof. By Proposition 2.1.2(ii), the k-subgroup $i_G(G)$ of $\mathrm{R}_{K/k}(L)$ is pseudo-simple and the subgroup $\mathscr{R}_u(i_G(G)_{\overline{k}}) \subset i_G(G)_{\overline{k}}$ has minimal field of definition K/k. The isomorphism property for $G \to i_G(G)$ when G is of minimal type with root system A_1 over k_s is part of Proposition 2.3.4, and to prove that i_G is surjective if $\text{char}(k) \neq 2$ it suffices to work over k_s (so G is pseudo-split).

We may now assume G is pseudo-split. Since $i_G(G)$ contains the split diagonal k-torus, by Proposition 3.1.7 it follows that if $\text{char}(k) \neq 2$ then there exists a subfield F of K containing k such that $i_G(G) \simeq \mathrm{R}_{F/k}(L_F)$ whereas if $\text{char}(k) = 2$ then $i_G(G)$ has the asserted form for a nonzero k-subspace $V \subset K$ that is a $k\langle V^2\rangle$-subspace. The minimal field of definition for the geometric unipotent radical of $i_G(G)$ is K/k, but this field of definition for the geometric unipotent radical of $\mathrm{R}_{F/k}(L_F)$ is F [CGP, Cor. A.5.16] and for both $H_{V,K/k}$ and $\mathrm{P}H_{V,K/k}$ it is $k\langle V\rangle$ (Proposition 3.1.4). We conclude that if $\mathrm{char}(k) \neq 2$ then $F = K$, settling (i), and if $\mathrm{char}(k) = 2$ then $k\langle V\rangle = K$. Thus, if $\mathrm{char}(k) = 2$ then V is a kK^2-subspace of K and the uniqueness of V up to $K^\times$-scaling is [CGP, Prop. 9.1.7] (as reviewed in the discussion preceding Definition 3.1.3). □

Proposition 3.1.9. *Let k be a field with $\mathrm{char}(k) \neq 2$ and let G be an absolutely pseudo-simple k-group with root system A_1 over k_s. Let K/k be the minimal field of definition for $\mathscr{R}_u(G_{\overline{k}}) \subset G_{\overline{k}}$ and define $G' = G_K/\mathscr{R}_{u,K}(G_K)$. The map $i_G : G \to \mathrm{R}_{K/k}(G')$ is an isomorphism, so G is standard and of minimal type.*

This is [CGP, Thm. 6.1.1] but the proof below is simpler.

Proof. We may assume $k = k_s$, so G' is K-isomorphic to SL_2 or PGL_2. By Proposition 3.1.8, i_G is surjective. Thus, by Proposition 2.3.4, $Z := \ker i_G$ is a central k-subgroup of G and if $G' = \mathrm{SL}_2$ then i_G is an isomorphism. Suppose instead that $G' = \mathrm{PGL}_2$ and consider the central extension

$$1 \to Z \to G \to \mathrm{R}_{K/k}(G') \to 1.$$

The central k-subgroup Z contains no nontrivial smooth connected k-subgroup since G is pseudo-reductive and $Z_{\overline{k}} \subset \mathscr{R}_u(G_{\overline{k}})$. We have to prove that Z is trivial.

Let $\widetilde{G}' = \mathrm{SL}_2$. For the étale isogeny $\pi : \widetilde{G}' \to G'$, the Weil restriction $f = \mathrm{R}_{K/k}(\pi)$ is also an étale isogeny. Hence, the f-pullback central extension

$$1 \to Z \to E \to \mathrm{R}_{K/k}(\widetilde{G}') \to 1$$

has middle term E étale over G, so E is smooth. Since $\widetilde{G}'$ is simply connected, the central extension E is uniquely split by [CGP, Prop. 5.1.3, Ex. 5.1.4]. The k-isomorphism $E = Z \times \mathrm{R}_{K/k}(\widetilde{G}')$ forces Z to be smooth. But Z contains no non-trivial smooth connected k-subgroup, so Z^0 is trivial. Hence, $E^0 = \mathrm{R}_{K/k}(\widetilde{G}')$.

The étale map $E \to G$ must carry E^0 onto G, so G is a quotient of $\mathrm{R}_{K/k}(\widetilde{G}')$ modulo a k-subgroup C of $\ker f = \mathrm{R}_{K/k}(\mu)$ where $\mu := \ker(\widetilde{G}' \to G') = Z_{\widetilde{G}'} \simeq \mu_2$ is finite étale of order 2 over K (as $\mathrm{char}(k) \neq 2$). Thus, $\mathrm{R}_{K/k}(\mu)$ is finite étale of order 2 since K/k is purely inseparable (see [CGP, Prop. A.5.13]). Clearly $C \neq 1$, so $C = \mathrm{R}_{K/k}(\mu) = \ker f$ and hence i_G is an isomorphism. □

3.2 Minimal field of definition for $\mathscr{R}_u(G_{\overline{k}})$

3.2.1. Let G be a *pseudo-split* pseudo-reductive group over an arbitrary field k, with K/k the minimal field of definition for the geometric unipotent radical of G, and define $G' := G_K^{\rm red} = G_K/\mathscr{R}_{u,K}(G_K)$. Let $i_G : G \to \mathrm{R}_{K/k}(G')$ be the natural homomorphism and $\pi : G_K \to G'$ the natural projection. Fix a split maximal k-torus T in G, and view $T' := T_K$ as a maximal K-torus in G', so $\Phi' := \Phi(G', T')$ is identified with the set of non-multipliable roots in $\Phi := \Phi(G, T)$ by [CGP, Thm. 2.3.10].

For each $c \in \Phi$, let U_c denote the associated root group of (G, T) as in [CGP, Def. 2.3.13], so by [CGP, Prop. 2.3.11] the Lie algebra $\mathrm{Lie}(U_c)$ is the c-weight space in $\mathrm{Lie}(G)$ when c is non-multipliable and is the span of the weight spaces for c and $2c$ when c is multipliable. By [CGP, Prop. 3.4.1], the k-subgroup $G_c := \langle U_c, U_{-c}\rangle$ is absolutely pseudo-simple with (split) maximal k-torus $T \cap G_c$ of dimension 1 and we have a useful alternative description of G_c in terms of the codimension-1 torus $T_c = (\ker c)^0_{\rm red}$ killed by c: if c is not divisible then $G_c = \mathscr{D}(Z_G(T_c))$ whereas if c is divisible (so k is imperfect of characteristic 2) then $G_c = \mathscr{D}(H_c)$ for the pseudo-reductive identity component H_c of the centralizer in $Z_G(T_c)$ of $(T \cap G_c)[2] \simeq \mu_2$.

In all cases $\pm c$ are the non-divisible roots in $\Phi(G_c, T \cap G_c)$ (see [CGP, Prop. 3.4.1(1)]), so $\Phi(G_c, T \cap G_c)$ is equal to $\{\pm c\}$ when c is not multipliable and it is equal to $\{\pm c, \pm 2c\}$ when c is multipliable. Define $c' = c$ when c is not multipliable and define $c' = 2c$ when c is multipliable, so $c \mapsto c'$ is a bijection from the set of non-multipliable roots of Φ onto the set Φ' and $\pi((U_c)_K)$ is the c'-root group of (G', T') (by [CGP, Cor. 2.1.9]). Thus, $\pi((G_c)_K)$ is equal to the subgroup $G'_{c'}$ of G' generated by the $\pm c'$-root groups.

3.2.2. For each $c \in \Phi$ we want to relate the minimal field of definition K/k for the geometric unipotent radical of G to the minimal field of definition K_c/k for the geometric unipotent radical of the k-subgroup G_c of rank 1. In view of the explicit description of G_c in 3.2.1, it is immediate from [CGP, Prop. A.4.8] (for non-divisible c) and [CGP, Prop. A.8.14(2)] (for divisible c) that $\mathscr{R}_u((G_c)_{\overline{k}}) = \mathscr{R}_u(G_{\overline{k}}) \cap (G_c)_{\overline{k}}$ for all $c \in \Phi$. Hence, for $G' := G_K^{\rm red}$ there is a natural identification of $(G_c)_{\overline{k}}^{\rm red}$ with $(G'_{c'})_{\overline{k}}$ and it also follows that $K_c \subset K$ for all $c \in \Phi$.

For each $c \in \Phi$ let $\mathscr{G}'_c := (G_c)_{K_c}/\mathscr{R}_{u,K_c}((G_c)_{K_c})$ be the K_c-descent of the maximal geometric reductive quotient of G_c, so naturally $(\mathscr{G}'_c)_K = G'_{c'}$ and hence there are natural inclusions:

$$\begin{aligned} \mathrm{R}_{K_c/k}(\mathscr{G}_c') \hookrightarrow \mathrm{R}_{K_c/k}(\mathrm{R}_{K/K_c}((\mathscr{G}_c')_K)) = \mathrm{R}_{K/k}((\mathscr{G}_c')_K) &= \mathrm{R}_{K/k}(G'_{c'}) \\ &\hookrightarrow \mathrm{R}_{K/k}(G'). \end{aligned}$$

Since i_{G_c} is the natural homomorphism $G_c \to \mathrm{R}_{K_c/k}(\mathscr{G}_c')$, so $i_G|_{G_c}$ and i_{G_c} are compatible via the universal property of Weil restriction, we obtain a natural identification of $i_{G_c}(G_c)$ with $i_G(G_c)$.

By Proposition 2.1.2(ii), the minimal field of definition over k for the geometric unipotent radical of $i_{G_c}(G_c)$ is K_c, so the minimal field of definition over k for the geometric unipotent radical of $i_G(G_c)$ is also K_c. This establishes the following useful lemma via Proposition 3.1.8:

Lemma 3.2.3. *If* $\mathrm{char}(k) \neq 2$ *or k is perfect then for every $c \in \Phi$ the k-group $i_G(G_c) = i_{G_c}(G_c)$ is isomorphic to* $\mathrm{R}_{K_c/k}(\mathrm{SL}_2)$ *or* $\mathrm{R}_{K_c/k}(\mathrm{PGL}_2)$. *If k is imperfect with* $\mathrm{char}(k) = 2$ *then for every $c \in \Phi$ there exists a nonzero kK_c^2-subspace $V_c \subset K_c$, unique up to $K_c^\times$-scaling, such that $k\langle V_c\rangle = K_c$ and $i_{G_c}(G_c)$ is isomorphic to $H_{V_c,K_c/k}$ or* $\mathrm{P}H_{V_c,K_c/k}$.

Remark 3.2.4. Assume in addition that G is absolutely pseudo-simple with a reduced root system (and continue to allow k to be arbitrary). If $\mathrm{char}(k) \neq 2$ or k is perfect then define $V_c = K_c$. If k is imperfect with characteristic 2 then define $V_c \subset K_c$ to be a member of the $K_c^\times$-homothety class of nonzero kK_c^2-subspaces of K_c determined by $i_{G_c}(G_c)$ as in Lemma 3.2.3.

The image $i_G(G)$ is perfect and hence is contained in $\mathscr{D}(\mathrm{R}_{K/k}(G'))$. The map i_G thereby induces a homomorphism

$$\xi_G : G \to \mathscr{D}(\mathrm{R}_{K/k}(G'))$$

identifying T with a split maximal k-torus of $\xi_G(G)$. The k-groups $\mathrm{R}_{K/k}(G')$ and $\mathscr{D}(\mathrm{R}_{K/k}(G'))$ have the same root groups with respect to T, and by [CGP, Lemma 3.1.5] these root groups generate $\mathscr{D}(\mathrm{R}_{K/k}(G'))$. By consideration of open cells it follows that ξ_G is surjective if and only if $i_G(U_c)$ exhausts the c-root group of $\mathrm{R}_{K/k}(G')$ for all $c \in \Phi$; i.e., if and only if $V_c = K$ for every c.

Here is another application of 3.2.2:

Proposition 3.2.5. *Assume G is pseudo-semisimple with a split maximal k-torus T. The extension K/k is generated by the subextensions K_c for $c \in \Phi(G,T)$.*

This is part of [CGP, Prop. 6.3.1(1)], but we provide a simpler proof.

Proof. We use the notation $T' = T_K \subset G'$ and identify T with the maximal k-torus $i_G(T)$ of the k-group $i_G(G)$ (note that $T \cap \ker i_G$ is trivial since $\ker i_G$ is unipotent). By Proposition 2.1.2, the k-group $i_G(G)$ is pseudo-semisimple and the minimal field of definition over k for its geometric unipotent radical is K. Since $i_G(G)$ contains T as a split maximal k-torus, by [CGP, Thm. 3.4.6] we may choose a Levi k-subgroup L of $i_G(G)$ containing T, so $\Phi(L,T) = \Phi(G',T') =: \Phi'$.

We naturally identify G' with L_K. For $c \in \Phi$ let $L_{c'}$ be the k-subgroup of L generated by the $\pm c'$-root groups of (L,T), so $L_{c'}$ is a Levi k-subgroup of G_c (as $(G_c)^{\text{red}}_{\overline{k}} = G'_{c'} \otimes_K \overline{k} = \mathscr{G}'_c \otimes_{K_c} \overline{k}$). Hence, the k-group $i_G(G_c) = i_{G_c}(G_c)$ is contained in $\mathrm{R}_{K_c/k}((L_{c'})_{K_c}) \subset \mathrm{R}_{K_c/k}(L_{K_c})$ inside $\mathrm{R}_{K/k}(L_K) = \mathrm{R}_{K/k}(G')$.

Let K' be the subfield of K generated over k by the subextensions K_c for $c \in \Phi$ and define $L' = L_{K'}$, so $i_G(G) \subset \mathrm{R}_{K'/k}(L')$ inside $\mathrm{R}_{K/k}(G')$. Hence,

$$L \subset i_G(G) \subset \mathrm{R}_{K'/k}(L'),$$

so by Proposition 2.1.2(i) and [CGP, Cor. A.5.16] the extension K'/k is a field of definition for the geometric unipotent radical of $i_G(G)$. But K/k is the minimal such extension, so the containment $K' \subset K$ over k is an equality. □

Proposition 3.2.6. *Let $f : H \to \overline{H}$ be a central quotient homomorphism between perfect smooth connected affine groups over a field k. The minimal fields of definition over k for the geometric unipotent radicals of H and $\overline{H}$ coincide.*

Proof. Since $f_{\overline{k}}$ carries $\mathscr{R}(H_{\overline{k}})$ onto $\mathscr{R}(\overline{H}_{\overline{k}})$, the induced map $H^{\text{ss}}_{\overline{k}} \to \overline{H}^{\text{ss}}_{\overline{k}}$ has central kernel. Hence, $f_{\overline{k}}$ induces an isomorphism $H^{\text{ad}}_{\overline{k}} \simeq \overline{H}^{\text{ad}}_{\overline{k}}$ between maximal geometric adjoint semisimple quotients (given as the maximal central quotients of the maximal geometric semisimple quotients). More specifically, if we define

$$R = \ker(H_{\overline{k}} \twoheadrightarrow H^{\text{ad}}_{\overline{k}}),\ \overline{R} = \ker(\overline{H}_{\overline{k}} \twoheadrightarrow \overline{H}^{\text{ad}}_{\overline{k}}),$$

then $R = f_{\overline{k}}^{-1}(\overline{R})$ and $(\ker f)_{\overline{k}} \subset R$ with $R/(\ker f)_{\overline{k}} = \overline{R}$ inside the $\overline{k}$-group $(H/\ker f)_{\overline{k}} = \overline{H}_{\overline{k}}$. Hence, if K is the minimal field of definition over k for $\overline{R} \subset \overline{H}_{\overline{k}}$ then the f_K-preimage of the K-descent of $\overline{R}$ is a K-subgroup of H_K that is a K-descent of R, so K is also a field of definition over k for $R \subset H_{\overline{k}}$. It must be minimal as such because if $F \subset K$ is a subfield over k such that R descends to an F-subgroup $R_0 \subset H_F$ then $(\ker f)_F \subset R_0$ and the F-subgroup $R_0/(\ker f)_F \subset \overline{H}_F$ is an F-descent of $\overline{R}$, forcing $F = K$.

We have established the equality over k of minimal fields of definition for projection onto maximal geometric adjoint semisimple quotients. It therefore remains to show that if G is a perfect smooth connected affine k-group then the minimal field of definition K/k for $\mathscr{R}_u(G_{\overline{k}}) = \ker(G_{\overline{k}} \twoheadrightarrow G_{\overline{k}}^{\mathrm{ss}})$ coincides with the minimal field of definition F/k for the kernel of the natural homomorphism $G_{\overline{k}} \twoheadrightarrow G_{\overline{k}}^{\mathrm{ss}}/Z_{G_{\overline{k}}^{\mathrm{ss}}}$. This equality is [CGP, Prop. 5.3.3], but for the convenience of the reader we now give a more elementary proof.

Clearly $F \subset K$, and to prove that this inclusion of purely inseparable extensions of k is an equality it suffices to check after scalar extension to k_s. By Galois descent, the formation of F and K commutes with such scalar extension. Thus, we may assume $k = k_s$. Let G_F^{ad} denote the quotient of G_F that descends the quotient $G_{\overline{k}}^{\mathrm{ss}}/Z_{G_{\overline{k}}^{\mathrm{ss}}}$ of $G_{\overline{k}}$. Since $F = F_s$, we see via consideration of the simply connected central cover of G_F^{ad} that every connected semisimple central cover of $G_F^{\mathrm{ad}} \otimes_F K$ uniquely descends to a connected semisimple central cover of G_F^{ad}. Hence, we get a central isogeny $G'_F \to G_F^{\mathrm{ad}}$ that descends the natural central isogeny $G_K^{\mathrm{ss}} \to G_F^{\mathrm{ad}} \otimes_F K$ between quotients of G_K, and we have to show that the K-homomorphism $q : G_K \twoheadrightarrow G_K^{\mathrm{ss}} = G'_F \otimes_F K$ descends to an F-homomorphism $G_F \to G'_F$ (i.e., q is "defined over F").

By perfectness, G_F is generated by its maximal F-tori [CGP, Prop. A.2.11]. For any maximal F-torus $T \subset G_F$, the image $q(T_K) \subset G_K^{\mathrm{ss}}$ is a maximal K-torus and so it is the preimage of its image in $G_F^{\mathrm{ad}} \otimes_F K$ (as for maximal tori relative to any central isogeny between connected semisimple groups). Thus, $q(T_K)$ arises from a maximal F-torus $T' \subset G'_F$. The resulting quotient map $T_K \twoheadrightarrow T'_K$ is defined over F since the F-tori T and T' are split (as $F = F_s$), so $q(T(F)) \subset G'_F(F)$. The subgroup $\Sigma \subset G(F)$ generated by the $T(F)$'s is Zariski-dense in G since such T's generate G. Since $q(\Sigma) \subset G'_F(F)$, it follows that q is defined over F. □

Proposition 3.2.7. *Let G be an absolutely pseudo-simple k-group whose root system over k_s is reduced. Then G is standard if and only if the natural map $\xi_G : G \to \mathscr{D}(\mathrm{R}_{K/k}(G_K^{\mathrm{ss}}))$ is surjective. If G_K^{ss} is simply connected and ξ_G is surjective then i_G is an isomorphism.*

This is the main content of [CGP, Thm. 5.3.8]; we now give a simpler proof.

Proof. The necessity of surjectivity of ξ_G is elementary (see [CGP, Rem. 5.3.6]), so now assume ξ_G is surjective. If G is of minimal type then $\ker \xi_G$ is trivial, so standardness is clear in such cases. If instead G_K^{ss} is simply connected then $\mathrm{R}_{K/k}(G_K^{\mathrm{ss}})$ is perfect by [CGP, Cor. A.7.11], so in such cases $\xi_G = i_G$ and (by Proposition 2.3.4) this map is an isomorphism when it is surjective.

It remains to show in the general case that G is standard when ξ_G is surjective. The formation of K/k is unaffected by passing to the central quotient $G/\mathscr{C}_G$ (by Proposition 3.2.6), so the formation of the target of ξ_G is unaffected by passing to $G/\mathscr{C}_G$. Hence, clearly $\xi_{G/\mathscr{C}_G}$ is surjective, so the settled minimal-type case ensures that the central quotient $G/\mathscr{C}_G$ is standard. We may conclude by applying Lemma 3.2.8 below. □

Lemma 3.2.8. *For a pseudo-reductive k-group $\overline{G}$ and central extension*

$$1 \to C \to G \to \overline{G} \to 1 \tag{3.2.8}$$

with a pseudo-reductive k-group G, if $\overline{G}$ is standard then so is G.

Proof. Standardness of G is equivalent to that of $\mathscr{D}(G)$ [CGP, Prop. 5.2.1], so we may replace G and $\overline{G}$ with their derived groups to arrange that each is perfect. As $\overline{G}$ is a standard pseudo-semisimple group, it is isomorphic to $\mathrm{R}_{k'/k}(G')/Z$ for a nonzero finite reduced k-algebra k', a semisimple k'-group G' whose fibers over the factor fields of k' are absolutely simple and simply connected, and a central k-subgroup $Z \subset \mathrm{R}_{k'/k}(G')$.

The pullback of (3.2.8) along $\pi : \mathrm{R}_{k'/k}(G') \twoheadrightarrow \mathrm{R}_{k'/k}(G')/Z = \overline{G}$ is a central extension E of $\mathrm{R}_{k'/k}(G')$ by C. Note that any smooth connected k-subgroup of C is unipotent, due to the finiteness of the center of the semisimple $G_{\overline{k}}^{\mathrm{red}}$, and hence is trivial since G is pseudo-reductive. Thus, the central extension E is (uniquely) split by [CGP, Prop. 5.1.3, Ex. 5.1.4]. The splitting provides a homomorphism $q : \mathrm{R}_{k'/k}(G') \to G$ with central kernel such that $C \times \mathrm{R}_{k'/k}(G') \to G$ is surjective, so the perfectness of $\mathrm{R}_{k'/k}(G')$ and G and the commutativity of C imply that q is surjective. This central quotient presentation of G can be upgraded to a standard presentation; see [CGP, Rem. 1.4.6]. □

3.3 Root field and applications

3.3.1. Let G be an absolutely pseudo-simple k-group with a *reduced* root system over k_s. Let K/k be the minimal field of definition for the geometric unipotent radical of G, and define $G' = G_K^{\mathrm{ss}}$.

For a maximal k-torus T in G, there is a unique T-stable linear complement $\mathfrak{g}(T)$ to the subspace $\mathfrak{g}^T$ of $\mathfrak{g} = \mathrm{Lie}(G)$. Since $\ker i_G$ is central by Proposition 2.3.4, so $\mathrm{Lie}(\ker i_G) \subset \mathrm{Lie}(Z_G(T)) = \mathfrak{g}^T$, $\mathrm{Lie}(i_G)$ defines a k-linear inclusion of $\mathfrak{g}(T)$ into $\mathrm{Lie}(\mathrm{R}_{K/k}(G')) = \mathrm{Lie}(G')$. Thus, it makes sense to consider the set F_T of elements $\lambda \in K$ such that multiplication by λ on $\mathrm{Lie}(G')$ preserves the k-subspace $\mathfrak{g}(T)$. This is an intermediate field between K and k, and the analogous such field for the pair (G_{k_s}, T_{k_s}) is $k_s \otimes_k F_T = (F_T)_s$. All maximal k_s-tori in G_{k_s} are $G(k_s)$-conjugate, so $(F_T)_s$ is independent of T and hence so is F_T. The following definition is therefore intrinsic to G (whose root system over k_s has been assumed to be reduced):

Definition 3.3.2. The *root field* of G is the subextension F of K/k that satisfies

$$F = \{\lambda \in K \mid \lambda \cdot \mathfrak{g}(T) \subset \mathfrak{g}(T) \text{ inside } \mathrm{Lie}(G')\}$$

for some (equivalently, every) maximal k-torus $T \subset G$.

In the pseudo-split case we can describe F in more explicit terms as follows. Let T be a split maximal k-torus and $\Phi := \Phi(G,T) = \Phi(G', T_K)$. For each $c \in \Phi$ the c-root group $U_c \subset G$ satisfies $U_c \cap \ker i_G = 1$ since $Z_G(U_c) = 1$ (as even U_c^T is trivial), so $i_G : U_c \to \mathrm{R}_{K/k}(U_c')$ has trivial kernel, where $U_c' \simeq \mathbf{G}_{\mathrm{a}}$ is the c-root group of G'. The T-action on U_c is compatible with the $\mathrm{R}_{K/k}(T_K)$-action on $\mathrm{R}_{K/k}(U_c')$, so U_c corresponds to a nonzero linear subgroup of the vector group $\mathrm{R}_{K/k}(U_c')$. Let G_c denote the pseudo-split absolutely pseudo-simple k-subgroup of G generated by U_c and U_{-c}, so G_c has split maximal k-torus $c^\vee(\mathrm{GL}_1)$ with weight spaces $\mathrm{Lie}(U_{\pm c})$ for its nontrivial weights.

The restriction $i_G|_{G_c}$ factors through i_{G_c}, and conjugation by any representative $n_c \in N_{G_c}(c^\vee(\mathrm{GL}_1))(k)$ of the reflection $r_c \in W(G_c, c^\vee(\mathrm{GL}_1))$ swaps c and $-c$ and acts K-linearly on $\mathrm{Lie}(G')$. Thus, the set F_c of elements $\lambda \in K$ whose scaling action on $\mathrm{R}_{K/k}(U_c')$ preserves U_c is contained in the minimal field of definition K_c/k for the geometric unipotent radical of G_c (as $\mathrm{Lie}(U_c)$ lies inside a K_c-line contained in the K-line $\mathrm{Lie}(U_c')$) and $F_{-c} = F_c$. In particular, F_c coincides with the root field of G_c. Clearly F_c as a subextension of K/k only depends on c through its orbit under $N_G(T)(k)/Z_G(T)(k) = W(\Phi)$, and

$$F = \bigcap_{c \in \Phi} F_c. \tag{3.3.2}$$

Remark 3.3.3. Let G be an absolutely pseudo-simple k-group with a reduced root system over k_s. Define K/k to be the minimal field of definition for its geometric unipotent radical and define $F \subset K$ to be the root field of G. Two useful properties of F are that $k' \otimes_k F$ is the root field of $G_{k'}$ for any separable extension k'/k and that the root field of any pseudo-reductive central quotient $\overline{G}$ of G coincides with F (as a purely inseparable extension of k). Let us establish these properties.

The behavior with respect to separable extension on k is immediate from the definition of F and the compatibility of the formation of both K/k and the quotient G' of G_K with such extension on k. The invariance under passage to a pseudo-reductive central quotient $\overline{G}$ of G reduces to two other invariance properties: (i) the minimal field of definition over k for the geometric unipotent radical of $\overline{G}$ coincides with K/k, and (ii) for a maximal k-torus T in G and its isogenous image $\overline{T}$ in $\overline{G}$, the natural map $\mathfrak{g} \to \overline{\mathfrak{g}}$ (which is generally neither injective nor surjective) induces an isomorphism $\mathfrak{g}(T) \simeq \overline{\mathfrak{g}}(\overline{T})$ (and so likewise for the induced central quotient map $G' \to \overline{G}'$ over K).

Assertion (i) follows from Proposition 3.2.6. To prove (ii) we may assume $k = k_s$, so T and $\overline{T}$ are split. By centrality and consideration of open cells we see that $\Phi := \Phi(G,T) = \Phi(\overline{G},\overline{T})$ inside $\mathrm{X}(T)_{\mathbf{Q}} = \mathrm{X}(\overline{T})_{\mathbf{Q}}$ and that $G \twoheadrightarrow \overline{G}$ restricts to an isomorphism between c-root groups for all $c \in \Phi$, so (ii) is clear.

The following two examples illustrate the root field in A_1-cases.

Example 3.3.4. Suppose G_{k_s} has root system A_1 and either $\mathrm{char}(k) \neq 2$ or k is perfect, so $G \simeq \mathrm{R}_{K/k}(G')$ (see Proposition 3.1.9). In particular, G is of minimal type and its root field is K. For a subfield $F \subset K$ over k, the natural map $G_F \to \mathrm{R}_{K/F}(G')$ is surjective with a smooth connected unipotent kernel by [CGP, Prop. A.4.8]. Thus, the maximal pseudo-reductive quotient G_F^{pred} of G_F is $\mathrm{R}_{K/F}(G')$. As the latter group is of minimal type, $G_F^{\mathrm{pred}} = G_F^{\mathrm{prmt}}$ (with notation as in (2.3.11)) and the natural map $G \to \mathrm{R}_{F/k}(G_F^{\mathrm{prmt}})$ is an isomorphism.

Example 3.3.5. Suppose G_{k_s} has root system A_1 and k is imperfect with characteristic 2. The central quotient $i_G(G)$ of G has the same root field as G (by Remark 3.3.3) and is of minimal type (as we saw in the discussion preceding Example 2.3.5), so also assume that G is of *minimal type*. Denote the root field of G as F, and define $\mathscr{G} := G_F^{\mathrm{prmt}}$ (notation as in (2.3.11)), so $\mathscr{G}$ is absolutely pseudo-simple with root system A_1 over F_s.

If G is pseudo-split over k then by Proposition 3.1.8 there is a nonzero kK^2-subspace $V \subset K$ such that $k\langle V\rangle = K$ and G is isomorphic to either $H_{V,K/k}$ or $\mathrm{P}H_{V,K/k}$, with V unique up to $K^{\times}$-scaling. Clearly

$$kK^2 \subset F = \{\lambda \in K \mid \lambda V \subset V\} \subset K,$$

so $\mathscr{G}$ is F-isomorphic to either $H_{V,K/F}$ or $\mathrm{P}H_{V,K/F}$ by Proposition 2.3.13.

In general, without a pseudo-split hypothesis on G, the following properties are straightforward to verify via reduction to the pseudo-split case over k_s:

- the root field of $\mathscr{G} = G_F^{\mathrm{prmt}}$ is F,
- K is the minimal field of definition over F for $\mathscr{R}_u(\mathscr{G}_{\overline{F}}) \subset \mathscr{G}_{\overline{F}}$,
- the natural map $G \to \mathscr{D}(\mathrm{R}_{F/k}\mathscr{G})$ is an isomorphism.

Likewise, for any subfield $F' \subset F$ over k, the natural k-homomorphism $G \to \mathscr{D}(\mathrm{R}_{F'/k}(G_{F'}^{\mathrm{prmt}}))$ is an isomorphism.

The isomorphisms built in Examples 3.3.4 and 3.3.5 are special cases of:

Proposition 3.3.6. *Let G be an absolutely pseudo-simple k-group of minimal type with a reduced root system over k_s. Let K/k be the minimal field of definition for the geometric unipotent radical of G, and let F' be a subfield of K containing k and contained in the root field F of G. The natural map $\theta : G \to \mathscr{D}(\mathrm{R}_{F'/k}(G_{F'}^{\mathrm{prmt}}))$ is an isomorphism.*

Proof. To prove that θ is an isomorphism, we may replace k with k_s (as the formation of the root field and $G_{F'}^{\mathrm{prmt}}$ is compatible with separable extension on k) so that k is separably closed. In particular, G is pseudo-split and so admits a Levi k-subgroup L by [CGP, Thm. 3.4.6]. Let $T \subset L$ be a maximal k-torus.

The inclusion $L \hookrightarrow G$ identifies L_K with $G' := G_K^{\mathrm{ss}}$. Since G is of minimal type and its root system is reduced, the homomorphism $i_G : G \to \mathrm{R}_{K/k}(G') = \mathrm{R}_{K/k}(L_K)$ has trivial kernel (Proposition 2.3.4) and so carries G isomorphically onto its image. In particular, $G_{F'}$ is naturally an F'-subgroup of $\mathrm{R}_{K/k}(L_K)_{F'}$. Let $\mathscr{G} \subset \mathrm{R}_{K/F'}(L_K)$ denote the image of $G_{F'}$ under the natural map

$$q : \mathrm{R}_{K/k}(L_K)_{F'} = \mathrm{R}_{F'/k}(\mathrm{R}_{K/F'}(L_K))_{F'} \to \mathrm{R}_{K/F'}(L_K).$$

Since the composite map $G_{F'} \to \mathrm{R}_{K/F'}(L_K)$ is visibly identified with $i_{G_{F'}}$, by Remark 2.3.12 the quotient map $G_{F'} \to \mathscr{G}$ identifies $\mathscr{G}$ with $G_{F'}^{\mathrm{prmt}}$.

By Proposition 2.1.2, $L_{F'}$ is a Levi F'-subgroup of $\mathscr{G}$ and the inclusion of $\mathscr{G}$ into $\mathrm{R}_{K/F'}(L_K)$ is $i_{\mathscr{G}}$; in particular, the reduced and irreducible root system $\Phi = \Phi(G,T)$ coincides with $\Phi(L,T)$ and $\Phi(\mathscr{G}, T_{F'})$. The inclusion of G into $\mathrm{R}_{K/k}(L_K) = \mathrm{R}_{F'/k}(\mathrm{R}_{K/F'}(L_K))$ clearly factors through $\mathrm{R}_{F'/k}(\mathscr{G})$, so we get an inclusion

$$G \hookrightarrow \mathscr{D}(\mathrm{R}_{F'/k}(\mathscr{G}))$$

that recovers θ. Hence, $\ker\theta = 1$ and we have to show that G exhausts the target. The root system $\Phi(\mathscr{D}(\mathrm{R}_{F'/k}(\mathscr{G})), T)$ coincides with $\Phi(\mathrm{R}_{F'/k}(\mathscr{G}), T) = \Phi(\mathscr{G}, T_{F'}) = \Phi$. Thus, since any pseudo-split pseudo-semisimple k-group is generated by its root groups relative to a split maximal k-torus, it suffices to show that for each $c \in \Phi$ the c-root groups of $\mathscr{D}(\mathrm{R}_{F'/k}(\mathscr{G}))$ and G coincide.

For $c \in \Phi$, let $G_c \subset G$ be the k-subgroup generated by the $\pm c$-root groups; define $L_c \subset L$ and $\mathscr{G}_c \subset \mathscr{G}$ similarly. The pseudo-semisimple $\mathscr{D}(\mathrm{R}_{F'/k}(\mathscr{G}_c))$ of rank 1 is generated by the $\pm c$-root groups of $\mathscr{D}(\mathrm{R}_{F'/k}(\mathscr{G}))$, so it suffices to show that the evident inclusion $j_c : G_c \subset \mathscr{D}(\mathrm{R}_{F'/k}(\mathscr{G}_c))$ is an equality for all c. The case of perfect k is trivial, so we now assume k is imperfect with characteristic p. By (3.3.2), the root field F of G is contained in the root field $F_c \subset K_c$ of G_c for all c, so $F' \subset F_c$ for all c. The compatibility of i_G and i_{G_c} implies that the inclusion i_G of G into $\mathrm{R}_{K/k}(L_K)$ carries G_c into $\mathrm{R}_{K_c/k}((L_c)_{K_c})$, so $\mathscr{G}_c = (G_c)_{F'}^{\mathrm{prmt}}$ by Remark 2.3.12 applied to $(G_c)_{F'}$. Thus, our problem is now intrinsic to the separate groups G_c (and the subextension F'/k of the root field F_c/k) for each c. That is, we may assume Φ is A_1. This in turn is handled by Examples 3.3.4 (for $p \neq 2$) and 3.3.5 (for $p = 2$). □

Definition 3.3.7. Let G be an absolutely pseudo-simple k-group with a reduced root system over k_s and minimal field of definition K/k for the geometric unipotent radical. Let T be a maximal k-torus of G, and define $\Phi = \Phi(G_{k_s}, T_{k_s})$.

The *long root field* of G is the unique subextension $F_> \subset K$ over k such that $k_s \otimes_k F_>$ is the common root field of $(G_{k_s})_a$ for all roots $a \in \Phi$ when all roots are of equal length, and for all long roots $a \in \Phi$ otherwise. (As $N_G(T)(k_s)$ acts transitively on the set of roots of a given length, $F_>$ exists.) Define the *short root field* $F_< \subset K$ over k similarly (so $F_< = F_>$ if Φ is simply laced).

It is clear that the subextensions $F_>$ and $F_<$ of K/k are independent of the choice of T. These root fields arise in the description of automorphisms of G that act trivially on a chosen Cartan subgroup; see Proposition 8.5.4.

By (3.3.2) the intersection of the root fields of the $(G_{k_s})_a$'s is F_s, where F/k is the root field of G. In general, if G is an absolutely pseudo-simple k-group then we claim that $F_> \subset F_<$, so $F = F_>$; this is a consequence of the following theorem that gives restrictions on field-theoretic and linear-algebraic data associated to pseudo-split absolutely pseudo-simple groups with a reduced root system.

Theorem 3.3.8. *Let G be a pseudo-split absolutely pseudo-simple group with rank $n \geqslant 2$ over an imperfect field k of characteristic p, and let $T \subset G$ be a split maximal k-torus. Assume the irreducible root system $\Phi := \Phi(G,T)$ is reduced.*

Let K/k be the minimal field of definition for the geometric (unipotent) radical of G, and for each $c \in \Phi$ let K_c/k be the minimal field of definition for the geometric (unipotent) radical of G_c. Let F be the root field of G, and let $F_>$ and $F_<$ respectively denote the long and short root fields.

(i) *Assume the Dynkin diagram of Φ does not have an edge of multiplicity p (as is automatic when $p \neq 2, 3$). Then $F = F_c = K_c = K$ for all $c \in \Phi$.*
(ii) *Assume the Dynkin diagram of Φ has an edge of multiplicity p (so either $p = 3$ with Φ of type G_2 or $p = 2$ with Φ of type B_n, C_n, or F_4). Then*

$K_c = K$ for all short roots c and there is a subfield $K_> \subset K$ containing kK^p such that $K_c = K_>$ for all long roots c. If $p = 3$ then $F_< = K$ and $F_> = K_>$ (so $F = F_>$).

(iii) *Assume $p = 2$, and let V_c be as in Lemma 3.2.3 for each $c \in \Phi$. If there is no edge of multiplicity 2 in the Dynkin diagram then $V_c = K$ for all c. Assuming that there is an edge of multiplicity 2 in the diagram, V_c is a nonzero $K_>$-subspace of K for c short and V_c is a nonzero kK^2-subspace of $K_>$ for c long, so*

$$kK^2 \subset F = F_> \subset K_> \subset F_<. \tag{3.3.8}$$

Moreover, if Φ is of type F_4 or B_n with $n \geqslant 3$ then V_c is a $K_>$-line for all long c (so $F_> = K_>$), and if Φ is of type F_4 or C_n with $n \geqslant 3$ then $V_c = K$ for all short c (so $F_< = K$).

This result includes [CGP, Prop. 6.3.2] as a special case (and has a simpler proof). It is used in the proof of Theorem 3.4.1, which classifies (in the spirit of the Isomorphism Theorem for split connected semisimple groups) pseudo-split absolutely pseudo-simple groups G of minimal type with a reduced root system of rank $\geqslant 2$. (The rank-1 case of that classification is handled in Propositions 3.1.8 and 3.1.9, as Proposition 2.3.4 ensures that $\ker i_G = 1$ for such G.)

Proof. By [CGP, Thm. 3.4.6], G has a Levi k-subgroup L containing T. As Φ is reduced, $\Phi(L,T) = \Phi$. The natural projection $G_K \to G' = G_K^{\mathrm{red}}$ maps L_K isomorphically onto G' and we identify the latter with the former. Via this identification, the homomorphism i_G is a homomorphism from G to $\mathrm{R}_{K/k}(L_K)$.

By Proposition 2.1.2(ii), the k-group $i_G(G)$ is absolutely pseudo-simple with K/k as the minimal field of definition for its geometric unipotent radical, and it is of minimal type (as we noted immediately above Example 2.3.5). Moreover, it has been observed in 3.2.2 that for $c \in \Phi$, $i_{G_c}(G_c)$ has a natural identification with $i_G(G_c)$ and the minimal field of definition for the geometric unipotent radical of $i_G(G_c)$ is K_c ($\subset K$). The root fields F_c of G_c and F of G are likewise unaffected by passing to the central quotient $i_G(G)$ (see Remark 3.3.3), so we can replace G with $i_G(G)$ to arrange that G is a k-subgroup of $\mathrm{R}_{K/k}(L_K)$ containing L (and the inclusion $G \hookrightarrow \mathrm{R}_{K/k}(L_K)$ is i_G).

To reduce our problem to the case where L is simply connected, let $\pi : \widetilde{L} \to L$ be the simply connected central cover and let $\widetilde{T} \subset \widetilde{L}$ be the unique k-torus over T. Via the identification of $\Phi(\widetilde{L}, \widetilde{T})$ with Φ, π restricts to an isomorphism between corresponding root groups. The same therefore holds for

$R_{K/k}(\pi_K)$, so in this way we may identify each root group of G with a k-subgroup of $R_{K/k}(\widetilde{L}_K)$ containing the corresponding root group of $\widetilde{L}$. Let $\widetilde{G}$ be the k-subgroup of $R_{K/k}(\widetilde{L}_K)$ generated by these root groups of G, so $\widetilde{G}$ contains $\widetilde{L}$ and admits G as a central quotient.

The k-group $\widetilde{G}$ lies between $R_{K/k}(\widetilde{L}_K)$ and $\widetilde{L}$, so it is pseudo-reductive by Proposition 2.1.2(ii), and it is generated by the root groups U_c of G. These U_c's are vector groups on which $\widetilde{T}$ acts linearly with only nontrivial weights, so each lies inside $\mathscr{D}(\widetilde{G})$. Hence, $\widetilde{G}$ is perfect. It follows from Proposition 3.2.6 that the minimal field of definition over k for the geometric unipotent radical of $\widetilde{G}$ coincides with the analogous such field K/k for its central quotient G. For each $c \in \Phi$, the centrality of $\widetilde{G}_c \twoheadrightarrow G_c$ implies that replacing G with $\widetilde{G}$ has no effect on the extension K_c/k or on the root field F_c/k, nor on the kK_c^2-subspace $V_c \subset K_c$ if $p = 2$. Thus, we may replace G with $\widetilde{G}$ to reduce to the case that L is simply connected.

Fix a basis Δ of Φ and a pinning of (L, T). For each $c \in \Delta$ the k-subgroup L_c of L generated by its $\pm c$-root groups is thereby identified with SL_2 carrying $T \cap L_c$ to the diagonal k-torus. Choose $c \in \Delta$. By Proposition 3.1.7, Proposition 3.1.8, and 3.2.2, if $p \neq 2$ then G_c is the subgroup $R_{K_c/k}((L_c)_{K_c})$ of $R_{K_c/k}(R_{K/K_c}((L_c)_K)) = R_{K/k}((L_c)_K) \subset R_{K/k}(L_K)$. By applying the same results if $p = 2$, in such cases there exists a kK_c^2-subspace V_c of K_c containing 1 such that $k[V_c] = K_c$ and (via the identification of L_c with SL_2 and the corresponding identification of $R_{K_c/k}((L_c)_{K_c})$ with the subgroup $R_{K_c/k}(SL_2)$ of $R_{K/k}(SL_2)$) we have $G_c = H_{V_c,K_c/k}$ inside $R_{K_c/k}(SL_2)$.

To make the subsequent arguments more uniform, we adopt the convention that if Φ is simply laced then all roots are considered to be both long and short. The action of $W(\Phi)$ is transitive on the set of roots of a given length, and the natural map $N_L(T)(k) \to W(\Phi)$ is surjective. Thus, the subfields $K_c \subset K$ for short roots c coincide with a common subfield $K_<$, and likewise the subfields $K_c \subset K$ for long roots c coincide with a common subfield $K_>$. The fields $K_>$ and $K_<$ generate K over k by Proposition 3.2.5, so in the simply laced case $K_> = K_< = K$. Likewise, if $p = 2$ then the k-subspaces $V_c \subset K_c$ for all long (resp. short) roots $c \in \Delta$ are equal to each other. For $p \neq 2$ we define $V_c = K_c$ for all $c \in \Delta$.

For any p we have $k[V_c] = K_c$ for all c and the k-group G_c is generated by the k-subgroups $U_{V_c}^{\pm}$ of the $\pm c$-root groups $R_{K/k}(\mathbf{G}_a)$ of $R_{K/k}(L_K)$ corresponding to the k-subspace $V_c \subset K$. For $c \in \Delta$, let $(V_c)^*_{K_c/k}$ be the Zariski closure in $R_{K_c/k}(GL_1)$ of the subgroup of $K_c^{\times}$ generated by $v \in V_c - \{0\}$. (If $p \neq 2$ then clearly $(V_c)^*_{K_c/k} = R_{K_c/k}(GL_1)$.) By Proposition 3.1.4 when $p = 2$, and directly for $p \neq 2$, the Cartan subgroup $Z_{G_c}(T \cap G_c)$ of G_c is the image of

$(V_c)^*_{K_c/k}$ under the map

$$\mathrm{R}_{K_c/k}(c^\vee_{K_c}) : \mathrm{R}_{K_c/k}(\mathrm{GL}_1) \to \mathrm{R}_{K_c/k}(T_{K_c}) \hookrightarrow \mathrm{R}_{K/k}(T_K).$$

For any $a, b \in \Delta$, the action of

$$Z_{G_a}(T \cap G_a)(k) = \mathrm{R}_{K_a/k}(a^\vee_{K_a})((V_a)^*_{K_a/k})(k) \subset T(K)$$

preserves the root groups $U_{\pm b}$ of (G,T), so it preserves the k-subspace V_b of K. In particular, for any nonzero $v \in V_a$, the $a^\vee_K(v)$-action preserves V_b, which is to say that scaling by $v^{\langle b, a^\vee \rangle} \in K^\times$ on K preserves V_b, so such powers lie in K_b. Similarly, for any nonzero $v \in V_b$, the $b^\vee_K(v)$-action preserves V_a, so scaling by $v^{\langle a, b^\vee \rangle} \in K^\times$ on K preserves V_a.

Let a and b be simple roots such that $\langle b, a^\vee \rangle = -1$ (so a and b correspond to adjacent vertices in the Dynkin diagram and either they are of same length or a is long and b is short). As $k[V_a] = K_a$, we see that V_b is a nonzero K_a-subspace of K_b inside K, so $K_a \subset K_b$. If Φ contains roots of unequal lengths, by choosing a to be long and b to be short we conclude that

$$K_> = K_a \subset F_< \subset K_b = K_<.$$

But as $K_>$ and $K_<$ generate K over k in all cases (even when Φ is simply laced), we infer that $K_< = K$ in general. Thus, the final assertion in (ii) is settled by the observation that if $p \neq 2$ then $F_< = K_<$ and $F_> = K_>$ by Proposition 3.1.9.

Suppose we can choose a and b of equal length; i.e., Φ is simply laced, or Φ is of type F_4 or B_n with $n \geqslant 3$ (so there exist non-orthogonal long simple roots), or Φ is of type F_4 or C_n with $n \geqslant 3$ (so there exist non-orthogonal short simple roots). Since $K_a = K_b$ and V_b is a nonzero K_a-subspace of K_b, we see that V_b is equal to K_b, so $F_b = K_b$. This settles the simply laced case (as part of (i)) due to (3.3.2) since $K_c = K_< = K$ for all c in such cases.

It remains to consider cases where Φ has roots of different lengths. Now we may choose a to be long and b to be short, so $\langle a, b^\vee \rangle = -m$ with $m \in \{2, 3\}$. Let $\mathcal{K}$ be the subfield of K generated over k by the mth powers of nonzero elements of V_b; note that $K/\mathcal{K}$ inherits pure inseparability from K/k. Since $K = K_< = K_b = k[V_b]$, if $m \neq p$ then the purely inseparable extension $K/\mathcal{K}$ is also separable and hence $\mathcal{K} = K$ in such cases. If instead $m = p$ then $\mathcal{K} = kK^p$.

For any nonzero $v \in V_b$ the scaling by $v^{-m} \in K^\times$ on K preserves V_a, so V_a is a $\mathcal{K}$-subspace of K. Hence, if $m \neq p$ then V_a is a nonzero K-subspace of K and so $V_a = K$. This implies that $K_> = K$ and $F_> = K$ if $m \neq p$ (so $F = K$

since $F_< \supset K_> = K$). We saw earlier that V_b is a nonzero $K_>$-subspace of K, so likewise $V_b = K$ if $m \neq p$. This settles (i) if Φ is not simply laced. Let us assume now that $m = p$, so $\mathcal{K} = kK^p$ and hence V_a is a kK^p-subspace of K_a. Thus, $kK^p \subset F_> \subset K_>$. We have proved all the assertions of the theorem. □

Corollary 3.3.9. *Let k be an imperfect field with characteristic $p \in \{2,3\}$. For any pair $(\mathscr{G}, K/k)$ as in Definition 2.2.3, the k-group $G = \mathrm{R}_{K/k}(\mathscr{G})$ determines $(\mathscr{G}, K/k)$ uniquely up to unique isomorphism.*

Proof. By the uniqueness of the desired isomorphism, it suffices to treat the case $k = k_s$. Writing $K = \prod K_i$ for fields K_i and letting $\mathscr{G}_i$ denote the K_i-fiber of $\mathscr{G}$, the k-subgroups $\mathrm{R}_{K_i/k}(\mathscr{G}_i)$ are precisely the pseudo-simple k-subgroups of G. Thus, it suffices to treat the case where K is a field (so K has no nontrivial k-automorphism and $\mathscr{G}$ is a basic exotic K-group). In such cases, since $\mathrm{R}_{K/k}(\mathscr{G})$ is perfect we see via Example 2.3.3 and the criterion in Proposition 2.3.13 that the natural map $G_K \to \mathscr{G}$ is the maximal pseudo-reductive quotient of minimal type. Hence, we just need to reconstruct K/k from G; we claim it is the root field of G.

Let $\mathscr{T} \subset \mathscr{G}$ be a maximal K-torus, and let T be the maximal k-torus of $\mathrm{R}_{K/k}(\mathscr{T})$, so T is a maximal k-torus of G. By parts (ii) and (iii) of Theorem 3.3.8, the root field of G coincides with that of G_c for long roots c in $\Phi(G,T) = \Phi(\mathscr{G}, \mathscr{T})$. For such c, the construction of basic exotic K-groups gives $\mathscr{G}_c \simeq \mathrm{SL}_2$ as K-groups (with $c_K^\vee(\mathrm{GL}_1)$ corresponding to the diagonal K-torus of SL_2), so $G_c = \mathscr{D}(\mathrm{R}_{K/k}(\mathscr{G}_c)) = \mathrm{R}_{K/k}(\mathrm{SL}_2)$. Hence, the root field of G_c is K. □

Proposition 3.3.10. *Let G be a pseudo-split pseudo-semisimple group of minimal type over a field k, and let T be a split maximal k-torus in G. Assume $\Phi := \Phi(G,T)$ is reduced. The k-isomorphism class of G is determined by $G_{\overline{k}}^{\mathrm{ss}}$ and the k-groups $\{G_c\}_{c \in \Phi}$ up to isomorphism.*

Note that the isomorphism class of G_c only depends on the orbit of c under $W(\Phi) = N_G(T)(k)/Z_G(T)(k)$, and $G_{\overline{k}}^{\mathrm{ss}}$ determines Φ.

Proof. Let K/k be the minimal field of definition for the geometric unipotent radical of G, so $G' := G_K/\mathscr{R}_{u,K}(G_K)$ is semisimple and split (with T_K a split maximal K-torus). By Proposition 2.3.4, the natural map $i_G : G \to \mathrm{R}_{K/k}(G')$ has trivial kernel. Let $L \subset G$ be a Levi k-subgroup containing T (as exists by [CGP, Thm. 3.4.6] since T is split), so $L_K \simeq G'$. By the Isomorphism Theorem for split connected semisimple groups, L is unique up to k-isomorphism since $L_{\overline{k}} \simeq G_{\overline{k}}^{\mathrm{ss}}$.

Let K_c/k be the minimal field of definition for the geometric unipotent radical of G_c, so K is generated over k by its subfields K_c/k (Proposition 3.2.5). Consider the k-subgroup $G_c \subset \mathrm{R}_{K/k}((L_c)_K)$ that contains L_c, so L_c is a Levi k-subgroup of G_c. Hence, $G_c \subset \mathrm{R}_{K_c/k}((L_c)_{K_c})$ inside $\mathrm{R}_{K/k}((L_c)_K)$. If $\mathrm{char}(k) \neq 2$ or k is perfect then $G_c = \mathrm{R}_{K_c/k}((L_c)_{K_c})$ for all c (see Proposition 3.1.9), so in such cases G is determined inside $\mathrm{R}_{K/k}(L_K)$ since it is generated by the k-subgroups G_c.

Now we may assume k is imperfect with characteristic 2, so G_c is determined up to k-isomorphism by L_c $(= \mathrm{SL}_2$ or $\mathrm{PGL}_2)$ and the $K_c^\times$-homothety class of a nonzero kK_c^2-subspace $V_c \subset K_c$ satisfying $k\langle V_c\rangle = K_c$: if $L_c = \mathrm{SL}_2$ (carrying $T \cap L_c$ to the diagonal k-torus D) then $G_c \simeq H_{V_c, K_c/k}$ and if $L_c = \mathrm{PGL}_2$ (carrying $T \cap L_c$ to the diagonal k-torus $\overline{D}$) then $G_c \simeq \mathrm{P}H_{V_c, K_c/k}$.

Let $\mathscr{G}$ be another pseudo-split pseudo-semisimple k-group of minimal type with a reduced root system such that $\mathscr{G}_{\overline{k}}^{\mathrm{ss}} \simeq G_{\overline{k}}^{\mathrm{ss}}$, so $\mathscr{G}$ has root system Φ. Upon choosing an isomorphism between the root systems, we may compatibly identify L as a Levi k-subgroup of $\mathscr{G}$, so $\mathscr{G}$ is a k-subgroup of $\mathrm{R}_{K/k}(L_K)$ containing L (as $\mathscr{G}$ is of minimal type and Φ is reduced). Assuming $\mathscr{G}_c \simeq G_c$ for all $c \in \Phi$, we will build an automorphism of $\mathrm{R}_{K/k}(L_K)$ that carries G onto $\mathscr{G}$.

Let Δ be a basis of Φ, so G is generated by $\{G_c\}_{c \in \Delta}$ (as $\{L_c\}_{c\in\Delta}$ generates L, and $N_L(T)(k)/T(k) = W(\Phi)$). Upon identifying the pair $(L_c, T \cap L_c)$ with (SL_2, D) or $(\mathrm{PGL}_2, \overline{D})$, $\mathscr{G}_c$ equals $H_{\mathcal{V}_c, K_c/k}$ or $\mathrm{P}H_{\mathcal{V}_c, K_c/k}$ respectively, where $\mathcal{V}_c$ is a nonzero kK_c^2-subspace of K_c such that $k\langle \mathcal{V}_c\rangle = K_c$.

The k-isomorphism $G_c \simeq \mathscr{G}_c$ implies that $\mathcal{V}_c = \lambda_c \cdot V_c$ for some $\lambda_c \in K_c^\times$. Since Δ is a basis of $\mathrm{X}(T/Z_L)$, there exists a unique $\overline{t} \in (T/Z_L)(K)$ such that $c(\overline{t}) = \lambda_c$ for all $c \in \Delta$. The action on $\mathrm{R}_{K/k}(L_K)$ by $\overline{t} \in \mathrm{R}_{K/k}((T/Z_L)_K)(k)$ carries G to a k-subgroup that contains $\mathscr{G}_c \supset L_c$ for all $c \in \Delta$ and hence contains L. As $\mathscr{G}$ is generated by $\{\mathscr{G}_c\}_{c\in\Delta}$, it is clear that $\overline{t}$ carries G onto $\mathscr{G}$. □

3.4 Application to classification results

Inspired by Theorem 3.3.8 and Proposition 3.3.10, we now establish (and use) a classification of the isomorphism classes of *pseudo-split* pseudo-simple groups G over an imperfect field k of characteristic p subject to the hypothesis that G is of *minimal type*. The associated irreducible root datum, which is sufficient to classify isomorphism classes in the semisimple case, needs to be supplemented with additional field-theoretic and (if $p = 2$) linear-algebraic data.

The possibilities for such G of minimal type with a *non-reduced* root system are fully described in terms of field-theoretic and linear-algebraic data in [CGP, Thm. 9.8.6, Prop. 9.8.9], so we shall focus on pseudo-split G of minimal type with a *reduced* (and irreducible) root system Φ. The case of type A_1 is settled by Propositions 3.1.8 and 3.1.9, so we assume Φ has rank $n \geqslant 2$.

Let L be the unique split connected absolutely simple k-group with the same root datum as $G^{\mathrm{ss}}_{\overline{k}}$, and let K/k be the minimal field of definition for the geometric unipotent radical of G (so K/k is a purely inseparable finite extension); note that the connected semisimple group $G' := G^{\mathrm{ss}}_K$ is simple and split. By [CGP, Thm. 3.4.6], L occurs as a Levi k-subgroup of G. Clearly $L_K \to G'$ is an isomorphism, so

$$L \subset G \subset \mathrm{R}_{K/k}(L_K)$$

(as $\ker i_G$ is trivial, by Proposition 2.3.4). Conversely, consider a perfect smooth connected k-subgroup $H \subset \mathrm{R}_{K/k}(L_K)$ containing L. By Proposition 2.1.2(i), H is pseudo-semisimple with L as a Levi k-subgroup and the minimal field of definition $\mathcal{K}/k$ for its geometric unipotent radical is a subextension of K/k. The given inclusion of H into $\mathrm{R}_{K/k}(L_K)$ is the composition of i_H and the inclusion $\mathrm{R}_{\mathcal{K}/k}(L_{\mathcal{K}}) \hookrightarrow \mathrm{R}_{K/k}(L_K)$. In particular, $\ker(i_H)$ is trivial, so H is of minimal type.

Upon specifying the root datum for $G^{\mathrm{ss}}_{\overline{k}}$ and a purely inseparable finite extension K/k, and letting L be the unique split connected absolutely simple k-group with the same root datum as $G^{\mathrm{ss}}_{\overline{k}}$, our task amounts to describing the k-isomorphism classes of perfect smooth connected k-subgroups G of $\mathrm{R}_{K/k}(L_K)$ that contain L and are "large enough" that the minimal field of definition for $\mathscr{R}_u(G_{\overline{k}}) \subset G_{\overline{k}}$ is K/k (rather than a proper subextension).

Theorem 3.4.1. *Let k be an imperfect field of characteristic p, and let K/k be a purely inseparable finite extension. Let L be an absolutely simple and split connected semisimple k-group whose root system Φ is of rank $n \geqslant 2$. The isomorphism classes of pseudo-split absolutely pseudo-simple k-groups G of minimal type with a reduced root system such that K/k is the minimal field of definition for the geometric unipotent radical and $G^{\mathrm{ss}}_K \simeq L_K$ are given by:*

(i) *If the Dynkin diagram of Φ does not have an edge of multiplicity p (as is automatic if $p \neq 2, 3$) then $\mathscr{D}(\mathrm{R}_{K/k}(L_K))$ is the unique such G. In particular, such a G is standard.*

(ii) *Assume $p = 3$ and Φ is of type G_2. Let $K_>$ be a subfield of K containing kK^3. Up to isomorphism there is exactly one G such that $K_>/k$ is the minimal field of definition for the geometric unipotent radical of G_c for long roots $c \in \Phi$. (The standard case is $K_> = K$, and the basic exotic case is $K_> = k$ with $K \neq k$.)*

(iii) *Assume* $p = 2$ *and* Φ *is of type* B_n, C_n, *or* F_4. *Let* $K_>$ *be a subfield of* K *containing* kK^2. *Let* V *be a nonzero* $K_>$*-subspace of* K *such that* $k\langle V\rangle = K$, *and let* $V_>$ *be a nonzero* kK^2*-subspace of* $K_>$ *such that* $k\langle V_>\rangle = K_>$. *Assume* $V = K$ *if* Φ *is of type* F_4 *or* C_n *with* $n \geqslant 3$, *and assume* $V_> = K_>$ *if* Φ *is of type* F_4 *or* B_n *with* $n \geqslant 3$.
Up to isomorphism there is exactly one G *such that: if* $c \in \Phi$ *is long then* $K_>/k$ *is the minimal field of definition over* k *for the geometric unipotent radical of* G_c *and the* $K_>^\times$*-homothety class of* $V_>$ *classifies the isomorphism class of* G_c *(via Proposition* 3.1.8*), and if* $c \in \Phi$ *is short then the* $K^\times$*-homothety class of* V *classifies the isomorphism class of* G_c.

Proof. Let T be a split maximal k-torus of L. We may and do identify any such G with a k-subgroup of $\mathrm{R}_{K/k}(L_K)$ containing L. First consider the case where the diagram does not have an edge of multiplicity p, so by Theorem 3.3.8(i) we have $K_c = K$ for all $c \in \Phi$ and the root field F is also equal to K. Thus, for each $c \in \Phi$ the c-root group U_c of (G,T) as a k-subgroup of the c-root group $\mathrm{R}_{K/k}(U_c')$ of $\mathrm{R}_{K/k}(L_K)$ corresponds to a nonzero k-subspace of K that is a K-subspace. Hence, the inclusion $G \subset \mathscr{D}(\mathrm{R}_{K/k}(L_K))$ between pseudo-semisimple groups is an equality on root groups and thus is an equality.

Now assume that the diagram has an edge of multiplicity p (so $p \in \{2,3\}$). Since $G_{\overline{K}}^{\mathrm{ss}} \simeq L_{\overline{K}}$, Proposition 3.3.10 shows that G is uniquely determined up to isomorphism by the isomorphism classes of the k-subgroups G_c for representatives c from the $W(\Phi)$-orbits on Φ.

Suppose that either $p = 3$ (so Φ is of type G_2) or that $p = 2$ with Φ of type F_4. Thus, L is simply connected, so $L_c = \mathrm{SL}_2$ for all c. In particular, $\mathrm{R}_{K_c/k}((L_c)_{K_c})$ is perfect for all c. By parts (ii) and (iii) of Theorem 3.3.8, we have $F_c = K_c$ for all c, so the inclusion $G_c \subset \mathrm{R}_{K_c/k}((L_c)_{K_c})$ between pseudo-semisimple groups (arising from the identification of L_c as a Levi k-subgroup of $G_c \subset \mathrm{R}_{K/k}((L_c)_K)$) is forced to be an equality on root groups and thus an equality. This shows that G is determined up to k-isomorphism by the subextension $K_>$ of K containing kK^p. If $K_> \neq K$ then applying $\mathrm{R}_{K_>/k}$ to a pseudo-split basic exotic $K_>$-group with root system Φ ensures existence for $K_>/k$, and $\mathrm{R}_{K/k}(L_K)$ settles existence for $K_> = K$.

It remains to address Φ of type B_n or C_n ($n \geqslant 2$), so $p = 2$. Since L has been specified and $(G_c)_{\overline{k}}^{\mathrm{ss}} \simeq (L_c)_{\overline{k}}$, the k-isomorphism class of G_c is determined by the $K_c^\times$-homothety class of a nonzero kK_c^2-subspace V_c of $K_c = k\langle V_c\rangle$ by Proposition 3.1.7. By Theorem 3.3.8(iii) we have $K_c = K$ for short c, so the V_c's for short c lie in a common $K^\times$-homothety class of nonzero $K_>$-subspaces V of K satisfying $k\langle V\rangle = K$. The V_c's for long c lie in a common $K_>^\times$-homothety class of nonzero kK^2-subspaces $V_>$ of $K_>$ that satisfy $k\langle V_>\rangle = K_>$. If $n \geqslant 3$ then Theorem 3.3.8(iii) ensures that $V_> = K_>$ for type B and $V = K$ for type C. Hence, every possibility for G gives rise to a triple $(K_>/k, V, V_>)$ that determines G up to k-isomorphism.

To show that every triple $(K_>/k, V, V_>)$ actually arises from some G, we may assume that both V and $V_>$ contain 1 by moving each of V and $V_>$ within their homothety class. In particular, $k[V_>] = K_>$. It is sufficient to settle existence for the simply connected central cover $\widetilde{L}$ of L (equipped with its split unique maximal k-torus $\widetilde{T}$ over T). Indeed, if $\widetilde{G} \subset \mathrm{R}_{K/k}(\widetilde{L}_K)$ is such a solution containing $\widetilde{L}$ then its image G in $\mathrm{R}_{K/k}(L_K)$ is a central quotient that contains L and has the desired root-space invariants because $\widetilde{L} \to L$ induces an isomorphism between root groups for $\widetilde{T}$ and T (and this G has the associated field invariants K/k and $K_>/k$ by Proposition 3.2.6).

Taking L to be simply connected, fix a basis Δ of $\Phi = \Phi(L,T)$ and a pinning of (L,T) to identify L_c with SL_2 carrying $T \cap L_c$ to the diagonal k-torus and the c-root group to the upper triangular unipotent subgroup for every $c \in \Delta$. Define the k-groups $H_c \subset \mathrm{R}_{K/k}((L_c)_K)$ for $c \in \Delta$ to be $H_{V,K/k}$ for short c and $H_{V_>,K_>/k}$ for long c, so $L_c \subset H_c$ since $1 \in V$ and $1 \in V_>$. A basis of $\mathrm{X}_*(T)$ is given by $\Delta^\vee$ since L is simply connected, so

$$C := \mathrm{R}_{K/k}(T_K) = \prod_{c \in \Delta} \mathrm{R}_{K/k}(c_K^\vee(\mathrm{GL}_1)).$$

By hypothesis, $V_> = K_>$ for type B with rank $\geqslant 3$ and $V = K$ for type C with rank $\geqslant 3$. Thus, since $V_>$ is a kK^2-subspace of $K_>$ and V is a $K_>$-subspace of K, it follows that H_c is normalized by C for all $c \in \Delta$. For $c \in \Delta$, the intersection $C_c := C \cap H_c$ coincides with the pseudo-reductive Cartan k-subgroup $Z_{H_c}(c^\vee(\mathrm{GL}_1))$ of H_c, so the k-group $C \cdot H_c = (H_c \rtimes C)/C_c$ is pseudo-reductive by [CGP, Prop. 1.4.3].

Consider the k-subgroup H of $\mathrm{R}_{K/k}(L_K)$ generated by the k-groups $C \cdot H_c$ for $c \in \Delta$. We have $L \subset H$ since L is generated by $\{L_c\}_{c\in\Delta}$, so H is pseudo-reductive with L as a Levi k-subgroup and $\Phi(H,T) = \Phi$. Since $Z_{C \cdot H_c}(T) = C$ for all $c \in \Delta$, it follows from [CGP, Thm. C.2.29(iii)] that H has c-root group equal to that of $C \cdot H_c$ for every $c \in \Delta$. The pseudo-semisimple derived group $G := \mathscr{D}(H)$ containing L has the same root system and root groups as H relative to T, so for each $c \in \Delta$ its c-root group inside that of $\mathrm{R}_{K_c/k}((L_c)_{K_c})$ corresponds to V when c is short and $V_>$ when c is long. □

We now obtain several nice consequences:

Theorem 3.4.2. *Let G be a pseudo-reductive group over a field k.*

(i) *If* $\mathrm{char}(k) \neq 2, 3$ *then G is standard.*
(ii) *Assume* $\mathrm{char}(k) = p \in \{2, 3\}$ *and that the root system Φ of G_{k_s} is reduced (automatic when $p = 3$) with Dynkin diagram having no edge of multiplicity p. If $p = 2$ then assume moreover that Φ is not* A_1. *The k-group G is standard.*

Proof. It suffices to check standardness over k_s (by [CGP, Cor. 5.2.3]), so we may and do assume $k = k_s$. By [CGP, Prop. 5.3.1] it is enough to prove standardness of the normal pseudo-simple k-subgroups of G, so we may assume G is (absolutely) pseudo-simple. Note that Φ is automatically reduced when $\mathrm{char}(k) \neq 2$, by [CGP, Thm. 2.3.10]. We may also assume k is imperfect with characteristic p or else there is nothing to do.

Our task is to show that a pseudo-simple group over $k = k_s$ with a reduced root system is standard except possibly when the Dynkin diagram either has an edge of multiplicity p or is a single point and $p = 2$. Proposition 3.1.9 settles the rank-1 cases away from $p = 2$, so assume Φ has rank $n \geqslant 2$. If G is of minimal type then we may conclude by Theorem 3.4.1(i). Hence, the central quotient $G/\mathscr{C}_G$ of minimal type is standard. By Lemma 3.2.8, standardness of $G/\mathscr{C}_G$ implies standardness of G. □

Proposition 3.4.3. *An absolutely pseudo-simple k-group G with root system* F_4 *or* G_2 *over k_s is either standard or exotic (see Definition* 2.2.3).

Proof. Let us first treat the case when $k = k_s$, so G is pseudo-split. Let K/k be the minimal field of definition for the geometric unipotent radical of G. By Theorem 3.4.1, G is standard except possibly if $\mathrm{char}(k) = p > 0$ with $p = 2$ for type F_4 and $p = 3$ for type G_2, and in these latter cases G is classified up to isomorphism by the subfield $K_> \subset K$ over kK^p that can be arbitrary. The cases with $K_> \neq K$ are obtained as Weil restrictions of (pseudo-split) basic exotic $K_>$-groups relative to $K/K_>$. The case $K_> = K$ corresponds to the standard case (i.e., $\mathrm{R}_{K/k}$ applied to a split K-group of type F_4 or G_2, depending on $p \in \{2, 3\}$). This settles the case when $k = k_s$.

Now consider general k, so G_{k_s} is either standard or exotic. If G_{k_s} is standard then G is standard by [CGP, Cor. 5.2.3]. Suppose instead that G_{k_s} is exotic, so k is imperfect with characteristic $p \in \{2, 3\}$ where $p = 2$ when the root system is F_4 and $p = 3$ when the root system is G_2. The preceding arguments construct a proper subfield $F' \subset K_s$ containing $k_s K_s^p$ and a basic exotic F'-group $\mathscr{G}'$ such that $G_{k_s} \simeq \mathrm{R}_{F'/k_s}(\mathscr{G}')$. More specifically, as we saw in the proof of Corollary 3.3.9, F'/k_s is the root field of G_{k_s} and the natural map $G_{F'} \to \mathscr{G}'$ is the maximal pseudo-reductive quotient of minimal type.

For any absolutely pseudo-simple k-group, the formation of its root field commutes with separable extension on k. Likewise, the formation of the maximal pseudo-reductive quotient of minimal type (for any smooth connected affine group) commutes with separable extension on the ground field (due to Lemma 2.3.6). Thus, the purely inseparable root field F/k of G satisfies $F \otimes_k k_s = F'$, and the quotient $\mathscr{G} := (G_F)^{\text{prmt}}$ of G_F satisfies $\mathscr{G}_{F'} = \mathscr{G}'$ as quotients of $G_{F'}$. But $\mathscr{G}'$ is basic exotic over $F' = F_s$, so $\mathscr{G}$ is basic exotic over F [CGP, Prop. 7.2.7(3)], and the natural map $G \to \mathrm{R}_{F/k}(\mathscr{G})$ is an isomorphism because over k_s it recovers the identification $G_{k_s} = \mathrm{R}_{F'/k_s}(\mathscr{G}')$. □

In [CGP, Thm. 5.1.1], standardness is characterized among pseudo-reductive groups in terms of dimensions of root spaces in the Lie algebra over k_s, assuming $[k : k^2] \leqslant 2$ if $\mathrm{char}(k) = 2$. (This is only interesting if k is imperfect of characteristic 2 or 3, as otherwise standardness always holds.) This degree restriction in characteristic 2 can now be eliminated:

Proposition 3.4.4. *Let G be a pseudo-reductive group over a field k, T a maximal k-torus of G, and $\Phi = \Phi(G_{k_s}, T_{k_s})$. The k-group G is standard if and only if the following conditions all hold: Φ is reduced, the k_s-dimension of the a-root space $(\mathfrak{g}_{k_s})_a$ in $\mathrm{Lie}(G)_{k_s}$ depends only on the irreducible component of a in Φ, and $\dim(\mathfrak{g}_{k_s})_a = [K_a : k_s]$ for isolated points a in the Dynkin diagram of Φ (where K_a/k_s is the minimal field of definition for the geometric unipotent radical of $(G_{k_s})_a$).*

The additional requirement on isolated points of the Dynkin diagram always holds if $\mathrm{char}(k) \neq 2$ (by Proposition 3.1.9) or if $\mathrm{char}(k) = 2$ with $[k : k^2] \leqslant 2$ (by [CGP, Prop. 9.2.4]), so it is only relevant when $\mathrm{char}(k) = 2$ with $[k : k^2] > 2$. In such cases this additional condition is necessary: for any such k we may choose $K \subset k^{1/2}$ of degree 4 over k and a nonzero k-subspace $V \subset K$ of dimension 3 (so $k\langle V \rangle = K$), and the non-standard absolutely pseudo-simple k-group $H_{V,K/k}$ of rank 1 satisfies all requirements except for the final one.

Proof. We may replace G with $\mathscr{D}(G)$ [CGP, Prop. 5.2.1], so G is perfect, and we may assume $G \neq 1$. Also, by [CGP, Cor. 5.2.3] we may assume $k = k_s$. Since standardness of G is equivalent to that of its pseudo-simple normal k-subgroups [CGP, Prop. 5.2.6, Prop. 5.3.1], we may also assume that G is pseudo-simple.

For a finite extension field k'/k and connected semisimple k'-group G', any pseudo-reductive central quotient of $\mathrm{R}_{k'/k}(G')$ is standard [CGP, Rem. 1.4.6].

Thus, if G is standard then so is any pseudo-reductive central quotient. Likewise, by Lemma 3.2.8, G is standard if its pseudo-simple central quotient $G/\mathscr{C}_G$ is standard. Since the root system and root groups of a pseudo-semisimple group are inherited by any pseudo-reductive central quotient (see [CGP, Prop. 2.3.15]), as is the minimal field of definition for the geometric unipotent radical (Proposition 3.2.6), we may replace G with $G/\mathscr{C}_G$ so that G is of minimal type.

Let K/k be the minimal field of definition for $\mathscr{R}_u(G_{\overline{k}}) \subset G_{\overline{k}}$. The root system is reduced in the standard case, and the standard pseudo-simple cases satisfy the dimension conditions on root spaces (as standard groups are a central quotient of $\mathrm{R}_{K/k}(G')$ for a simple semisimple K-group G').

It remains to prove the sufficiency of the given set of conditions, so we can assume Φ is reduced. The possibilities for (minimal type and pseudo-simple) G with rank $\geqslant 2$ are classified up to isomorphism in Theorem 3.4.1. We shall check that the dimension conditions on root spaces are not satisfied in the non-standard cases on that list. Case (i) in Theorem 3.4.1 is standard. In cases (ii) and (iii), V is a nonzero $K_>$-subspace of K and $V_> \subset K_>$, so the dimension condition on root spaces forces $V_> = K_>$ and V to be a $K_>$-line. But then $K = k\langle V\rangle = K_>$, so all root spaces have dimension $[K:k]$. That forces the inclusion $G \hookrightarrow \mathscr{D}(\mathrm{R}_{K/k}(L_K))$ between pseudo-simple k-groups to be an equality on root groups and thus an equality of k-groups.

Finally, we can assume that Φ is of type A_1 and (by Proposition 3.1.9) that $\mathrm{char}(k) = 2$. Then G is either $H_{V,K/k}$ or $\mathrm{P}H_{V,K/k}$ with $V \subset K$ a nonzero kK^2-subspace satisfying $k\langle V\rangle = K$ (by Proposition 3.1.8(ii) and Proposition 2.3.4). The dimension condition on root spaces forces $V = K$, so G is either $\mathrm{R}_{K/k}(\mathrm{SL}_2)$ or $\mathscr{D}(\mathrm{R}_{K/k}(\mathrm{PGL}_2))$, each of which is standard. □

4

Central extensions and groups locally of minimal type

4.1 Central quotients

The following general lemma on the behavior of the scheme-theoretic center with respect to the formation of central quotient maps between pseudo-reductive groups generalizes a familiar fact in the connected reductive case and will frequently be useful in our subsequent work.

Lemma 4.1.1. *For any central extension of k-group schemes*

$$1 \to \mathscr{Z} \to \mathscr{G} \xrightarrow{f} \overline{\mathscr{G}} \to 1$$

with pseudo-reductive $\mathscr{G}$ and $\overline{\mathscr{G}}$, the inclusion $Z_{\mathscr{G}} \subset f^{-1}(Z_{\overline{\mathscr{G}}})$ is an equality.

Recall that $Z_{\mathscr{G}}$ (resp. $Z_{\overline{\mathscr{G}}}$) denotes the scheme-theoretic center of $\mathscr{G}$ (resp. $\overline{\mathscr{G}}$).

Proof. We may assume $k = k_s$. Let T be a maximal k-torus in $\mathscr{G}$, so its image $\overline{T} \subset \overline{\mathscr{G}}$ is a maximal k-torus. Let $C = Z_{\mathscr{G}}(T)$ and $\overline{C} = Z_{\overline{\mathscr{G}}}(\overline{T})$ denote the associated Cartan k-subgroups, which are commutative. The map $C \to \overline{C}$ is surjective, hence faithfully flat, and $\mathscr{Z} \subset C$, so the scheme-theoretic preimage $f^{-1}(\overline{C})$ is equal to C. Thus, $f^{-1}(Z_{\overline{\mathscr{G}}}) \subset C$. Our problem is therefore to show that for any k-algebra A, any $g \in C(A)$ whose image in $\overline{\mathscr{G}}(A)$ centralizes $\overline{\mathscr{G}}_A$ also centralizes $\mathscr{G}_A$.

Since $\mathscr{G} = C \cdot \mathscr{D}(\mathscr{G})$ with C commutative, and $\mathscr{D}(\mathscr{G})$ is generated by the T-root groups [CGP, Lemma 3.1.5], $Z_{\mathscr{G}}$ represents the functor of points of C centralizing all T-root groups in $\mathscr{G}$. By [CGP, Cor. 2.1.9] and the centrality of $\mathscr{Z}$, each T-root group in $\mathscr{G}$ maps isomorphically onto a $\overline{T}$-root group in $\overline{\mathscr{G}}$. Thus, for points of C we can detect the centralizing property against a given root group for $(\mathscr{G}, T)$ by passing to $(\overline{\mathscr{G}}, \overline{T})$ (as C normalizes every T-root group of $\mathscr{G}$). □

4.1.2. Let G be a smooth connected affine group over a field k. Let T be a maximal k-torus in G, with $C := Z_G(T)$ the associated Cartan subgroup of G. To study the automorphism functor of G, two useful functors on k-algebras are:

$$\underline{\mathrm{Aut}}_{G,C} : A \rightsquigarrow \{f \in \mathrm{Aut}_A(G_A) \mid f|_{C_A} = \mathrm{id}_{C_A}\},$$

$$\underline{\mathrm{Aut}}_{G,T} : A \rightsquigarrow \{f \in \mathrm{Aut}_A(G_A) \mid f|_{T_A} = \mathrm{id}_{T_A}\}.$$

The following properties of these functors are known by [CGP, Thm. 2.4.1, Cor. 2.4.4]: the functor $\underline{\mathrm{Aut}}_{G,C}$ is represented by an affine finite type k-group scheme $\mathrm{Aut}_{G,C}$, if G is generated by tori (for example, if G is perfect) then the functor $\underline{\mathrm{Aut}}_{G,T}$ is represented by an affine finite type k-group scheme $\mathrm{Aut}_{G,T}$ for which the natural map $\mathrm{Aut}_{G,C} \to \mathrm{Aut}_{G,T}$ is a closed immersion, and if moreover G is pseudo-reductive (and generated by tori) then the maximal smooth closed k-subgroups of $\mathrm{Aut}_{G,C}$ and $\mathrm{Aut}_{G,T}$ coincide.

We denote the maximal smooth closed k-subgroup of $\mathrm{Aut}_{G,C}$ by $Z_{G,C}$. By [CGP, Thm. 2.4.1], if G is pseudo-reductive then $Z_{G,C}$ is commutative with pseudo-reductive identity component. Assuming that G is pseudo-reductive, we will prove later (see Proposition 6.1.4) that $Z_{G,C}$ is in fact connected.

Proposition 4.1.3. *The central quotient G/Z_G of a pseudo-reductive group G is pseudo-reductive with trivial center. In particular, it is of minimal type.*

Proof. The natural conjugation action of C on G extends the identity on C, and so is classified by a k-homomorphism $C \to Z^0_{G,C}$. Thus, we can form the noncommutative pushout $\mathscr{G} = (G \rtimes Z^0_{G,C})/C$ as a central quotient of $G \rtimes Z^0_{G,C}$ (using the central anti-diagonal inclusion of C), and by [CGP, Prop. 1.4.3] this pushout is pseudo-reductive. The image of G in $\mathscr{G}$ is normal, hence also pseudo-reductive, and it is the quotient of G modulo $\ker(C \to Z^0_{G,C}) = Z_G$, so G/Z_G is pseudo-reductive. Lemma 4.1.1 now implies that G/Z_G has trivial center. □

Corollary 4.1.4. *For a pseudo-reductive k-group G and closed k-subgroup scheme $Z \subset Z_G$, the central quotient G/Z is pseudo-reductive if and only if Z_G/Z does not contain a nontrivial smooth connected unipotent k-subgroup.*

Proof. The quotient G/Z_G of G/Z modulo Z_G/Z is pseudo-reductive by Proposition 4.1.3, so the smooth connected k-group $\mathscr{R}_{u,k}(G/Z)$ is contained in

Z_G/Z. Since any smooth connected unipotent k-subgroup of Z_G/Z is central in G/Z (and so is contained in $\mathscr{R}_{u,k}(G/Z)$), we are done. □

Note that in contrast with the reductive case, a pseudo-reductive group with trivial center can fail to be perfect (i.e., it may not be pseudo-semisimple). For example, if k'/k is a nontrivial purely inseparable finite extension of fields with characteristic $p > 0$ then $G = \mathrm{R}_{k'/k}(\mathrm{PGL}_p)$ has trivial center by [CGP, Prop. A.5.15(1)] but is not perfect (due to [CGP, Prop. 1.3.4, Ex. 1.3.5]).

Proposition 4.1.5. *Let $f : H \to \overline{H}$ be a central quotient homomorphism between pseudo-semisimple groups over a field k, and let K/k be the common minimal field of definition for their geometric unipotent radicals. The homomorphism $\xi_H : H \to \mathscr{D}(\mathrm{R}_{K/k}(H_K^{\mathrm{ss}}))$ is surjective if and only if $\xi_{\overline{H}}$ is surjective.*

The extension K/k is the same for H and $\overline{H}$ due to Proposition 3.2.6.

Proof. We may assume $k = k_s$. Let T be a maximal k-torus of H and $\overline{T}$ its image in $\overline{H}$. By consideration of open cells and centrality of the kernel we see that $\Phi(H,T) = \Phi(\overline{H},\overline{T})$ via the inclusion $\mathrm{X}(\overline{T}) \hookrightarrow \mathrm{X}(T)$. Let $H' = H_K^{\mathrm{ss}}$ and $\overline{H}' = \overline{H}_K^{\mathrm{ss}}$, so $\overline{H}'$ is a central quotient of H' and the K-tori $T' := T_K$ and $\overline{T}' := \overline{T}_K$ are maximal in H' and $\overline{H}'$ respectively.

The root system $\Phi' := \Phi(H',T')$ coincides with $\Phi(\overline{H}',\overline{T}')$ via the inclusion $\mathrm{X}(\overline{T}') \hookrightarrow \mathrm{X}(T')$ and they both equal the set of non-multipliable roots in $\Phi(H,T)$. By [CGP, Ex. 2.3.2], Φ' naturally coincides with the root systems of $\mathrm{R}_{K/k}(H')$ and $\mathrm{R}_{K/k}(\overline{H}')$ relative to the maximal k-tori $\mathscr{T} \subset \mathrm{R}_{K/k}(T')$ and $\overline{\mathscr{T}} \subset \mathrm{R}_{K/k}(\overline{T}')$, and the associated root groups are given by applying $\mathrm{R}_{K/k}$ to root groups of (H',T') and $(\overline{H}',\overline{T}')$. Hence, by consideration of open cells, centrality of the kernel of $H' \to \overline{H}'$ implies that each $\mathscr{T}$-root group of $\mathrm{R}_{K/k}(H')$ maps isomorphically onto the corresponding $\overline{\mathscr{T}}$-root group of $\mathrm{R}_{K/k}(\overline{H}')$.

Every pseudo-semisimple k-group is generated by the root groups relative to a maximal k-torus (recall that $k = k_s$), so in the commutative diagram

$$\begin{array}{ccc} H & \xrightarrow{\xi_H} & \mathscr{D}(\mathrm{R}_{K/k}(H')) \\ {\scriptstyle f}\downarrow & & \downarrow{\scriptstyle f'} \\ \overline{H} & \xrightarrow[\xi_{\overline{H}}]{} & \mathscr{D}(\mathrm{R}_{K/k}(\overline{H}')) \end{array}$$

the map f' with central kernel is surjective. (The map analogous to f' between Weil restrictions, without the intervention of derived groups, can fail to be surjective.) Thus, if ξ_H is surjective then so is $\xi_{\overline{H}}$. Conversely, assuming that $\xi_{\overline{H}}$ is surjective, it follows that the natural k-homomorphism

$$H \times \ker(f') \to \mathscr{D}(\mathrm{R}_{K/k}(H'))$$

is surjective on $\overline{k}$-points. Thus, by passing to derived groups of $\overline{k}$-points we see that ξ_H is surjective (as H and $\mathscr{D}(\mathrm{R}_{K/k}(H'))$ are perfect). □

Proposition 4.1.6. *Let $\mathscr{G}$ and G be pseudo-semisimple k-groups. Assume that $\mathscr{G}$ is of minimal type and $G_{\overline{k}}^{\mathrm{ss}}$ is simply connected. Let $f : \mathscr{G} \to G$ be a central quotient homomorphism. Then f is an isomorphism and G is of minimal type.*

Proof. Let K/k be the minimal field of definition for $\mathscr{R}_u(G_{\overline{k}}) \subset G_{\overline{k}}$. By Proposition 3.2.6, K/k is also the minimal field of definition for the geometric unipotent radical of $\mathscr{G}$. The central quotient homomorphism $f_K^{\mathrm{ss}} : \mathscr{G}_K^{\mathrm{ss}} \to G_K^{\mathrm{ss}}$ is an isomorphism since G_K^{ss} is simply connected.

Let $\mathscr{T}$ be a maximal k-torus in $\mathscr{G}$ and define $\mathscr{C} = Z_{\mathscr{G}}(\mathscr{T})$. Since f is a central quotient homomorphism we have $\ker f \subset \mathscr{C}$, and since $\mathscr{G}$ is of minimal type the map $i_{\mathscr{G}}|_{\mathscr{C}}$ has trivial kernel. In the commutative diagram

$$\begin{array}{ccc} \mathscr{G} & \xrightarrow{i_{\mathscr{G}}} & \mathrm{R}_{K/k}(\mathscr{G}_K^{\mathrm{ss}}) \\ {\scriptstyle f}\downarrow & & \downarrow{\scriptstyle \mathrm{R}_{K/k}(f_K^{\mathrm{ss}})} \\ G & \xrightarrow[i_G]{} & \mathrm{R}_{K/k}(G_K^{\mathrm{ss}}) \end{array}$$

the right map is an isomorphism, so the triviality of $\ker(i_{\mathscr{G}}|_{\mathscr{C}})$ implies that $\ker(f|_{\mathscr{C}})$ is trivial. But $\ker(f|_{\mathscr{C}}) = \ker(f)$, so f is an isomorphism. □

4.2 Beyond the quadratic case

Let k be an imperfect field of characteristic 2. There are several phenomena that arise when $[k : k^2] > 2$ but not when $[k : k^2] = 2$. We now describe four of them, as each underlies a problem that we have to overcome.

4.2.1. Let G be absolutely pseudo-simple over k, with K/k the minimal field of definition for the geometric unipotent radical $\mathscr{R}_u(G_{\overline{k}}) \subset G_{\overline{k}}$, and assume G_{k_s} has root system A_1. Let $G' = G_K/\mathscr{R}_{u,K}(G_K)$ (a K-form of SL_2 or PGL_2).

Suppose $[k : k^2] = 2$. By [CGP, Prop. 9.2.4], the k-group G is standard and if moreover $G'_{K_s} \simeq \mathrm{SL}_2$ (rather than PGL_2) then $G \simeq \mathrm{R}_{K/k}(G')$. Due to the role of

SL_2 in the structure of connected semisimple groups, this standardness for type A_1 underlies the proof (when $[k:k^2]=2$) that any pseudo-reductive k-group with a *reduced* root system over k_s admits a "generalized standard" form using exotic constructions for types F_4, B_n, and C_n ($n \geqslant 2$) [CGP, Thm. 10.2.1(2)].

This breaks down when $[k:k^2]>2$: there are many *non-standard* pseudo-split absolutely pseudo-simple k-groups G with a reduced root system such that $G'_{K_s} \simeq SL_2$. Fortunately, such G of minimal type can be classified (see Proposition 3.1.8). The abundance of these rank-1 groups G leads to the classes of "generalized basic exotic" and rank-2 "basic exceptional" k-groups in Chapter 8 that go beyond the "basic exotic" k-groups in [CGP, Ch. 7–8].

4.2.2. If $[k:k^2] \geqslant 16$ then there exist non-standard pseudo-split absolutely pseudo-simple k-groups $\overline{G}$ of minimal type with root system A_1 such that $\overline{G}^{ss}_{\overline{k}} \simeq SL_2$ and $\overline{G}$ admits a smooth connected central extension G by α_2 or $\mathbf{Z}/(2)$ (so G is absolutely pseudo-simple) with G *not* of minimal type. (By Proposition B.3.1, no such examples exist if $[k:k^2] \leqslant 8$.) In Examples B.4.1 and B.4.3 we give analogous examples over k with root system BC_n ($n \geqslant 1$) whenever $[k:k^2]>2$.

To build such G with root system A_1, choose $K \subset k^{1/2}$ with k-degree 16 and a 2-basis $\{e_1, e_2, e_3, e_4\}$. Let $v = e_1e_2 + e_3e_4$ (resp. $v = e_1e_2e_3e_4$). Define

$$V = k \oplus ke_1 \oplus ke_2 \oplus ke_3 \oplus ke_4 \oplus kv \subset K. \tag{4.2.2}$$

The subfield $F = \{\lambda \in K \mid \lambda V \subset V\}$ over k is equal to k. Indeed, since $\dim_k V = 6$ the only other possibility is $[F:k]=2$. Since $F \subset V$ (as $1 \in V$), if $[F:k]=2$ then $F = k \oplus kx$ where $x = \sum a_i e_i + av$ for $a_i, a \in k$ not all 0. But $xe_j \in V$, so the 2-basis property of $\{e_1, e_2, e_3, e_4\}$ yields a contradiction. Hence, $F = k$. In terms of Definitions 3.1.2 and 3.3.2, this says that $H_{V,K/k}$ has root field k.

Let $q: V \hookrightarrow K \to k$ be the squaring map and $C = \mathrm{Sym}(V)/(v \cdot v - q(v))$, so C is a finite local k-algebra with residue field K. Since $\mathrm{char}(k)=2$, the definition of C shows that all squares in C lie inside k; i.e., $C^2 \subset k$.

Consider the k-subgroup $G = H_{V,C/k} \subset R_{C/k}(SL_2)$ generated by the k-subspace $V \subset C$ viewed inside both the upper and lower triangular unipotent k-subgroups. By [CGP, 9.1.9–9.1.11], G is a pseudo-split absolutely pseudo-simple central extension of $\overline{G} := H_{V,K/k}$ by α_2 (resp. $\mathbf{Z}/(2)$). Since $\overline{G}^{ss}_{\overline{k}} \simeq SL_2$, by Proposition 4.1.6 the absolutely pseudo-simple k-group G is not of minimal type and does not admit an absolutely pseudo-simple central extension of minimal type. In §B.1–§B.2 we construct higher-rank pseudo-split absolutely pseudo-simple k-groups (with root system B_n or C_n for any $n \geqslant 2$) that admit no absolutely pseudo-simple central extension of minimal type.

In view of such examples, when $[k : k^2] > 2$ we will give a structure theorem only for pseudo-reductive k-groups that are locally minimal type in the sense of Definition 4.3.1 (also see Remark 4.3.2 and Proposition 5.3.3).

4.2.3. Consider a pseudo-split absolutely pseudo-simple k-group G whose root system is BC_n with $n \geqslant 1$. The classification of all such G *of minimal type* with a specified minimal field of definition K/k for the geometric unipotent radical is the main result of [CGP, Ch. 9] (see [CGP, Thm. 9.8.6, Prop. 9.8.9]) and involves nonzero K^2-subspaces of K satisfying nontrivial conditions.

If $[k : k^2] = 2$ then (without assuming G to be of minimal type at the outset) G admits no nontrivial pseudo-semisimple central extension (see [CGP, Prop. 9.9.1]). For such k there are very few possibilities for the relevant K^2-subspaces of K (see [CGP, Cor. 9.8.11]). This implies that if $[k : k^2] = 2$ and G is absolutely pseudo-simple over k with G_{k_s} having root system BC_n then by [CGP, Thm. 9.9.3(1)] two properties hold: G is *necessarily* pseudo-split and of minimal type over k, and G is determined up to k-isomorphism by n and the minimal field of definition K/k of its geometric (unipotent) radical (which can be any nontrivial purely inseparable finite extension of k). These conclusions about G generally fail when $[k : k^2] > 2$.

4.2.4. Let G be a pseudo-reductive k-group. If $[k : k^2] = 2$ then by [CGP, Prop. 10.1.6, Thm. 10.2.1(1)] there is a unique decomposition $G = G_1 \times G_2$ where $(G_1)_{k_s}$ has a reduced root system and G_2 is pseudo-semisimple with root system over k_s having only non-reduced irreducible components. This direct product structure without a pseudo-semisimplicity hypothesis on G rests on the fact, specific to the case $[k : k^2] = 2$, that any pseudo-simple k-group $\mathscr{G}$ with root system BC_n over k_s satisfies $\mathrm{Aut}_k(\mathscr{G}) = \mathscr{G}(k)$ (via conjugation). Without restriction on $[k : k^2]$ but assuming $\mathscr{G}$ is of minimal type and pseudo-split, $\mathrm{Aut}_k(\mathscr{G})$ is computed by Proposition 6.2.2 and [CGP, Prop. 9.8.15] and the natural map $\mathscr{G}(k) \to \mathrm{Aut}_k(\mathscr{G})$ is generally not a surjection when $[k : k^2] > 2$ (see [CGP, Ex. 9.8.16]) though it is injective since $Z_{\mathscr{G}} = 1$ by [CGP, Prop. 9.4.9].

4.3 Groups locally of minimal type

In our study of the structure and classification of pseudo-reductive groups, we want to assume less than that G is of minimal type, since over any imperfect field there are standard absolutely pseudo-simple groups that are not of minimal type (Example 2.3.5). It is also too strong to assume every $(G_{k_s})_c$ is of minimal type, as this also fails, even over k of characteristic 2 for which $[k:k^2]=2$ (see Example 2.3.5 with $p=2$). The hypothesis we shall use is:

Definition 4.3.1. A pseudo-reductive group G over a field k is *locally of minimal type* if, for a maximal torus of G_{k_s}, the pseudo-simple k_s-subgroup of rank 1 generated by any pair of opposite root groups is a central quotient of an absolutely pseudo-simple k_s-group of minimal type (necessarily of rank 1).

This definition is of most interest when k is imperfect of characteristic 2 because if either k is perfect or $\mathrm{char}(k) \neq 2$ then G is necessarily locally of minimal type. Indeed, the case of perfect k is obvious (as then G is reductive), so suppose $\mathrm{char}(k) \neq 2$. We may assume $k = k_s$ and that G is pseudo-simple of rank 1. The root system is reduced since $\mathrm{char}(k) \neq 2$, so it is A_1. Hence, G is of minimal type by Proposition 3.1.9.

In general, the description of i_{G_c} in 3.2.2 implies that a pseudo-reductive group of minimal type is locally of minimal type. It is a tautology that every commutative pseudo-reductive k-group is locally of minimal type (as the root system is empty in such cases), or this could be regarded as a convention.

Remark 4.3.2. Assume G is not locally of minimal type. As we saw above, $\mathrm{char}(k) = 2$ and k is imperfect, so $[k:k^2] \geqslant 2$. We claim that $[k:k^2] > 2$, and that if moreover G_{k_s} has a reduced root system then $[k:k^2] \geqslant 16$. To prove these claims we may assume $k = k_s \neq \overline{k}$ and G is pseudo-simple of rank 1.

If the root system is non-reduced (so BC_1) and $[k:k^2]=2$ then G is of minimal type by [CGP, Thm. 9.9.3(1)]; counterexamples whenever $[k:k^2]>2$ are given in §B.4. If the root system is reduced and $[k:k^2] \leqslant 8$ then G is locally of minimal type by Proposition B.3.1.

When $[k:k^2] \geqslant 16$, there exist pseudo-split absolutely pseudo-simple k-groups G not locally of minimal type that have root system B_n or C_n for any $n \geqslant 2$ (we saw the same for $n=1$ in 4.2.2). Indeed, for any subfield $K \subset k^{1/2}$ with degree 16 over k, in Appendix B we give a pseudo-split absolutely pseudo-simple k-group G with root system B_n or C_n and minimal field of definition K/k for $\mathscr{R}_u(G_{\overline{k}}) \subset G_{\overline{k}}$ such that for some root a the subgroup G_a is of the form $H := H_{V,C/k}$ as in 4.2.2 that is not locally of minimal type, where V is a certain 6-dimensional subspace of K and C is the Clifford algebra of V relative to its squaring map into k. Thus, G is not locally of minimal type.

Proposition 4.3.3. *Let G be pseudo-semisimple and $T \subset G$ a maximal k-torus.*

(i) *If T is split and Δ is a basis of $\Phi(G,T)$ then $C := Z_G(T)$ is generated by $\{C_a\}_{a \in \Delta}$ for $C_a := G_a \cap Z_G(T) = Z_{G_a}(T \cap G_a)$.*

(ii) *If G is locally of minimal type and $G_{\overline{k}}^{\mathrm{ss}}$ is simply connected then G is of minimal type, and if moreover T is split then $\prod_{a\in\Delta} C_a \twoheadrightarrow C = Z_G(T)$ is an isomorphism for any basis Δ of $\Phi(G,T)$.*

The equality $G_a \cap Z_G(T) = Z_{G_a}(T \cap G_a)$ for $a \in \Phi(G,T)$ follows from the description of G_a provided by [CGP, Prop. 3.4.1(2),(3)] and the behavior of Cartan subgroups under intersection against derived groups in [CGP, Lemma 1.2.5(ii)]. The product structure in (ii) on $Z_G(T)$ generalizes the familiar fact that coroots associated to a basis of the root system form a $\mathbf{Z}$-basis of the cocharacter lattice of a split maximal torus in a simply connected semisimple group.

Proof. We may assume $k = k_s$. Let Δ be a basis of $\Phi = \Phi(G,T)$, and let $\mathscr{C} \subset C$ be the k-subgroup generated by $\{C_a\}_{a\in\Delta}$, so $\mathscr{C}$ normalizes G_a for all $a \in \Delta$. For each $a \in \Delta$, the k-group $H_a = \mathscr{C} \cdot G_a$ contains T and $Z_{H_a}(T) = \mathscr{C}$. By [CGP, Thm. C.2.29(i)], for the k-subgroup $H \subset G$ generated by $\{H_a\}_{a\in\Delta}$ we have $Z_H(T) = \mathscr{C}$. But $N_H(T)(k)$ contains representatives of reflections that generate $W(\Phi)$, so H contains G_a for every $a \in \Phi$ and thus $H = G$. Hence, $C = Z_G(T) = \mathscr{C}$, proving (i).

Now we assume that G is locally of minimal type and $G_{\overline{k}}^{\mathrm{ss}}$ is simply connected. Letting K/k be the minimal field of definition for $\mathscr{R}_u(G_{\overline{k}}) \subset G_{\overline{k}}$, the K-group $G' := G_K^{\mathrm{ss}}$ is simply connected and semisimple with maximal torus $T' := T_K$. To prove that G is of minimal type, we need to show that $(\ker i_G) \cap C$ is trivial for the natural map $i_G : G \to \mathrm{R}_{K/k}(G')$.

By [CGP, Prop. 3.2.10, Thm. 2.3.10], $\Phi' := \Phi(G', T')$ coincides with the set Φ_{nm} of non-multipliable roots in Φ. For each $a \in \Phi$ let a' denote a_K or $2a_K$ depending on whether a is non-multipliable or multipliable respectively, so the quotient map $G_K \twoheadrightarrow G'$ carries $(U_a)_K$ onto $U'_{a'}$. The set $\Delta' = \{a'\}_{a\in\Delta}$ is a basis of Φ', so since G' is simply connected we see that the natural map $\mathrm{GL}_1^{\Delta} \to T'$ defined by $(t_a) \mapsto \prod a'^{\vee}(t_a)$ is a K-isomorphism. In particular, the coroots associated to Δ' constitute a basis of $\mathrm{X}_*(T')$. Applying $\mathrm{R}_{K/k}$ then implies that the Cartan k-subgroup $\mathrm{R}_{K/k}(T')$ of $\mathrm{R}_{K/k}(G')$ is the direct product of its intersections $\mathrm{R}_{K/k}(a'^{\vee}(\mathrm{GL}_1))$ with the pseudo-simple k-subgroups $\mathrm{R}_{K/k}(G'_{a'}) \simeq \mathrm{R}_{K/k}(\mathrm{SL}_2)$ generated by the $\pm a'$-root groups for $a' \in \Delta'$.

By 3.2.2, the map i_G carries C_a into $\mathrm{R}_{K/k}(a'^{\vee}(\mathrm{GL}_1)) = \mathrm{R}_{K/k}(\mathrm{GL}_1)$ via the composition of i_{G_a} with $\mathrm{R}_{K_a/k}(\mathrm{GL}_1) \hookrightarrow \mathrm{R}_{K/k}(\mathrm{GL}_1)$. The composite map

$$\prod_{a\in\Delta} C_a \to C \xrightarrow{i_G} \mathrm{R}_{K/k}(T') = \prod_{a\in\Delta} \mathrm{R}_{K/k}(a'^{\vee}(\mathrm{GL}_1))$$

is the direct product of the maps $i_{G_a}|_{C_a}$ (post-composed with inclusions). By 3.2.2, for every $a \in \Phi$ we have $(G_a)^{\mathrm{ss}}_{\overline{k}} = (G'_{a'})_{\overline{K}}$, and this subgroup is SL_2 since G' is simply connected. Thus, by Proposition 4.1.6, G_a is of minimal type for every a (due to the hypothesis that G is locally of minimal type). Hence, the restriction $i_{G_a}|_{C_a}$ has trivial kernel for all $a \in \Delta$, so $i_G|_{C_a}$ has trivial kernel for all $a \in \Delta$. This implies that the multiplication map

$$m : \prod_{a \in \Delta} C_a \to C$$

has trivial kernel. We know from (i) that m is onto, so it is an isomorphism. Therefore, $(\ker i_G) \cap C$ is trivial and so G is of minimal type. □

Remark 4.3.4. In [CGP, §9.8] the pseudo-split absolutely pseudo-simple k-groups of minimal type with a non-reduced root system are classified over any imperfect field of characteristic 2. Proposition 4.3.3(ii) shows that in this classification there is no effect if we relax the "minimal type" hypothesis to "locally of minimal type".

5

Universal smooth k-tame central extension

5.1 Construction of central extensions

In this chapter we construct canonical central extensions that are analogues for perfect smooth connected affine k-groups of the simply connected central cover of a connected semisimple k-group.

Definition 5.1.1. Let k be a field. A commutative affine k-group scheme of finite type is *k-tame* if it does not contain a nontrivial unipotent k-subgroup scheme. For an affine k-group scheme G of finite type, a central extension

$$1 \to Z \to E \to G \to 1$$

with affine E of finite type is *k-tame* if Z is k-tame.

Since Z might not be smooth or connected, the absence of nontrivial unipotent k-subgroups is much weaker than the condition that Z is of multiplicative type. As an example, let K/k be a finite extension of fields and $q : \widetilde{G}' \to G'$ the simply connected central cover of a connected semisimple K-group G'. For the finite k-group $\mu := \ker q$ of multiplicative type, we obtain a central extension

$$1 \to \mathrm{R}_{K/k}(\mu) \to \mathrm{R}_{K/k}(\widetilde{G}') \to \mathscr{D}(\mathrm{R}_{K/k}(G')) \to 1 \qquad (5.1.1)$$

by [CGP, Prop. 1.3.4]. There is no nontrivial unipotent k-subgroup $j : U \hookrightarrow \mathrm{R}_{K/k}(\mu)$ since j corresponds to a K-homomorphism $U_K \to \mu$ that must be trivial as there are no nontrivial homomorphisms between unipotent and multiplicative type group schemes over a field [SGA3, XVII, Prop. 2.4]. Thus, (5.1.1) is a k-tame central extension. If K/k is not separable and μ is not K-étale then $\mathrm{R}_{K/k}(\mu)$ has positive dimension and is *not* of multiplicative type.

Proposition 5.1.2. *Let Z be a commutative affine k-group scheme of finite type and k'/k a separable extension field. Then Z is k-tame if and only if $Z_{k'}$ is k'-tame.*

Proof. For any commutative affine k-group scheme Z of finite type we claim that there exists a unipotent k-subgroup scheme Z^{unip} of Z containing all others and that for any separable extension k'/k the inclusion $(Z^{\mathrm{unip}})_{k'} \subset (Z_{k'})^{\mathrm{unip}}$ is an equality.

Since Z is commutative and an extension of a unipotent k-group scheme by a unipotent k-group scheme is unipotent [SGA3, XVII, Prop. 2.2(iv)], by choosing a unipotent k-subgroup scheme $U \subset Z$ with maximal dimension we may pass to Z/U^0 so that all unipotent k-subgroup schemes of Z are 0-dimensional.

If $\mathrm{char}(k) = 0$ (so Z is smooth) then $Z^0 = T \times U$ for a k-torus T and (smooth) connected unipotent k-group U, so necessarily $U = 1$ and Z^0 is a torus. Since all unipotent groups in characteristic 0 are connected, it follows that in such cases $Z^{\mathrm{unip}} = 1$ and likewise after any extension on k.

Now we may and do assume $\mathrm{char}(k) = p > 0$. To prove the existence of Z^{unip} as a *finite* k-subgroup of Z it suffices to show that for any commutative affine k-group scheme Z of finite type and sequence of unipotent k-subgroup schemes $U_1 \subset U_2 \subset U_3 \subset \dots$, the closed subscheme $U \subset Z$ whose ideal is the intersection of those of the U_i's (denoted as $\bigcup_i U_i$ and called the *schematic union* of the U_i's) is a unipotent k-subgroup scheme of Z. (Indeed, then $\dim U > 0$ if $\{U_i\}$ is strictly increasing, yielding a contradiction if all unipotent k-subgroups of Z are finite.)

It is elementary to check (or see [EGA, IV_3, 11.9.7.1]) that $U \times U_{i_0} = \bigcup_i (U_i \times U_{i_0})$ for every i_0 and $U \times U = \bigcup_i (U \times U_i)$, so $U \times U = \bigcup_i (U_i \times U_i)$. Hence, U is a k-subgroup scheme of Z. For the purpose of proving U is unipotent, we may replace Z with U so that $Z = \bigcup_i U_i$. For any closed k-subgroup scheme $Z' \subset Z$, the faithful flatness of $Z \to Z/Z'$ implies that $Z/Z' = \bigcup_i (U_i/(U_i \cap Z'))$, so Z/Z' cannot be of multiplicative type unless it is trivial. Setting Z' to be the kernel $\ker(F_{Z/k,n})$ of the n-fold relative Frobenius morphism of Z, by [SGA3, $\mathrm{VII_A}$, Prop. 8.3] if we take n sufficiently large then Z/Z' is smooth. This commutative *smooth* affine k-group quotient of Z has no nontrivial quotient of multiplicative type, so it must be unipotent.

For the $\overline{k}$-groups $\mathbf{G}_{\mathrm{a}}$, α_p, and $\mathbf{Z}/p\mathbf{Z}$, the relative Verschiebung morphism (in the sense of [SGA3, $\mathrm{VII_A}$, 4.2–4.3]) vanishes. Thus, there are integers $m, m' > 0$ such that $V_{(Z/Z')/k,m}$ vanishes and every *unipotent* k-subgroup of the k-finite Z' has vanishing m'-fold relative Verschiebung morphism, so $V_{U/k,m+m'} = 0$ for every unipotent k-subgroup $U \subset Z$. For example, $V_{U_i/k,m+m'} = 0$ for every

i. Thus, $V_{Z/k,m+m'} : Z^{(p^{m+m'})} \to Z$ kills $U_i^{(p^{m+m'})}$ for all i. But $Z^{(p^r)} = \bigcup_i U_i^{(p^r)}$ for any $r \geqslant 0$, so $V_{Z/k,m+m'} = 0$. Hence, $Z_{\overline{k}}$ cannot contain μ_ℓ for any prime ℓ whatsoever (treat $\ell = p$ separately), so Z is unipotent by [SGA3, XVII, Thm. 4.6.1(vi)]. This completes the construction of Z^{unip} in general.

It remains to show that the inclusion $(Z^{\mathrm{unip}})_{k'} \subset (Z_{k'})^{\mathrm{unip}}$ is always an equality (so Z^{unip} is trivial if and only if $(Z_{k'})^{\mathrm{unip}}$ is trivial). By Galois descent, $(Z^{\mathrm{unip}})_{k_s} = (Z_{k_s})^{\mathrm{unip}}$ and $(Z_{k'}^{\mathrm{unip}})_{k'_s} = (Z_{k'_s})^{\mathrm{unip}}$, so it is equivalent to check that the inclusion $(Z_{k_s}^{\mathrm{unip}})_{k'_s} \subset (Z_{k'_s})^{\mathrm{unip}}$ is an equality. Hence, we may replace k'/k with k'_s/k_s to reduce to the case that $k = k_s$. By passing to Z/Z^{unip} we may arrange that Z^{unip} is trivial and need to rule out the existence of a nontrivial unipotent k'-subgroup $U' \subset Z_{k'}$ when $k = k_s$. Assume there is such a U'. We may descend to the case where k' is finitely generated over k, so by separability k' is the fraction field of a smooth k-algebra A of finite type. By standard "spreading out" arguments, at the cost of shrinking $S := \mathrm{Spec}(A)$ around its generic point $\eta = \mathrm{Spec}(k')$ we may find a closed flat S-subgroup scheme $\mathscr{U}' \subset Z \times S$ with generic fiber U'. If $\mathscr{U}'$ has nontrivial unipotent fibers over a dense open subset $\Omega \subset S$ then since $\Omega(k)$ is non-empty (as $k = k_s$ and S is k-smooth) we could specialize at such a k-point to obtain a nontrivial unipotent k-subgroup of Z, a contradiction.

To complete the proof of the second assertion of the lemma we shall prove that a flat finite type S-group scheme $\mathscr{U}'$ with nontrivial unipotent η-fiber has nontrivial unipotent fibers over all points in a dense open subset of S. By [SGA3, XVII, Thm. 3.5], $\mathscr{U}'_\eta$ has a finite composition series $\{1 = U'_0 \subset \cdots \subset U'_m = \mathscr{U}'_\eta\}$ over η with $m \geqslant 1$ and U'_i a normal η-subgroup scheme of U'_{i+1} such that U'_{i+1}/U'_i is isomorphic to $\mathbf{G}_{\mathrm{a}}$, α_p, or a nontrivial p-torsion commutative finite étale k'-group. This provides a faithfully flat homomorphism from U'_{i+1} onto $\mathbf{G}_{\mathrm{a}}$, α_p, or a nontrivial p-torsion commutative finite étale group such that the kernel of the homomorphism is U'_i. By shrinking S around η, we may arrange that this extends to such a composition series $\{\mathscr{U}'_j\}$ over S. Every fiber $\mathscr{U}'_s$ therefore admits such a composition series and so is unipotent and nontrivial. □

Let G be a perfect smooth connected affine k-group, and consider two central extensions

$$1 \to Z \to E \to G \to 1, \quad 1 \to \mathscr{Z} \to \mathscr{E} \to G \to 1,$$

with E and $\mathscr{E}$ perfect smooth connected affine k-groups. There is at most one k-homomorphism $f : \mathscr{E} \to E$ over G since any two are related through multiplication against a k-homomorphism $h : \mathscr{E} \to Z$ and the only such h is the trivial homomorphism since $\mathscr{E}/(\ker h)$ is a perfect smooth connected affine k-group that is commutative and hence trivial. Moreover, such an f is surjective since $f(\mathscr{E}) \times Z \to E$ is visibly surjective on $\overline{k}$-points yet $E(\overline{k})$ is perfect. When f exists we say that $\mathscr{E}$ *dominates* E (as central extensions of G).

Theorem 5.1.3. *Let G be a perfect smooth connected affine group over a field k, with K/k the minimal field of definition for $\mathscr{R}_u(G_{\overline{k}}) \subset G_{\overline{k}}$. Let $G' := G_K^{\mathrm{ss}}$ and let $\widetilde{G}' \to G'$ be its simply connected central cover over K. Consider a k-tame central extension $1 \to Z \to E \to G \to 1$ with E a perfect smooth connected affine k-group.*

(i) *The minimal field of definition over k for $\mathscr{R}_u(E_{\overline{k}}) \subset E_{\overline{k}}$ is K, and the natural map $E_K^{\mathrm{ss}} =: E' \to G'$ is a central isogeny between connected semisimple K-groups.*
(ii) *If $1 \to \mathscr{Z} \to \mathscr{E} \to G \to 1$ is a k-tame central extension with $\mathscr{E}$ a perfect smooth connected affine k-group then $\mathscr{E}$ dominates E over G if and only if $\mathscr{E}' := \mathscr{E}_K^{\mathrm{ss}}$ dominates E' over G' (in which case $\mathscr{E} \to E$ is unique).*
(iii) *The functor $E \rightsquigarrow E'$ is an equivalence from the category of perfect smooth connected k-tame central extensions of G onto the category of connected semisimple central extensions of G' over K.*

In particular, up to unique isomorphism there is a unique k-tame central extension

$$1 \to Z \to \widetilde{G} \to G \to 1$$

such that $\widetilde{G}$ is a perfect smooth connected affine k-group and $\widetilde{G}_K^{\mathrm{ss}}$ is simply connected. The formation of this central extension is functorial with respect to isomorphisms in G and separable extension of the ground field.

Proof. The determination of the minimal field of definition over k in (i) is a special case of Proposition 3.2.6 since E is perfect. The quotient map $E' \to G'$ between connected semisimple K-groups has kernel that is a quotient of the central K-subgroup scheme $\ker(E_K \to G_K) \subset E_K$ (since $\mathscr{R}(E_{\overline{k}}) \to \mathscr{R}(G_{\overline{k}})$ is onto), so $E' \to G'$ is a central isogeny. Part (i) is thereby proved.

In the setting of (ii), if there is a k-homomorphism $f : \mathscr{E} \to E$ over G then f is surjective due to the perfectness of E since $E(\overline{k}) = f(\mathscr{E}(\overline{k}))Z(\overline{k})$ with central $Z(\overline{k}) \subset E(\overline{k})$. Such an f is unique since the perfect smooth $\mathscr{E}$ has no nontrivial k-homomorphism to a commutative k-group scheme (such as the central $Z \subset E$). By surjectivity, f induces a K-homomorphism $f' : \mathscr{E}' \to E'$ as connected semisimple central covers of G'.

The main task is to show that if f' is given over G' then there exists an f over G giving rise to f'. This is where we will use the k-tameness of the central extensions. The idea is to give a functorial reconstruction of E from E' (as central extensions). More specifically, there is an evident k-homomorphism

$$\theta_E : E \to G \times_{\mathscr{D}(\mathrm{R}_{K/k}(G'))} \mathscr{D}(\mathrm{R}_{K/k}(E')) =: \mathscr{G}(E')$$

over G. In general, $\mathscr{G}(E')$ might not be smooth, even if G is absolutely pseudo-simple. (The difficulty arises from Cartan subgroups; see Example 5.1.6.) Recall that for any finite type k-group scheme H, there is a smooth closed k-subgroup $H^{\mathrm{sm}} \subset H$ that contains all others and is final among smooth k-schemes equipped with a k-morphism to H [CGP, Lemma C.4.1]; note that this "maximal smooth closed k-subgroup" is generally smaller than the underlying reduced scheme of H (see [CGP, Rem. C.4.2]). Using this, the k-group scheme $\mathscr{G}(E')$ reconstructs E from E' as follows (completing the proof of (ii)):

Lemma 5.1.4. *The map θ_E carries E isomorphically onto $\mathscr{D}(\mathscr{G}(E')^{\mathrm{sm}})$.*

A refinement of the proof below shows that E represents the sheafified commutator of the (not necessarily smooth) fppf group sheaf $\mathscr{G}(E')$, but we do not need this and so do not discuss it.

Proof. As a first step, we check that the scheme-theoretic kernel of θ_E is trivial (so θ_E is a closed immersion). This kernel is contained inside $\ker(E \to G) = Z$, so it has no nontrivial unipotent k-subgroup due to the *k-tameness* hypothesis on E as a central extension of G. But $\ker \theta_E$ is also contained in the kernel of the natural map

$$h : E \to \mathscr{D}(\mathrm{R}_{K/k}(E')),$$

so if $\ker h$ is unipotent then $\ker \theta_E$ is unipotent and hence trivial.

The subgroup $\mathscr{R}_u(\mathrm{R}_{K/k}(E')_{\overline{k}}) \subset \mathrm{R}_{K/k}(E')_{\overline{k}}$ is defined over K by [CGP, Prop. A.5.11(1),(2)], so by [CGP, Prop. A.4.8] the geometric unipotent radical of $\mathscr{D}(\mathrm{R}_{K/k}(E'))$ is defined over K and

$$\mathscr{R}_{u,K}(\mathscr{D}(\mathrm{R}_{K/k}(E'))_K) = \mathscr{D}(\mathrm{R}_{K/k}(E'))_K \cap \mathscr{R}_{u,K}(\mathrm{R}_{K/k}(E')_K).$$

This yields a composite map

$$\begin{aligned} E_K \stackrel{h_K}{\to} \mathscr{D}(\mathrm{R}_{K/k}(E'))_K \twoheadrightarrow \mathscr{D}(\mathrm{R}_{K/k}(E'))_K^{\mathrm{red}} = \mathscr{D}(\mathrm{R}_{K/k}(E')_K^{\mathrm{red}}) &= \mathscr{D}(E') \\ &= E' \end{aligned}$$

that is easily checked to be the natural projection whose kernel is $\mathscr{R}_{u,K}(E_K)$. Thus, $\ker h_K$ is unipotent, so $\ker h$ is unipotent. This completes the proof that $\ker\theta_E = 1$. We may now view E as a k-subgroup scheme of $\mathscr{G}(E')$, and hence as a k-subgroup scheme of $\mathscr{D}(\mathscr{G}(E')^{\text{sm}})$.

To show that $\mathscr{D}(\mathscr{G}(E')^{\text{sm}}) \subset E$, it clearly suffices to show that the commutator subgroup of $\mathscr{G}(E')(\overline{k})$ lies inside $E(\overline{k})$. But $E(\overline{k}) \to G(\overline{k})$ is surjective and the k-group scheme

$$\begin{aligned}\mathscr{Z} &:= \ker(\mathrm{pr}_1 : \mathscr{G}(E') \to G)\\ &= \ker(\mathscr{D}(\mathrm{R}_{K/k}(E')) \to \mathscr{D}(\mathrm{R}_{K/k}(G'))) \subset \mathrm{R}_{K/k}(\ker(E' \to G'))\end{aligned}$$

is visibly central in $\mathscr{G}(E')$, so $\mathscr{G}(E')(\overline{k}) = E(\overline{k})\mathscr{Z}(\overline{k})$ with $\mathscr{Z}(\overline{k})$ central in $\mathscr{G}(E')(\overline{k})$. The commutator subgroup of $\mathscr{G}(E')(\overline{k})$ therefore coincides with $E(\overline{k})$. □

5.1.5. Completion of the proof of Theorem 5.1.3. Since part (ii) of Theorem 5.1.3 has been proved, to establish part (iii) it suffices to show that for any connected semisimple central extension $E' \to G'$ over K there is a k-tame central extension E of G with E a perfect smooth connected affine k-group such that $E_K^{\text{red}} \simeq E'$ as central extensions of G'. We will show that the smooth connected k-group $E := \mathscr{D}((\mathscr{G}(E')^{\text{sm}})^0)$ works.

As a first step, we show that E is perfect and a k-tame central extension of G via pr_1. By [CGP, Prop. 1.3.4], $\mathscr{D}(\mathrm{R}_{K/k}(E')) = \mathrm{R}_{K/k}(\widetilde{G}')/\mathrm{R}_{K/k}(\mu_{E'})$ for $\mu_{E'} := \ker(\widetilde{G}' \twoheadrightarrow E')$, and similarly for G' in place of E' (with $\mu_{G'} \supset \mu_{E'}$). (Recall that $\mathrm{R}_{K/k}(\widetilde{G}')$ is perfect since $\widetilde{G}'$ is simply connected [CGP, Cor. A.7.11].) Thus,

$$\mathscr{G}(E') = G \times_{\mathrm{R}_{K/k}(\widetilde{G}')/\mathrm{R}_{K/k}(\mu_{G'})} (\mathrm{R}_{K/k}(\widetilde{G}')/\mathrm{R}_{K/k}(\mu_{E'})),$$

so $\mathscr{G}(E') \to G$ is the quotient modulo the *central* closed k-subgroup $1 \times \mathscr{Z}$ where

$$\mathscr{Z} := \mathrm{R}_{K/k}(\mu_{G'})/\mathrm{R}_{K/k}(\mu_{E'}) \subset \mathrm{R}_{K/k}(\mu_{G'}/\mu_{E'})$$

does not contain any nontrivial unipotent k-subgroup scheme (since $\mu_{G'}/\mu_{E'}$ is of multiplicative type). Hence, if we show that $\mathscr{G} := \mathscr{G}(E')^{\text{sm}}$ is mapped onto G by pr_1, it would follow that $\mathscr{G}^0$ is a smooth connected k-tame central extension of G and that $\mathscr{G}^0 = \mathscr{D}(\mathscr{G}^0) \cdot Z_{\mathscr{G}^0}$. An immediate consequence of this would be that the k-group $E = \mathscr{D}(\mathscr{G}^0)$ is perfect and is a k-tame central extension of G.

To show that $\mathscr{G}$ is mapped onto G by pr_1, we may replace k by k_s and seek a smooth connected k-subgroup H of $\mathscr{G}(E')$ which maps onto G. To construct such an H, we shall use the "open cell" structure provided by the dynamic method for rather general smooth affine k-groups in [CGP, §2.1]. To this end, let $T \subset G$ be a (split) maximal k-torus and let $\widetilde{T}'$ be the unique maximal K-torus in $\widetilde{G}'$ lifting the maximal K-torus $T' := T_K$ in the central quotient G' of $\widetilde{G}'$. The k-group $i_G(T)$ is a maximal k-torus of $\mathrm{R}_{K/k}(\widetilde{G}')/\mathrm{R}_{K/k}(\mu_{G'})$ since it is the image of a (unique) maximal k-torus $\widetilde{T}$ in $\mathrm{R}_{K/k}(\widetilde{G}')$. (Explicitly, $\widetilde{T}$ is the maximal k-torus in $\mathrm{R}_{K/k}(\widetilde{T}')$.)

For any cocharacter $\lambda \in \mathrm{X}_*(T)$, if we multiply it by a sufficiently divisible positive integer then we can arrange that the map

$$\mathrm{X}_*(T)_{\mathbf{Q}} \twoheadrightarrow \mathrm{X}_*(i_G(T))_{\mathbf{Q}} = \mathrm{X}_*(\widetilde{T})_{\mathbf{Q}} = \mathrm{X}_*(\widetilde{T}')_{\mathbf{Q}}$$

carries λ into $\mathrm{X}_*(\widetilde{T}')$. Choose such a $\lambda \in \mathrm{X}_*(T)$ that does not annihilate any nonzero T-weights that occur in $\mathrm{Lie}(G)$, so $Z_G(\lambda) = Z_G(T)$. Thus, we have an open subscheme

$$\Omega = U_G(-\lambda) \times Z_G(T) \times U_G(\lambda) \subset G$$

via multiplication (see [CGP, Prop. 2.1.8]), with $U_G(\pm\lambda)$ a unipotent smooth connected k-group.

The k-groups $U_G(\lambda)$ and $U_G(-\lambda)$ generate G. Indeed, the smooth closed k-subgroup N of G generated by $U_G(\lambda)$ and $U_G(-\lambda)$ is normalized by the open cell Ω and hence is normal in G. The composite homomorphism $Z_G(T) \to G/N$ is dominant and thus surjective, so G/N is a perfect smooth connected affine k-group with a central maximal k-torus. Such a group is trivial, so $N = G$ as claimed.

To simplify the notation, define the central quotient

$$Q_{E'} = \mathrm{R}_{K/k}(\widetilde{G}')/\mathrm{R}_{K/k}(\mu_{E'})$$

and likewise for $Q_{G'}$. There is a natural central quotient map $q : Q_{E'} \to Q_{G'}$, and by [CGP, Cor. 2.1.9] the induced maps $U_{Q_{E'}}(\pm\lambda) \to U_{Q_{G'}}(\pm\lambda)$ are isomorphisms. The restrictions of their inverses to the k-subgroups $i_G(U_G(\pm\lambda)) \subset U_{Q_{G'}}(\pm\lambda)$ have associated graph morphisms that define homomorphisms

$$j_{\pm} : U_G(\pm\lambda) \hookrightarrow G \times_{Q_{G'}} Q_{E'} = \mathscr{G}(E')$$

whose compositions with pr_1 are the natural inclusions $U_G(\pm\lambda) \hookrightarrow G$ that generate G. Thus, the smooth closed k-subgroup H of $\mathscr{G}(E')$ generated by

$j_{\pm}(U_G(\pm\lambda))$ maps onto G.

It remains to build an isomorphism $E_K^{\mathrm{red}} \simeq E'$ as connected semisimple central extensions of G'. Via the inclusion θ_E of E into the fiber product $\mathscr{G}(E')$ in Lemma 5.1.4, we obtain a K-homomorphism

$$E_K \to \mathscr{D}(\mathrm{R}_{K/k}(E'))_K \twoheadrightarrow E' \tag{5.1.5}$$

over $G_K \twoheadrightarrow G'$. Thus, surjectivity of $E \to G$ and the isogeny property for $E' \to G'$ imply that $E_K \to E'$ is surjective (since E' is connected), so we obtain a K-homomorphism $f : E_K^{\mathrm{red}} \to E'$ over G'. Since f is a map between connected semisimple central extensions of G', $\ker f$ must be finite of multiplicative type. To prove f is an isomorphism, it suffices to show that $\ker f$ is unipotent.

It is equivalent to prove that $E_K \to E'$ has unipotent kernel. Since the second map in (5.1.5) has unipotent kernel, it suffices to show that the first map in (5.1.5) has unipotent kernel. In other words, for $\mathrm{pr}_2 : \mathscr{G}(E') \to E'$ we want $\mathrm{pr}_2|_E$ to have unipotent kernel. But $\ker(\mathrm{pr}_2) = \ker i_G$ is unipotent. This completes the proof of Theorem 5.1.3. □

Example 5.1.6. Here are examples for which the $\mathscr{G}(\cdot)$-construction in the proof of Theorem 5.1.3 is not smooth. Let k be imperfect of characteristic p, and choose $t \in k - k^p$. Define $k' = k(t^{1/p})$ and $K = k(t^{1/p^2})$. Let

$$G = \mathrm{R}_{K/k}(\mathrm{SL}_p)/\mathrm{R}_{k'/k}(\mu_p).$$

As explained in [CGP, Ex. 5.3.7] more generally, this is absolutely pseudo-simple and its geometric unipotent radical has minimal field of definition over k equal to K. Clearly $G' := G_K^{\mathrm{ss}}$ is equal to PGL_p, so $\widetilde{G}' = \mathrm{SL}_p$. Since $\mathrm{R}_{K/k}(\mathrm{SL}_p)$ is perfect, we have

$$\mathscr{G}(\widetilde{G}') = G \times_{\mathrm{R}_{K/k}(\mathrm{PGL}_p)} \mathrm{R}_{K/k}(\mathrm{SL}_p).$$

This is isomorphic to $Q \times \mathrm{R}_{K/k}(\mathrm{SL}_p)$ for $Q := \mathrm{R}_{K/k}(\mu_p)/\mathrm{R}_{k'/k}(\mu_p)$. The k-group Q has positive dimension and no nontrivial k_s-points, so it is not smooth. (See [CGP, Ex. 5.3.7] for a detailed discussion of a generalization of Q, including a proof that $Q(k_s)$ is trivial.)

5.2 A universal construction

For a perfect smooth connected affine group G over a field k, we call the canonical central extension $\widetilde{G}$ of G in Theorem 5.1.3 the *universal smooth k-tame* *central extension.* To see that this terminology is reasonable, consider a k-tame central extension

$$1 \to Z \to E \to G \to 1$$

with E a smooth affine k-group. The k-group $\mathscr{D}^j(E^0)$ is another such extension of G for any $j \geqslant 0$. For sufficiently large j, $\mathscr{D}^j(E^0)$ stabilizes and thus is perfect, so it receives a unique map from $\widetilde{G}$ as k-tame central extensions of G. This establishes universality for $\widetilde{G}$ among all smooth affine k-tame central extensions of G.

Remark 5.2.1. If G is a connected semisimple group then $\widetilde{G}$ is its simply connected central cover. If G is pseudo-semisimple then $\widetilde{G}$ is pseudo-semisimple. Indeed, the unipotent smooth connected normal k-subgroup $\mathscr{R}_{u,k}(\widetilde{G})$ of $\widetilde{G}$ must have trivial image in the pseudo-reductive quotient G, so $\mathscr{R}_{u,k}(\widetilde{G}) \subset Z := \ker(\widetilde{G} \to G)$. By k-tameness, this forces $\mathscr{R}_{u,k}(\widetilde{G})$ to be trivial.

Proposition 5.2.2. *Let G be a perfect smooth connected affine group over a field k, and let E be a perfect smooth connected k-tame central extension of G. If G admits a Levi k-subgroup L then E admits a unique Levi k-subgroup $L(E)$ over L.*

Proof. Let K/k be the minimal field of definition for the geometric unipotent radical of G, so it is also the minimal such field for E (Theorem 5.1.3(i)). Let $E' = E_K^{\mathrm{red}}$, so E' is a connected semisimple group that is a central extension of $G' = G_K^{\mathrm{red}}$. By Lemma 5.1.4, the natural map $E \to \mathscr{D}(\mathscr{G}(E')^{\mathrm{sm}})$ is an isomorphism, where

$$\mathscr{G}(E') := G \times_{\mathscr{D}(\mathrm{R}_{K/k}(G'))} \mathscr{D}(\mathrm{R}_{K/k}(E')).$$

The natural map $L_K \to G'$ is an isomorphism (as L is a Levi k-subgroup of G), so L is a k-descent of G'.

Let $\widetilde{G} \to G$ be the universal smooth k-tame central extension, so there is a unique homomorphism $f : \widetilde{G} \to E$ over G and the induced map $\widetilde{G}' := \widetilde{G}_K^{\mathrm{red}} \to E'$ is the simply connected central cover (Theorem 5.1.3(iii)). The simply connected semisimple central cover $\widetilde{L} \to L$ is uniquely identified (over L) as a k-descent of $\widetilde{G}'$. Note that $C := \ker(\widetilde{G}' \to G')$ is of multiplicative type with k-descent $\ker(\widetilde{L} \to L)$, and all K-subgroups of C descend uniquely to k-subgroups of $\widetilde{L}$ since it suffices to check the analogue for the étale Cartier duals (and the purely inseparable extension K/k induces an isomorphism between absolute Galois groups). Hence, the intermediate central cover $E' \to G'$ uniquely descends to a central cover $L(E) \to L$. In particular, $L(E)$ is a Levi k-subgroup of $\mathscr{D}(\mathrm{R}_{K/k}(L(E)_K)) = \mathscr{D}(\mathrm{R}_{K/k}(E'))$.

Likewise, L is identified as a Levi k-subgroup of $\mathscr{D}(\mathrm{R}_{K/k}(G'))$ in an analogous way, yielding a k-subgroup inclusion

$$L(E) = L \times_L L(E) \hookrightarrow G \times_{\mathscr{D}(\mathrm{R}_{K/k}(G'))} \mathscr{D}(\mathrm{R}_{K/k}(E')) = \mathscr{G}(E'),$$

so $L(E) \subset \mathscr{D}(\mathscr{G}(E')^{\mathrm{sm}}) = E$. By construction, the resulting composite map $L(E)_K \to E_K \twoheadrightarrow E_K^{\mathrm{red}} = E'$ is the isomorphism which defines $L(E)$ as a k-descent of E', so $L(E)$ is a Levi k-subgroup of E. Moreover, by design $E \to G$ carries $L(E)$ into L via a map $L(E) \to L$ that descends $E' \twoheadrightarrow G'$, so $L(E)$ maps onto L. This completes the existence proof.

It remains to show that $L(E)$ is the only Levi k-subgroup of E over L. If $N \subset E$ is a Levi k-subgroup over L then the resulting map $N \to L$ must be a k-descent of the quotient map $E' \to G'$ due to the Levi subgroup property, so the simply connected central cover of N is uniquely identified with $\widetilde{L}$ over L. Since $\ker(\widetilde{L} \to N)$ is a k-descent of $\ker(\widetilde{G}' \to E')$, it equals $\ker(\widetilde{L} \to L(E))$. Hence, uniquely $N \simeq L(E)$ over L and this isomorphism is as quotients of $\widetilde{L}$, so via the identification of E' as N_K and as $L(E)_K$ we see that $N = L(E)$ inside $\mathscr{D}(\mathrm{R}_{K/k}(E'))$. The perfect smooth N and $L(E)$ are each carried onto L under the central quotient map $E \to G$, so $N = L(E)$ inside $\mathscr{D}(\mathscr{G}(E')^{\mathrm{sm}}) = E$. □

5.3 Properties and applications of $\widetilde{G}$

We begin by recalling the following useful result (which is [CGP, Prop. 2.2.12]).

Proposition 5.3.1. *Let $f : G \to \overline{G}$ be a surjective homomorphism between smooth affine groups over a field k, and assume* $\ker f$ *is central in G.*

(i) *For every maximal k-torus $\overline{T} \subset \overline{G}$, the schematic preimage $f^{-1}(\overline{T})$ is commutative and contains a unique maximal k-torus T. This k-torus is maximal in G, and $f(T) = \overline{T}$. The map $\overline{T} \mapsto T$ defines a bijection between the sets of maximal k-tori of $\overline{G}$ and G, with inverse $T \mapsto f(T)$.*

(ii) *Assume G is connected and perfect. The schematic center $Z_{\overline{G}}$ coincides with $Z_G/(\ker f)$. In particular, if $\overline{G} \to \overline{\overline{G}}$ is a central quotient map then the composite surjection $G \to \overline{\overline{G}}$ has central kernel and G/Z_G has trivial center.*

(iii) *If G and $\overline{G}$ are pseudo-reductive then $P \mapsto f(P)$ is a bijection between the sets of pseudo-parabolic k-subgroups of G and pseudo-parabolic k-subgroups of $\overline{G}$, with inverse $\overline{P} \mapsto f^{-1}(\overline{P})$.*

To establish good properties of the universal smooth tame central extension by means of Proposition 5.3.1, we begin with a lemma in the rank-1 case:

Lemma 5.3.2. *Let G be an absolutely pseudo-simple k-group of rank 1 over k_s such that G is a central quotient of an absolutely pseudo-simple k-group H of minimal type. Then the universal smooth k-tame central extension $\widetilde{G}$ of G is of minimal type and $\widetilde{G}$ is pseudo-split if G is pseudo-split.*

Proof. By Proposition 5.3.1(i) the k-ranks of G and $\widetilde{G}$ coincide, so if G is pseudo-split (i.e., its k-rank is 1) then $\widetilde{G}$ is pseudo-split. For the rest of the proof we may and do assume $k = k_s$. If $G^{\mathrm{ss}}_{\overline{k}} \simeq \mathrm{SL}_2$ (as occurs whenever the root system of G is non-reduced) then by Proposition 4.1.6 we have $H = G$, so G is of minimal type, and $\widetilde{G} = G$ as $G^{\mathrm{ss}}_{\overline{k}}$ is simply connected. Thus, it remains to consider the case $G^{\mathrm{ss}}_{\overline{k}} \simeq \mathrm{PGL}_2$, so the root system of G is A_1.

By Proposition 3.2.6, G and H have the same minimal field of definition K over k for their geometric unipotent radicals. The root systems of G and H coincide, by [CGP, Prop. 2.3.15]. By Proposition 3.1.7, if $\mathrm{char}(k) \neq 2$ then either (i) $H \simeq \mathrm{R}_{K/k}(\mathrm{SL}_2)$ or (ii) $H \simeq \mathrm{R}_{K/k}(\mathrm{PGL}_2)$ whereas if $\mathrm{char}(k) = 2$, there is a nonzero kK^2-subspace $V \subset K$ such that $k\langle V\rangle = K$ and either (i′) $H \simeq H_{V,K/k}$ or (ii′) $H \simeq \mathrm{P}H_{V,K/k}$. Consider cases (i) and (i′), so the central quotient map $H \to G$ has kernel contained in $Z_H \subset \mathrm{R}_{K/k}(\mu_2)$. Thus, this kernel is k-tame and H is uniquely dominated by $\widetilde{G}$ as central extensions of G. But $H^{\mathrm{ss}}_{\overline{k}} \simeq \mathrm{SL}_2$, so $\widetilde{G} = H$ and therefore $\widetilde{G}$ is of minimal type. In cases (ii) and (ii′) we have $Z_H = 1$, so $G = H$ and the k-group $\widetilde{G}$ coincides with $\mathrm{R}_{K/k}(\mathrm{SL}_2)$ or $H_{V,K/k}$, giving once again that $\widetilde{G}$ is of minimal type. □

Proposition 5.3.3. *Let G be a pseudo-semisimple group over a field k. Then G is locally of minimal type if and only if $\widetilde{G}$ is of minimal type.*

Proof. The property "locally of minimal type" is inherited by pseudo-reductive central quotients of pseudo-reductive groups, so if $\widetilde{G}$ is of minimal type (hence locally of minimal type) then its central quotient G is locally of minimal type. Conversely, assuming G is locally of minimal type, it suffices to show that $\widetilde{G}$ is locally of minimal type (due to Proposition 4.3.3). We may and do assume $k = k_s$. Let $\widetilde{T} \subset \widetilde{G}$ be a maximal k-torus, and $T \subset G$ the maximal k-torus image of $\widetilde{T}$ in G. By [CGP, Prop. 2.3.15], the inclusion $\mathrm{X}(T) \hookrightarrow \mathrm{X}(\widetilde{T})$ identifies $\Phi(G,T)$ with $\Phi(\widetilde{G},\widetilde{T})$. For each $a \in \Phi(G,T)$, let $G_a \subset G$ and $\widetilde{G}_a \subset \widetilde{G}$ be the k-subgroups generated by the $\pm a$-root groups. Root group considerations show that $\widetilde{G}_a$ maps onto G_a, thereby making $\widetilde{G}_a$ a k-tame central extension of G_a.

Applying 3.2.2 to the pseudo-semisimple $\widetilde{G}$ we see that $(\widetilde{G}_a)^{\rm ss}_{\overline{k}}$ can be identified with the subgroup of $\widetilde{G}^{\rm ss}_{\overline{k}}$ generated by a pair of opposite $\widetilde{T}_{\overline{k}}$-root groups (relative to a in the non-multipliable case, and $2a$ in the multipliable case). Any such pair of opposite root groups generates SL_2 since $\widetilde{G}^{\rm ss}_{\overline{k}}$ is simply connected, so $\widetilde{G}_a$ is the universal smooth k-tame central extension of G_a. By the very definition of G being locally of minimal type, it then follows from Lemma 5.3.2 that $\widetilde{G}_a$ is of minimal type, so $\widetilde{G}$ is locally of minimal type. □

The preservation of centrality under composition in Proposition 5.3.1(ii) underlies the centrality of the composite map f in the following result.

Proposition 5.3.4. *Let G be a pseudo-semisimple group locally of minimal type over a field k, $\widetilde{G}$ its universal smooth k-tame central extension, and $\{G_i\}$ the finite set of pseudo-simple normal k-subgroups of G. The composite central quotient map*

$$f : \prod \widetilde{G}_i \to \prod G_i \to G$$

uniquely lifts to an isomorphism $\prod \widetilde{G}_i \simeq \widetilde{G}$.

See [CGP, Prop. 3.1.8] for the proof that the G_i pairwise commute and $\prod G_i \to G$ is a central quotient map.

Proof. The construction (or universal property) of $\widetilde{G}$ implies that its formation is compatible with separable extension of the ground field, so it is elementary to reduce to the case $k = k_s$. Each $\widetilde{G}_i$ is absolutely pseudo-simple since $k = k_s$, and each is of minimal type by Proposition 5.3.3. Hence, any Cartan k-subgroup of $\widetilde{G}_i$ lies inside a k-group of the form $\mathrm{R}_{K_i/k}(T_i)$ for the minimal field of definition K_i/k for the geometric unipotent radical of $\widetilde{G}_i$ and a maximal K_i-torus $T_i \subset \widetilde{G}_i$. In particular, $Z_{\widetilde{G}_i}$ is contained inside such a Weil restriction, so $Z_{\widetilde{G}_i}$ contains no nontrivial unipotent k-subgroup. Thus, $\prod \widetilde{G}_i$ is a perfect smooth connected k-tame central extension of G, and its geometric maximal semisimple quotient is the direct product $\prod (\widetilde{G}_i)^{\rm ss}_{\overline{k}}$ that is simply connected. By the unique characterization of $\widetilde{G}$ near the end of Theorem 5.1.3, it follows that $\prod \widetilde{G}_i$ is uniquely isomorphic to $\widetilde{G}$ over G. □

Although inseparable Weil restriction does not generally preserve perfectness, the formation of the universal smooth k-tame central extension interacts well with *derived groups* of Weil restrictions:

Proposition 5.3.5. *Let k be a field, G an absolutely pseudo-simple k-group, and $\pi : \widetilde{G} \to G$ the universal smooth k-tame central extension of G. Let K/k*

be the common minimal field of definition for the geometric unipotent radicals of G *and* $\widetilde{G}$, *define* $G' = G_K^{\mathrm{ss}}$, *and let* $\widetilde{G}' \to G'$ *be the simply connected central cover of* G' *over* K.

(i) *The* k*-group* G *is standard if and only if* $\widetilde{G}$ *is standard. In such cases* $\widetilde{G} \simeq \mathrm{R}_{K/k}(\widetilde{G}')$ *over the canonical central quotient map* $\mathrm{R}_{K/k}(\widetilde{G}') \twoheadrightarrow G$.

(ii) *Let* $k_0 \subset k$ *be a subfield over which* k *is purely inseparable of finite degree. Consider the absolutely pseudo-simple* k_0*-groups* $G_0 := \mathscr{D}(\mathrm{R}_{k/k_0}(G))$ *and* $\widetilde{G}_0 := \mathscr{D}(\mathrm{R}_{k/k_0}(\widetilde{G}))$. *The map* $\widetilde{G}_0 \to G_0$ *is the universal smooth* k_0*-tame central extension of* G_0, *and if* G *is of minimal type then so are* G_0 *and* $\widetilde{G}_0$.

Proof. If the pseudo-semisimple $\widetilde{G}$ is standard then its central quotient G is standard by [CGP, Rem. 1.4.6]. Conversely, any pseudo-reductive central extension of a standard pseudo-reductive group is standard (Lemma 3.2.8). Since K/k is the minimal field of definition for the maximal geometric reductive quotient of $\widetilde{G}$, and $\widetilde{G}_K^{\mathrm{ss}} = \widetilde{G}'$, the desired description of $\widetilde{G}$ in the standard case in (i) follows from Proposition 3.2.7 (applied to $\widetilde{G}$).

For (ii), if G is of minimal type then so is $\mathrm{R}_{k/k_0}(G)$ (Example 2.3.9) and hence so is its normal derived group G_0 (by Lemma 2.3.10). Let $\pi : \widetilde{G} \to G$ be the natural central quotient map, and define $\pi_0 = \mathscr{D}(\mathrm{R}_{k/k_0}(\pi))$. Clearly π_0 has central kernel, and it is also surjective (since to prove surjectivity we may assume $k = k_s$, over which surjectivity is verified via consideration of root groups).

To prove that π_0 is the universal smooth k_0-tame central extension of G_0, it is harmless to extend the ground field k_0 to a separable closure, so $k = k_s$. Let μ denote the k-tame $\ker \pi$, so $\mu_0 := \ker \pi_0$ is contained in the k_0-group $\mathrm{R}_{k/k_0}(\mu)$ that is k_0-tame since μ is k-tame. It follows that μ_0 is k_0-tame, so $\widetilde{G}_0$ is a perfect smooth connected k_0-tame central extension of G_0. By [CGP, Prop. A.5.11(2), Prop. A.4.8], the natural map $(\widetilde{G}_0)_k \to \widetilde{G}$ induces an isomorphism between the maximal geometric semisimple quotients, so $\widetilde{G}_0$ has simply connected maximal geometric semisimple quotient. Thus, $\widetilde{G}_0$ is the universal smooth k_0-tame central extension of G_0, so Proposition 5.3.3 implies that if G_0 is of minimal type then so is $\widetilde{G}_0$. This proves (ii). □

6

Automorphisms, isomorphisms, and Tits classification

6.1 Isomorphism Theorem

The proofs of our main classification results (Theorems 6.3.11, 8.4.5, and 9.2.1) require understanding isomorphisms among pseudo-reductive groups, so we now establish a version of the Isomorphism Theorem for pseudo-split pseudo-reductive groups. A pseudo-reductive variant of the Isogeny Theorem for split connected semisimple groups is proved in Appendix A. The key to both proofs is a technique in [CGP, Thm. C.2.29] to construct pseudo-reductive subgroups of an ambient smooth affine group.

Theorem 6.1.1 (Isomorphism Theorem). *Let k be a field and let the pairs (H, S) and (H', S') be pseudo-split pseudo-reductive k-groups equipped with split maximal k-tori. Suppose there is given an isomorphism $f : Z_H(S) \simeq Z_{H'}(S')$ restricting to an isomorphism $f_S : S \simeq S'$ carrying a basis Δ' of $\Phi' := \Phi(H', S')$ to a basis Δ of $\Phi := \Phi(H, S)$. Assume in addition the following:*

(i) *For every $a \in \Delta$ and the corresponding $a' \in \Delta'$ via f_S, an isomorphism $f_a : H_a \simeq H'_{a'}$ is given that is equivariant via f_S for the actions of the k-tori S and S', where $H_a = \langle U_a, U_{-a}\rangle$ and $H'_{a'} = \langle U'_{a'}, U'_{-a'}\rangle$.*
(ii) *For every $a \in \Delta$, $f_S \circ a^\vee = a'^\vee$.*

Then there is a unique k-isomorphism $H \simeq H'$ extending f and $\{f_a\}_{a\in\Delta}$.

Since any k-automorphism of SL_2 or PGL_2 that restricts to the identity on the diagonal torus is easily checked to arise from the action of a unique diagonal element of $PGL_2(k)$, the interested reader may verify without difficulty that the Isomorphism Theorem for pinned split connected reductive groups is a consequence of Theorem 6.1.1.

Proof. Our proof will use [CGP, Thm. C.2.29], which was suggested by a result of Steinberg that he used to give a new proof of the Isomorphism Theorem in the reductive case. The maximality of S and S' as k-tori forces $Z_H(S)$ and $Z_{H'}(S')$ to be commutative. By [CGP, Thm. C.2.29], it suffices to show that f_a and f glue (necessarily uniquely) to an isomorphism $H_a \cdot Z_H(S) \simeq H'_{a'} \cdot Z_{H'}(S')$ for all matching pairs (a, a').

Since $Z_a := H_a \cap Z_H(S)$ is the centralizer of the S-action on H_a, it is smooth and connected and inherits pseudo-reductivity from $Z_H(S)$. Similarly, the centralizer $Z'_{a'}$ of the S'-action on $H'_{a'}$ is pseudo-reductive. The hypothesis in (i) gives S-equivariance of f_a, so $f_a(Z_a) = Z'_{a'}$. This isomorphism $f_a : Z_a \simeq Z'_{a'}$ between commutative pseudo-reductive k-groups carries the unique maximal k-torus $S \cap H_a$ of Z_a onto the unique maximal k-torus $S' \cap H'_{a'}$ of $Z'_{a'}$. By hypothesis (ii), f_S also carries $a^\vee(\mathrm{GL}_1) = S \cap H_a$ onto $a'^\vee(\mathrm{GL}_1) = S' \cap H'_{a'}$. Now the S-equivariance of f_a implies that for all $s \in S \cap H_a$, $f_S(s)^{-1} f_a(s)$ belongs to the center of $H'_{a'}$. But since $H'_{a'}$ is pseudo-simple, we conclude using [CGP, Lemma 1.2.1] that $f_S(s)^{-1} f_a(s) = 1$ for all $s \in S \cap H_a$. Hence, $f|_{S \cap H_a} = f_S|_{S \cap H_a} = f_a|_{S \cap H_a}$. Since $Z_{H'}(S')$ is a commutative pseudo-reductive group, the two homomorphisms $f|_{Z_a}$ and $f_a|_{Z_a}$ that carry Z_a into $Z_{H'}(S')$ and coincide on the maximal torus $S \cap H_a$ of Z_a are equal by [CGP, Prop. 1.2.2]. Hence, f carries Z_a into $Z'_{a'}$ via $f_a : Z_a \simeq Z'_{a'}$.

The anti-diagonal embedding of Z_a into $H_a \rtimes Z_H(S)$ has central image, so we have a central quotient presentation

$$(H_a \rtimes Z_H(S))/Z_a \simeq H_a \cdot Z_H(S)$$

and similarly for (H', S', a'). Since f and f_a restrict to the same isomorphism $Z_a \simeq Z'_{a'}$, to glue them to an isomorphism $H_a \cdot Z_H(S) \simeq H'_{a'} \cdot Z_{H'}(S')$ it remains to check that f_a is equivariant with respect to f (i.e., $f_a(zhz^{-1}) = f(z) f_a(h) f(z)^{-1}$ for $h \in H_a$ and $z \in Z_H(S)$). We may assume $k = k_s$. It suffices to show for $z \in Z_H(S)(k)$ that the map $H_a \to H'_{a'}$ defined by

$$h \mapsto f(z)^{-1} f_a(zhz^{-1}) f(z)$$

coincides with f_a. By pseudo-reductivity of $H'_{a'}$ and [CGP, Prop. 1.2.2], such an equality of k-homomorphisms is equivalent to equality for the induced isomorphisms $(H_a)^{\mathrm{red}}_{\overline{k}} \simeq (H'_{a'})^{\mathrm{red}}_{\overline{k}}$ between maximal geometric reductive quotients. The map $(H_a)^{\mathrm{red}}_{\overline{k}} \to (H'_{a'})^{\mathrm{red}}_{\overline{k}}$ induced by f_a is equivariant with respect to $(f_S)_{\overline{k}}$, so it suffices to show that the action of $Z_H(S)_{\overline{k}}$ on $(H_a)^{\mathrm{red}}_{\overline{k}}$ factors through its maximal torus quotient that is also the image of $S_{\overline{k}}$, and likewise for (H', S', a').

By [CGP, Prop. A.4.8], the quotient $(H_a)^{\mathrm{red}}_{\overline{k}}$ of $(H_a)_{\overline{k}}$ is the image of $(H_a)_{\overline{k}}$ in $H^{\mathrm{red}}_{\overline{k}}$, so it remains to observe that the image of $Z_H(S)_{\overline{k}} = Z_{H_{\overline{k}}}(S_{\overline{k}})$ under the quotient map $H_{\overline{k}} \twoheadrightarrow H^{\mathrm{red}}_{\overline{k}}$ is the centralizer of $S_{\overline{k}}$, and that this centralizer is $S_{\overline{k}}$ since $H^{\mathrm{red}}_{\overline{k}}$ is connected reductive. □

Let G be a pseudo-reductive group over a field k, and let C be a Cartan k-subgroup of G. We shall now use the Isomorphism Theorem for pseudo-split pseudo-reductive groups to prove in general the connectedness of the (commutative) maximal smooth closed k-subgroup $Z_{G,C}$ of $\mathrm{Aut}_{G,C}$ (see 4.1.2). This requires the following useful lemma:

Lemma 6.1.2. *Let G be a pseudo-reductive group over a field k and $Z \subset G$ a closed central k-subgroup such that $\overline{G} := G/Z$ is pseudo-reductive. Let $C \subset G$ be a Cartan k-subgroup, and $\overline{C} = C/Z$ its Cartan k-subgroup image in $\overline{G}$.*

(i) *If $C' := C \cap \mathscr{D}(G)$ is the associated Cartan k-subgroup of $\mathscr{D}(G)$ then the natural map $Z_{G,C} \to Z_{\mathscr{D}(G),C'}$ is an isomorphism.*
(ii) *For the natural map $\alpha : \mathrm{Aut}_{G,C} \to \mathrm{Aut}_{\overline{G},\overline{C}}$, the induced map $Z_{G,C} \to Z_{\overline{G},\overline{C}}$ is an isomorphism.*

By [CGP, Lemma 1.2.5(iii), Prop. 1.2.6] the map $C \mapsto C \cap \mathscr{D}(G)$ is a bijection between the sets of Cartan k-subgroups of G and $\mathscr{D}(G)$. An interesting special case of (ii) is the central quotient $\overline{G} = G/\mathscr{C}_G$ as in §2.3.

Proof. Since $G = (\mathscr{D}(G) \rtimes C)/C'$ (using the central anti-diagonal inclusion of C'), it is clear that the restriction map $Z_{G,C} \to Z_{\mathscr{D}(G),C'}$ between smooth affine k-groups induces a bijection between points valued in every separable extension of k. Hence, this map is an isomorphism (as explained in the proof of [CGP, Prop. 8.2.6]), so (i) is proved.

By (i), the natural map of k-group schemes

$$\theta : \mathrm{Aut}_{\mathscr{D}(G),C'} \to \mathrm{Aut}_{G,C}$$

has restriction $\theta^{\mathrm{sm}} : Z_{\mathscr{D}(G),C'} \to Z_{G,C}$ between maximal smooth closed k-subgroups that is an isomorphism. (In fact, θ is an isomorphism of k-group schemes because $\mathscr{D}(G)$ represents the fppf sheafification of the commutator subfunctor of G on the category of k-schemes, but we do not need this stronger property of θ.) The same holds compatibly for the pair $(\overline{G}, \overline{C})$, so for the proof of (ii) we may replace G with $\mathscr{D}(G)$ to reduce to the case that G is pseudo-semisimple. Hence, $Z(k_s)$ is finite.

Since k is arbitrary and the formation of α commutes with separable extension on k, to prove (ii) it suffices to prove α is bijective on k-points. By Galois descent we may assume $k = k_s$, so the maximal k-torus T of C is split. It remains to show that any k-automorphism $\overline{f}$ of $\overline{G}$ restricting to the identity on $\overline{C}$ uniquely lifts to a k-automorphism f of G restricting to the identity on C.

Let $\overline{T}$ be the image of T under the central quotient map $q : G \to \overline{G}$, so q induces an isomorphism between the root systems $\Phi(G,T)$ and $\Phi(\overline{G},\overline{T})$ [CGP, Prop. 2.3.15]. If $\lambda \in X_*(T) \subset X_*(\overline{T})$ does not annihilate any roots then q carries $C = Z_G(\lambda)$ onto $\overline{C} = Z_{\overline{G}}(\lambda)$ and carries $U_G(\pm\lambda)$ isomorphically onto $U_{\overline{G}}(\pm\lambda)$. Since any k-automorphism of G restricting to the identity on C must restrict to an automorphism of each root group, and likewise for $(\overline{G},\overline{C})$, the only possibility for f is a k-automorphism of G extending the automorphism f' of $\Omega := U_G(-\lambda)\times C \times U_G(\lambda)$ given by $uzu' \mapsto \overline{f}(u)z\overline{f}(u')$ via the isomorphism $q : U_G(\pm\lambda) \simeq U_{\overline{G}}(\pm\lambda)$. To construct f extending f', we just have to check that f' is a rational homomorphism; i.e., the map $h : (\Omega\times\Omega)\cap m_G^{-1}(\Omega) \to G$ given by $(\omega_1,\omega_2) \mapsto f'(\omega_1\omega_2)f'(\omega_2)^{-1}f'(\omega_1)^{-1}$ is the constant map equal to 1. Obviously $q \circ h = 1$, so h is valued in $\ker(q) = Z$. But $Z(k)$ is finite whereas $(\Omega\times\Omega)\cap m_G^{-1}(\Omega)$ is connected with a Zariski-dense set of k-points, so h is the constant map equal to 1. □

As a preliminary step towards a description of $Z_{G,C}$ in general, first we need to understand the case where G_{k_s} has a rank-1 root system. The BC_1-case is part of [CGP, Prop. 9.8.15], so now we address the A_1-case:

Lemma 6.1.3. *Let G be absolutely pseudo-simple with minimal field of definition K/k for its geometric unipotent radical, and assume G_{k_s} has root system A_1. Let T be a maximal k-torus in G and define $C = Z_G(T)$, $T^{\rm ad} = T/(T\cap Z_G)$, and $G' = G_K^{\rm ss}$. Let $F \subset K$ be the root field of G.*

The natural action on $R_{K/k}(G')$ by the k-group $R_{F/k}(T_F^{\rm ad}) \subset R_{K/k}(T_K^{\rm ad})$ uniquely lifts through i_G to an action on G that is trivial on C, and the k-homorphism $f : R_{F/k}(T_F^{\rm ad}) \to Z_{G,C}$ classifying this action is an isomorphism.

Proof. The formation of Z_G is compatible with passing to central quotients by Proposition 5.3.1(ii), and likewise for the formation of root fields by Remark 3.3.3. Thus, by Lemma 6.1.2(ii) we can replace G with $G/\mathscr{C}_G$ (and T with its image in this quotient) to arrange that G is of minimal type. We can then replace G with its universal smooth k-tame central extension $\widetilde{G}$ and replace T with the corresponding maximal k-torus $\widetilde{T} \subset \widetilde{G}$ via Proposition 5.3.1(i) so that $G_K^{\rm ss} \simeq SL_2$.

In view of the uniqueness assertion, by Galois descent we may now assume $k = k_s$, so T is split. By [CGP, Thm. 3.4.6] there exists a Levi k-subgroup of G containing T, and we may identify it with SL_2 over k carrying T onto the diagonal k-torus. Thus, via Proposition 3.1.8, if $\mathrm{char}(k) \neq 2$ then i_G identifies G with $\mathrm{R}_{K/k}(\mathrm{SL}_2)$ (carrying T to the diagonal k-torus) whereas if $\mathrm{char}(k) = 2$ then i_G identifies G with $H_{V,K/k} \subset \mathrm{R}_{K/k}(\mathrm{SL}_2)$ for a nonzero kK^2-subspace $V \subset K$ containing 1 such that $k[V] = K$ and T is identified with the diagonal k-torus in $H_{V,K/k}$. Define $V = K$ if $\mathrm{char}(k) \neq 2$, so in all cases the T-root groups of $G = i_G(G) \subset \mathrm{R}_{K/k}(\mathrm{SL}_2)$ are the subgroups $U_V^{\pm}$ corresponding to $V \subset K$ inside the (upper and lower triangular) root groups of $\mathrm{R}_{K/k}(\mathrm{SL}_2)$.

Under the natural action of $\mathrm{R}_{K/k}(\mathrm{PGL}_2)$ on $\mathrm{R}_{K/k}(\mathrm{SL}_2)$, the diagonal k-subgroup $\mathrm{R}_{F/k}(\mathrm{GL}_1) \subset \mathrm{R}_{K/k}(\mathrm{PGL}_2)$ preserves the root groups $U_V^{\pm}$ of $i_G(G)$ that generate $i_G(G)$. This defines an action of $\mathrm{R}_{F/k}(\mathrm{GL}_1)$ on $i_G(G) = G$ that is trivial on C and compatible with the natural action on $\mathrm{R}_{K/k}(\mathrm{SL}_2)$. This is exactly an action of $\mathrm{R}_{F/k}(T_F^{\mathrm{ad}})$ on G extending the identity on C and equivariant through i_G with the natural action on $\mathrm{R}_{K/k}(G')$, and it is unique as such because uniqueness holds on k-points (since any k-automorphism of G is uniquely determined by the induced K-automorphism of G' [CGP, Prop. 1.2.2]). We have proved existence and uniqueness of the desired $f : \mathrm{R}_{F/k}(T_F^{\mathrm{ad}}) \to Z_{G,C}$ and need to prove that f is an isomorphism.

Recall the general fact (reviewed in the proof of [CGP, Prop. 8.2.6]) that a k-homomorphism between smooth k-groups of finite type is an isomorphism if it is bijective on points valued in every separable extension k'/k, and that we may restrict attention to separably closed k'. The formation of f and F/k commutes with separable extension on k, so we may rename such k' as k to reduce to checking that f is bijective on k-points. In other words, it suffices to show that for any k-automorphism θ of G that is trivial on C, the induced automorphism of $\mathrm{R}_{K/k}(G')$ arises from a unique element of $T^{\mathrm{ad}}(F)$.

Since $\theta|_C = \mathrm{id}_C$, the induced K-automorphism θ' of G' restricts to the identity on the image T_K of C_K and so is induced by a unique $t' \in T^{\mathrm{ad}}(K)$; i.e., t' is diagonal in $\mathrm{PGL}_2(K) = \mathrm{Aut}_K(G')$. Writing $t' = \mathrm{diag}(\lambda, 1)$ for $\lambda \in K^{\times}$, the effect of t' on the T_K-root lines in $\mathfrak{sl}_2(K)$ is multiplication by $\lambda^{\pm 1}$, and this has to preserve the subspaces arising from the T-root groups of G. Thus, multiplication by $\lambda^{\pm 1}$ on K preserves V; i.e., $\lambda \in F$. Hence, $t' \in T^{\mathrm{ad}}(F)$. □

By [CGP, Thm. 2.4.1], if G is a pseudo-reductive k-group and C is a Cartan k-subgroup then $Z_{G,C}^0$ is pseudo-reductive. Here is an improvement:

Proposition 6.1.4. *The k-group $Z_{G,C}$ is connected. In particular, $Z_{G,C}$ is pseudo-reductive. Moreover, $Z_{G,C}$ is of minimal type (equivalently, $Z_{G,C}$ does* *not contain a nontrivial unipotent k-subgroup scheme), and if G is pseudo-split with basis Δ of $\Phi(G,T)$ for a split maximal k-torus $T \subset G$ then naturally*

$$Z_{G,C} \simeq \prod_{a\in\Delta} Z_{G_a,C_a} \tag{6.1.4}$$

for $G_a := \langle U_a, U_{-a}\rangle$, $T_a := T \cap G_a$ a split maximal k-torus in G_a, and $C_a := Z_{G_a}(T_a)$ a Cartan k-subgroup of G_a.

Proof. Everything is obvious if G is commutative (equivalently, $G = C$), so we may and do assume G is non-commutative. By Lemma 6.1.2(i) the formation of $Z_{G,C}$ is unaffected by replacing G with $\mathscr{D}(G)$, so we may assume G is pseudo-semisimple. By Lemma 6.1.2(ii) with $Z = \mathscr{C}_G$, we may assume G is of minimal type. Likewise, we can pass to the universal smooth k-tame central extension of G so that $G^{\rm ss}_{\overline{k}}$ is simply connected. It suffices to treat the case when the maximal k-torus T in C is split (via scalar extension to the splitting field of T). The idea is to reduce to the case where G is pseudo-semisimple of rank 1, and then to carry out a direct analysis in the rank-1 cases.

Let Δ be a basis of $\Phi := \Phi(G,T)$, so for $a \in \Delta$ a k-automorphism of G_a restricts to the identity on C_a if and only if it restricts to the identity on T_a (as C_a is commutative and pseudo-reductive). For any k-algebra A and $f \in \mathrm{Aut}_{G,C}(A)$, the functorial characterization of root groups over k-algebras implies that f carries each $(U_a)_A$ into itself, hence isomorphically onto itself (by using f^{-1}). Thus, likewise f carries $(G_a)_A$ isomorphically onto itself, so we obtain a natural k-homomorphism $\mathrm{Aut}_{G,C} \to \prod_{a\in\Delta} \mathrm{Aut}_{G_a,C_a}$ and hence a k-homomorphism as in (6.1.4) between maximal smooth closed k-subgroups. To prove that this latter homomorphism is an isomorphism it suffices to do so on rational points over all separable extensions of k (as explained in the proof of [CGP, Prop. 8.2.6]), and by Galois theory we may restrict attention to such extensions that are separably closed. Renaming such an extension as k reduces the isomorphism property for (6.1.4) to its bijectivity on k-points. This in turn is an immediate special case of the Isomorphism Theorem (Theorem 6.1.1) using the identity map on C since $C = \prod_{a\in\Delta} C_a$ by Proposition 4.3.3.

To show that $Z_{G,C}$ is conncted and of minimal type, it now suffices to treat the pairs (G_a, C_a) separately, so we may assume that G is absolutely pseudo-simple of minimal type and Φ has rank 1. The k-group $Z_{G,C}$ is described explicitly in Lemma 6.1.3 when $\Phi = \mathrm{A}_1$ and in [CGP, Prop. 9.8.15] when $\Phi = \mathrm{BC}_1$. In both cases, we conclude by inspection. □

Example 6.1.5. As an application of the connectedness (and hence the pseudo-reductivity) of $Z_{G,C}$ in Proposition 6.1.4, we may use the canonical map $f : C \to Z_{G,C}$, composed with inversion, to obtain a central quotient

$$\mathscr{G} := (G \rtimes Z_{G,C})/C \tag{6.1.5}$$

of $G \rtimes Z_{G,C}$ that is pseudo-reductive (by [CGP, Prop. 1.4.3]) with Cartan k-subgroup $Z_{G,C}$. Since $\ker(f) = Z_G$, the image of G in $\mathscr{G}$ is G/Z_G. (In Proposition 6.2.4 we will see that if $G = \mathscr{D}(G)$ then $\mathscr{G}$ is the identity component of the maximal smooth k-subgroup of the automorphism scheme of G.)

This construction leads to some instructive examples (not used in later proofs) over imperfect fields k of characteristic 2. To explain the context, recall that if $[k : k^2] = 2$ then any pseudo-reductive k-group $\mathscr{H}$ has the unique decomposition $\mathscr{H}_1 \times \mathscr{H}_2$ where $\mathscr{H}_1$ has a reduced root system (over k_s) and $\mathscr{H}_2$ is pseudo-semisimple with root system (over k_s) whose irreducible components are non-reduced [CGP, Prop. 10.1.6]. If $[k : k^2] > 2$ then such a unique decomposition exists when $\mathscr{H}$ is perfect of minimal type, as we may check over k_s by using both the central quotient map $\prod H_i \to \mathscr{H}_{k_s}$ for the pseudo-simple normal k_s-subgroups H_i of $\mathscr{H}_{k_s}$ [CGP, Prop. 3.1.8, Prop. 3.2.10] and the triviality of the center of *minimal type* pseudo-split pseudo-simple groups whose irreducible root system is non-reduced [CGP, Prop. 9.4.9]. However, generally there is no pseudo-reductive direct factor $\mathscr{H}_2$ of $\mathscr{H}$ whose root system (over k_s) is the union of the non-reduced irreducible components of the root system of $\mathscr{H}_{k_s}$ when $\mathscr{H}$ is either

(i) perfect but not of minimal type,
(ii) non-perfect of minimal type.

Examples as in (i) can be made over any k satisying $[k : k^2] \geqslant 16$ as follows. Let G_1 be a pseudo-split absolutely pseudo-simple k-group such that $\mathscr{C}_{G_1} = \alpha_2$ and the root system is reduced (see §B.1, §B.2, or 4.2.2). Let G_2 be a pseudo-split absolutely pseudo-simple k-group such that $\mathscr{C}_{G_2} \supset \alpha_2$ and the root system is non-reduced (see Example B.4.1, Remark B.4.2, and Example B.4.3). Pasting G_1 and G_2 along the central α_2 gives a k-group G that is an extension of $G_1/\mathscr{C}_{G_1}$ by G_2. The k-group G is clearly pseudo-semisimple, and it is not of minimal type since its normal k-subgroup G_2 is not of minimal type.

It is more interesting to give examples as in (ii), assuming $[k : k^2] > 2$. We will do this via the pushout construction (6.1.5). For $m \equiv 2 \bmod 4$, define $G_1 = \mathscr{D}(\mathrm{R}_{K/k}(\mathrm{PGL}_m))$ with a split maximal k-torus denoted as T_1, so $Z_{G_1} = 1$ and the equality $G_1 = \mathrm{R}_{K/k}(\mathrm{SL}_m)/\mathrm{R}_{K/k}(\mu_m)$ identifies T_1 with the maximal k-

torus in $C_1 := \mathrm{R}_{K/k}(T)/\mathrm{R}_{K/k}(\mu_m)$ for a unique (split) maximal K-torus T of SL_m. The map $C_1 \hookrightarrow Z_{G_1,C_1}$ and natural inclusion $j : \mathrm{R}_{K/k}(T)/\mathrm{R}_{K/k}(\mu_m) \hookrightarrow \mathrm{R}_{K/k}(T/\mu_m)$ are thereby identified. Since $[m](\mathrm{R}_{K/k}(\mathrm{GL}_1)) = \mathrm{R}_{kK^2/k}(\mathrm{GL}_1)$ (as $m \equiv 2 \bmod 4$), the description of $\mu_m = Z_{\mathrm{SL}_m}$ in terms of T-valued coroots identifies j with $\mathrm{R}_{kK^2/k}(\mathrm{GL}_1) \times \mathrm{R}_{K/k}(\mathrm{GL}_1)^{m-2} \to \mathrm{R}_{K/k}(\mathrm{GL}_1)^{m-1}$.

The pseudo-split pseudo-simple k-groups of minimal type with root system BC_n ($n \geqslant 1$) are classified in [CGP, Thm. 9.8.6, Prop. 9.8.9] (these groups all have *trivial center*). Let (G_2, T_2) be such a group as constructed in [CGP, Ex. 9.8.16]; the minimal field of definition K/k for its geometric unipotent radical may be any nontrivial purely inseparable finite extension of k, and kK^2 is then necessarily a proper subfield of K. Since we assume $[k : k^2] \geqslant 4$, we may choose K so that $[K : kK^2] \geqslant 4$ (e.g., let $K = k(u^{1/2}, v^{1/2})$ for $\{u, v\}$ part of a 2-basis of k). The inclusion $C_2 := Z_{G_2}(T_2) \hookrightarrow Z_{G_2,C_2}$ is identified with the natural map $\mathrm{R}_{K/k}(\mathrm{GL}_1)^n \hookrightarrow \mathrm{R}_{F_\natural/k}(\mathrm{GL}_1) \times \mathrm{R}_{K/k}(\mathrm{GL}_1)^{n-1}$ for a finite extension $F_\natural/K$ contained in $K^{1/2}$ due to [CGP, 9.6.8, (9.7.6), Thm. 9.8.1(2), (9.8.6)]. Explicitly, G_2 rests on an arbitrary choice of nonzero kK^2-subspace $V' \subset K$ and a nonzero finite-dimensional K^2-subspace $V^{(2)} \subset K$ such that $V' \cap V^{(2)} = 0$ (with the additional requirement that $k\langle V^{(2)} + V'\rangle = K$ when $n = 1$), and $F_\natural^2$ is the K^2-finite subfield of elements of K whose multiplication action preserves V' and $V^{(2)} \oplus V'$.

In [CGP, Ex. 9.8.16] it is explained how to choose $V^{(2)}$ and V' so that $F_\natural^2$ strictly contains K^2, and that property underlies the construction of G_2. But we shall require more: $F_\natural^2$ should also not be contained in kK^2. (This would be impossible to arrange if $[K : kK^2] = 2$, as then V' would have to be a kK^2-line, forcing $F_\natural^2 \subset kK^2$.) Since $[K : kK^2] \geqslant 4$, we can choose a proper subfield E of K that strictly contains kK^2. Let $V' = E$, pick $\alpha \in E - kK^2$, and let $V^{(2)} \oplus V'$ correspond to the $K^2(\alpha)$-span of an E-basis of the quotient K/V' (so $k\langle V^{(2)} + V'\rangle = K$). Thus, $F_\natural^2$ contains the field $K^2(\alpha)$ that strictly contains K^2 and is not contained in kK^2.

Define $G := G_1 \times G_2$ (so $Z_G = 1$), $C := C_1 \times C_2$, and consider a smooth connected k-subgroup

$$\mathscr{C} \subset Z_{G,C} = Z_{G_1,C_1} \times Z_{G_2,C_2}$$

(equality by Proposition 6.1.4) with $C \subset \mathscr{C}$. The smooth connected k-subgroup

$$\mathscr{H} := (G \rtimes \mathscr{C})/C \subset (G \rtimes Z_{G,C})/C$$

is normal and hence pseudo-reductive, with $\mathscr{D}(\mathscr{H}) = G$ (since $Z_G = 1$). If $\mathscr{H} = \mathscr{H}_1 \times \mathscr{H}_2$ (so each $\mathscr{H}_j$ is pseudo-reductive and pseudo-split) where $\mathscr{H}_2$ has root system that is the non-reduced component of that of $\mathscr{H}$ then $\mathscr{D}(\mathscr{H}_j) = G_j$ and the Cartan k-subgroup $\mathscr{C}$ of $\mathscr{H}$ must have the form $\mathscr{C}_1 \times \mathscr{C}_2$ inside $Z_{G,C}$ for $\mathscr{C}_j$ lying between Z_{G_j,C_j} and C_j. Hence, to rule out such a direct product decomposition for $\mathscr{H}$ it suffices to choose $\mathscr{C}$ that does not have the form $\mathscr{C}_1 \times \mathscr{C}_2$. Equivalently, we seek a smooth connected k-subgroup of

$$Z_{G,C}/C = (\mathrm{R}_{K/k}(\mathrm{GL}_1)/\mathrm{R}_{kK^2/k}(\mathrm{GL}_1)) \times (\mathrm{R}_{F_\natural/k}(\mathrm{GL}_1)/\mathrm{R}_{K/k}(\mathrm{GL}_1))$$

that is not a direct product of k-subgroups of the two factors. The squaring map from the second factor into the first is nontrivial since $F_\natural^2 \not\subset kK^2$, so its graph Γ is such a k-subgroup.

We will now prove that for any $\mathscr{C} \subset Z_{G,C}$ containing C, the corresponding pseudo-reductive group $\mathscr{H}$ is of minimal type. It suffices to show that $\mathscr{H}$ does not contain a nontrivial central unipotent k-subgroup scheme (since $\mathscr{C}_{\mathscr{H}}$ is a central unipotent k-subgroup scheme). For this, first note that the natural projection $G \rtimes \mathscr{C} \to (G \rtimes \mathscr{C})/C = \mathscr{H}$ carries $\mathscr{C}$ isomorphically onto a Cartan k-subgroup of $\mathscr{H}$. The center of $\mathscr{H}$ is contained in any Cartan k-subgroup, and $\mathscr{C}$ is a k-subgroup of $Z_{G,C}$. But $Z_{G,C}$ does not contain a nontrivial unipotent k-subgroup scheme since $Z_{G,C} = Z_{G_1,C_1} \times Z_{G_2,C_2}$. This proves that $\mathscr{H}$ is of minimal type.

6.1.6. As an application of the Isomorphism Theorem and Lemma 6.1.2, we next address the behavior of $Z_{G,C}$ with respect to Weil restriction in the pseudo-reductive case. To formulate this, we first review some constructions.

Let k' be a nonzero finite reduced algebra over a field k and G' a smooth affine k'-group with pseudo-reductive fibers over the factor fields of k'. Let $G = \mathrm{R}_{k'/k}(G')$ and define $Z_{G',C'}$ to be the commutative fiberwise maximal smooth k'-subgroup of the automorphism scheme $\mathrm{Aut}_{G',C'}$ over k' (built in the evident manner, working separately over the factor fields of k'). By [CGP, Prop. A.5.15(3)], $C' \mapsto \mathrm{R}_{k'/k}(C')$ is a bijection from the set of Cartan k'-subgroups of G' to the set of Cartan k-subgroups of $G := \mathrm{R}_{k'/k}(G')$. Moreover, $C \mapsto C \cap \mathscr{D}(G)$ is a bijection between the sets of Cartan k-subgroups of G and $\mathscr{D}(G)$ due to [CGP, Lemma 1.2.5(iii), Prop. 1.2.6]. Hence, each Cartan k-subgroup $\mathscr{C}$ of $\mathscr{D}(G)$ has the form $\mathrm{R}_{k'/k}(C') \cap \mathscr{D}(G)$ for a unique Cartan k'-subgroup $C' \subset G'$ and all such intersections are Cartan k-subgroups of $\mathscr{D}(G)$.

Fix a C in G, and consider the associated C' in G' and $\mathscr{C} := C \cap \mathscr{D}(G)$ in $\mathscr{D}(G)$. By Lemma 6.1.2(i) $Z_{\mathscr{D}(G),\mathscr{C}} \simeq Z_{G,C}$. To relate $\mathrm{R}_{k'/k}(Z_{G',C'})$ and

$Z_{G,C}$, observe that for any k-algebra A and $A_{k'}$-automorphism φ of $G'_{A_{k'}}$ that restricts to the identity on $C'_{A_{k'}}$, the map $\mathrm{R}_{A_{k'}/A}(\varphi)$ is an A-automorphism of $\mathrm{R}_{A_{k'}/A}(G'_{A_{k'}}) = G_A$ that restricts to the identity on C_A. This defines a homomorphism $\mathrm{R}_{k'/k}(\mathrm{Aut}_{G',C'}) \to \mathrm{Aut}_{G,C}$ of k-group schemes that must carry the commutative k-smooth closed subgroup $\mathrm{R}_{k'/k}(Z_{G',C'})$ into the commutative maximal k-smooth closed subgroup $Z_{G,C}$, yielding k-homomorphisms

$$\mathrm{R}_{k'/k}(Z_{G',C'}) \to Z_{G,C} \to Z_{\mathscr{D}(G),\mathscr{C}}. \tag{6.1.6}$$

Proposition 6.1.7. *The maps in* (6.1.6) *are isomorphisms. Moreover, using $\mathscr{C}$-conjugation on $\mathscr{D}(G)$, the resulting k-homomorphism $\mathscr{C} \to Z_{\mathscr{D}(G),\mathscr{C}}$ between commutative pseudo-reductive k-groups restricts to an isogeny between their unique maximal k-tori.*

Proof. By Lemma 6.1.2(i), the second map in (6.1.6) is an isomorphism. By the same reasoning as in the proof of Lemma 6.1.3 (making an arbitrary separable extension on the ground field, and renaming the new ground field as k), the first map in (6.1.6) is an isomorphism if it induces a bijection on k-points in general, and we may moreover assume $k = k_s$ (so all factor fields of k' are separably closed). By left-exactness of Weil restriction, G and its k-subgroup $\mathrm{R}_{k'/k}(\mathscr{D}(G'))$ have the same derived groups, so we may assume $G' = \mathscr{D}(G')$.

If $\{k'_i\}$ is the set of factor fields of k' and (G'_i, C'_i) is the k'_i-fiber of (G', C') then $G = \prod G_i$ and $C = \prod C_i$ for $G_i := \mathrm{R}_{k'_i/k}(G'_i)$ and $C_i = \mathrm{R}_{k'_i/k}(C'_i)$. Let $T' \subset C'$ be the unique maximal k'-torus, so if T'_i is its k'_i-fiber then the maximal k-torus T of C is $\prod T_i$ where T_i is the maximal k-torus of $\mathrm{R}_{k'_i/k}(T'_i)$. For the isomorphism assertion, our task is to show that every k-automorphism f of G restricting to the identity on C has the form $\mathrm{R}_{k'/k}(f')$ for a unique k'-automorphism f' of G' restricting to the identity on C'. Since we have arranged that each G'_i is perfect and hence is generated by its T'_i-root groups, any k'-automorphism of G' restricting to the identity on C' (equivalently, on T') must arise from a collection of automorphisms $f'_i \in Z_{G'_i,C'_i}(k'_i)$. Likewise, for the Cartan k-subgroups $\mathscr{C} := C \cap \mathscr{D}(G)$ and $\mathscr{C}_i := C_i \cap \mathscr{D}(G_i)$ we have

$$Z_{G,C}(k) = Z_{\mathscr{D}(G),\mathscr{C}}(k) = \prod Z_{\mathscr{D}(G_i),\mathscr{C}_i}(k) = \prod Z_{G_i,C_i}(k),$$

so for both the isomorphism and torus isogeny assertions it suffices to treat each triple $(k'_i/k, G'_i, C'_i)$ separately; i.e., we can assume k' is a field.

By Lemma 6.1.2(ii) it is harmless to replace G' with the pseudo-reductive central quotient G'/Z' for $Z' = \mathscr{C}_{G'}$ since for the central k-subgroup scheme

$Z = G \cap \mathrm{R}_{k'/k}(Z')$ in G the inclusion $G/Z \hookrightarrow \mathrm{R}_{k'/k}(G'/Z')$ induces an equality on derived groups (as we see by considerations of root groups). Thus, we can assume that G' is of minimal type. Likewise, we can replace G' with its universal smooth k'-tame central extension so that its root datum is simply connected. The root systems of the pairs (G,T) and (G',T') are naturally identified and the corresponding root groups are related through Weil restriction (see [CGP, Ex. 2.3.2]), so by the Isomorphism Theorem and Proposition 6.1.4 we are reduced to the rank-1 case. We now separately consider when the common root system Φ of (G,T) and (G',T') is A_1 or BC_1.

Suppose Φ is A_1, and let K'/k' be the minimal field of definition for the geometric unipotent radical of G'. Identify $(G'_{K'})^{\mathrm{ss}}$ with SL_2 carrying $T'_{K'}$ to the diagonal K'-torus and carrying a chosen coroot to $t' \mapsto \mathrm{diag}(t', 1/t')$. By Proposition 3.1.7, if $\mathrm{char}(k) \neq 2$ then $G' = \mathrm{R}_{K'/k'}(\mathrm{SL}_2)$ whereas if $\mathrm{char}(k) = 2$ then $G' = H_{V',K'/k'}$ for a nonzero $k'K'^2$-subspace $V' \subset K'$ satisfying $k'\langle V' \rangle = K'$. Letting $F' \subset K'$ be the root field of G' (so $k' \subset F'$, and $F' = K'$ if $\mathrm{char}(k) \neq 2$), Lemma 6.1.3 gives that $Z_{G',C'} = \mathrm{R}_{F'/k'}(\mathrm{GL}_1)$ with its action on G' induced by the diagonal of $\mathrm{R}_{K'/k'}(\mathrm{PGL}_2)$ via $t' \mapsto \mathrm{diag}(t', 1)$.

If $\mathrm{char}(k) \neq 2$ then $G = \mathrm{R}_{K'/k}(\mathrm{SL}_2)$ is perfect. If $\mathrm{char}(k) = 2$ then (3.1.6) gives $\mathscr{D}(G) = H_{V',K'/k}$ with $k\langle V' \rangle = k'\langle V' \rangle = K'$. In both cases, by inspection the root field over k is also equal to F' and $Z_{G,C}$ is identified with $\mathrm{R}_{F'/k}(\mathrm{GL}_1)$, under which the map of interest $\mathrm{R}_{k'/k}(Z_{G',C'}) \to Z_{G,C}$ is easily seen to be the natural isomorphism $\mathrm{R}_{k'/k}(\mathrm{R}_{F'/k'}(\mathrm{GL}_1)) \simeq \mathrm{R}_{F'/k}(\mathrm{GL}_1)$ via calculation on k-points.

Next consider the BC_1-case, so k is imperfect with characteristic 2. Once again let K'/k' be the minimal field of definition for the geometric unipotent radical of G'. An explicit description of (G',T') as the derived group of a construction in terms of certain linear algebra data relative to K'/k' is given in [CGP, Thm. 9.8.6(i)], and its compatibility with passage to $\mathscr{D}(G) = \mathscr{D}(\mathrm{R}_{k'/k}(G'))$ is given in [CGP, Prop. 9.8.13] using "the same" linear algebra data viewed relative to K'/k. Upon choosing a basis of the root system, an identification of $Z_{G',C'}$ with $\mathrm{R}_{F'_\natural/k'}(\mathrm{GL}_1)$ for a subfield $F'_\natural \subset K'^{1/2}$ over K' is given in [CGP, (9.8.7)]. The definition of $F'_\natural$ in terms of the linear algebra data shows that it is also the field (now viewed over k) which arises in the description of $Z_{G,C}$, and the construction of these identifications is easily seen to identify the map of interest $\mathrm{R}_{k'/k}(Z_{G',C'}) \to Z_{G,C}$ with the natural isomorphism $\mathrm{R}_{k'/k}(\mathrm{R}_{F'_\natural/k'}(\mathrm{GL}_1)) \simeq \mathrm{R}_{F'_\natural/k}(\mathrm{GL}_1)$. □

6.2 Automorphism schemes

Away from the reductive case, the consideration of automorphism schemes for pseudo-reductive groups has to be restricted to the pseudo-semisimple case because commutative pseudo-reductive groups that are not tori generally have a non-representable automorphism functor:

Example 6.2.1. Let k be an imperfect field of characteristic $p > 0$, and let $G = \mathrm{R}_{k'/k}(\mathrm{GL}_1)$ for a nontrivial finite extension k'/k satisfying $k'^p \subset k$. Since $G_{k'} = \mathrm{GL}_1 \times \mathbf{G}_{\mathrm{a}}^n$ for $n = [k' : k] - 1$, if $\underline{\mathrm{Aut}}_{G/k}$ is represented by a k-scheme $\mathrm{Aut}_{G/k}$ then $\underline{\mathrm{Aut}}_{\mathbf{G}_{\mathrm{a}}^n/k'}$ is represented by a closed subscheme of $(\mathrm{Aut}_{G/k})_{k'}$. Thus, non-representability of $\underline{\mathrm{Aut}}_{G/k}$ is reduced to the non-representability of $\underline{\mathrm{Aut}}_{\mathbf{G}_{\mathrm{a}}^n/k'}$ (on the category of k'-algebras). If this latter functor were representable then the representing scheme would be locally of finite type over k', due to the functorial criterion in [EGA, IV_3, 8.14.2], so the nilradical would locally be killed by a single exponent. Hence, by considering additive polynomials with nilpotent coefficients, we see that representability cannot hold.

Proposition 6.2.2. *Let G be a pseudo-semisimple k-group. The functor on k-algebras $\underline{\mathrm{Aut}}_{G/k} : A \rightsquigarrow \mathrm{Aut}_A(G_A)$ is represented by an affine k-group scheme $\mathrm{Aut}_{G/k}$ of finite type.*

Proof. Let T be a maximal k-torus of G. Let $\underline{\mathrm{Stab}}_{\underline{\mathrm{Aut}}_{G/k}}(T)$ be the subfunctor of $\underline{\mathrm{Aut}}_{G/k}$ assigning to any k-algebra A the set of A-group automorphisms of G_A that carry T_A into itself, hence isomorphically onto itself by [EGA, IV_4, 17.9.6]; this is visibly a subgroup functor. Since G/Z_G is identified with a normal subgroup functor of $\underline{\mathrm{Aut}}_{G/k}$ that meets $\underline{\mathrm{Stab}}_{\underline{\mathrm{Aut}}_{G/k}}(T)$ in $N_G(T)/Z_G$, we can form a non-commutative pushout subgroup functor as in [CGP, Rem. 1.4.5]:

$$(G \rtimes \underline{\mathrm{Stab}}_{\underline{\mathrm{Aut}}_{G/k}}(T))/N_G(T) \subset \underline{\mathrm{Aut}}_{G/k}. \tag{6.2.2}$$

But this inclusion of fppf group sheaves on the category of k-algebras is an *equality* due to the fppf-local conjugacy of maximal tori in smooth affine group schemes [SGA3, XI, Cor. 5.4], so it suffices to prove the representability of $\underline{\mathrm{Stab}}_{\underline{\mathrm{Aut}}_{G/k}}(T)$ by an affine k-group scheme of finite type.

There is a natural homomorphism of group functors

$$\underline{\mathrm{Stab}}_{\underline{\mathrm{Aut}}_{G/k}}(T) \to \underline{\mathrm{Aut}}_{T/k}$$

whose kernel is $\underline{\mathrm{Aut}}_{G,T}$. Since $\underline{\mathrm{Aut}}_{T/k}$ is represented by an étale k-scheme $\mathrm{Aut}_{T/k}$ and $\underline{\mathrm{Aut}}_{G,T}$ is represented by an affine k-group scheme $\mathrm{Aut}_{G,T}$ of finite type, we have a left-exact sequence of group functors

$$1 \to \mathrm{Aut}_{G,T} \to \underline{\mathrm{Stab}}_{\underline{\mathrm{Aut}}_{G/k}}(T) \xrightarrow{f} \mathrm{Aut}_{T/k}$$

on the category of k-algebras. Viewing the target as the automorphism scheme of the Galois lattice $X = \mathrm{X}(T_{k_s})$, the map f lands inside the automorphism scheme of the root system $(\Phi, X_{\mathbf{Q}})$ of $(G,T)_{k_s}$. (Note that Φ spans $X_{\mathbf{Q}}$ because G is pseudo-semisimple.) But this latter automorphism scheme is *finite*, so by affineness of $\mathrm{Aut}_{G,T}$ and calculations over a finite Galois splitting field of T we can represent $\underline{\mathrm{Stab}}_{\underline{\mathrm{Aut}}_{G/k}}(T)$ by the Galois descent of a disjoint union of finitely many $\mathrm{Aut}_{G,T}$-torsors. □

The k-group $\mathrm{Aut}_{G/k}$ is generally not smooth when the pseudo-semisimple G is not semisimple, due to the existence of $\overline{k}/k$-forms of G that are not pseudo-reductive:

Example 6.2.3. Let k be an imperfect field of characteristic $p > 0$ and $G = \mathrm{R}_{k'/k}(\mathscr{G}_{k'})$ for a purely inseparable extension k'/k of degree p and a nontrivial simply connected semisimple k-group $\mathscr{G}$. For $A = k[x]/(x^p)$, the smooth connected k-group $H = \mathrm{R}_{A/k}(\mathscr{G}_A)$ is an extension of $\mathscr{G}$ by a nontrivial smooth connected unipotent k-group. Thus, H is not pseudo-reductive but $H_{\overline{k}} \simeq G_{\overline{k}}$ because $k' \otimes_k \overline{k} \simeq A \otimes_k \overline{k}$ as $\overline{k}$-algebras. By descent theory with the affine Aut-scheme of G, there is an Isom-scheme $\mathrm{Isom}(G,H)$ that is a right fppf $\mathrm{Aut}_{G/k}$-torsor. This non-empty Isom-scheme has no k_s-points since H is not pseudo-reductive, so it is not smooth. But it is an fppf $\mathrm{Aut}_{G/k}$-torsor, so $\mathrm{Aut}_{G/k}$ is not smooth.

We are mainly interested in the maximal smooth closed k-subgroup scheme $\mathrm{Aut}^{\mathrm{sm}}_{G/k}$ whose group of points valued in any separable extension K/k coincides with $\mathrm{Aut}_K(G_K)$. (See [CGP, Lemma C.4.1] for a general discussion of the maximal smooth closed k-subgroup of a finite type k-group scheme.)

Proposition 6.2.4. *Let G be a pseudo-semisimple k-group, T a maximal k-torus of G, and $C = Z_G(T)$. Then $(G \rtimes Z_{G,C})/C \to (\mathrm{Aut}^{\mathrm{sm}}_{G/k})^0$ is an isomorphism and $(\mathrm{Aut}^{\mathrm{sm}}_{G/k})^0$ is pseudo-reductive with derived group $\mathscr{D}((\mathrm{Aut}^{\mathrm{sm}}_{G/k})^0) = G/Z_G$. Moreover, if T is k-split then $(G(k) \rtimes Z_{G,C}(k))/C(k) = (\mathrm{Aut}^{\mathrm{sm}}_{G/k})^0(k)$.*

Proof. By Proposition 6.1.4, $Z_{G,C}$ is pseudo-reductive. Since $Z_{G,C}$ is the maximal smooth closed k-subgroup of $\mathrm{Aut}_{G,T}$ and moreover $Z_G(C) = C$, the construction of $\mathrm{Aut}_{G/k}$ identifies the smooth connected $(G \rtimes Z_{G,C})/C$ with a (necessarily closed) k-subgroup scheme of $\mathrm{Aut}_{G/k}$, so this k-subgroup scheme lies inside $(\mathrm{Aut}^{\mathrm{sm}}_{G/k})^0$. The pseudo-reductivity of this k-subgroup is a special case of [CGP, Prop. 1.4.3] (using that $Z_{G,C}$ is pseudo-reductive), and to prove that it exhausts the identity component of $\mathrm{Aut}^{\mathrm{sm}}_{G/k}$ we just need to check that its set of k_s-points has finite index in $\mathrm{Aut}^{\mathrm{sm}}_{G/k}(k_s) = \mathrm{Aut}_{k_s}(G_{k_s})$.

Let f be a k_s-automorphism of G_{k_s}. Since $f(T_{k_s})$ is $G(k_s)$-conjugate to T_{k_s}, modifying f by $G(k_s)$-conjugation brings us to the case where f preserves T_{k_s}. In other words,

$$f \in \underline{\mathrm{Stab}}_{\underline{\mathrm{Aut}}_{G/k}}(T)(k_s),$$

so it suffices to show that the subgroup $Z_{G,C}(k_s) = \mathrm{Aut}_{G,T}(k_s)$ of this stabilizer group has finite index. But the image of f in $\mathrm{Aut}(T_{k_s})$ lands inside the *finite* automorphism group of the based root datum of $(G,T)_{k_s}$, so the finite-index property is established. This completes the determination of $(\mathrm{Aut}^{\mathrm{sm}}_{G/k})^0$, and the computation of its derived group is immediate from the commutativity of $Z_{G,C}$.

It remains to describe $(\mathrm{Aut}^{\mathrm{sm}}_{G/k})^0(k)$ when T is k-split. The transitivity of the $G(k)$-action on the set of split maximal k-tori [CGP, Thm. C.2.3] and the surjectivity of $N_G(T)(k)/Z_G(T)(k) \to W(\Phi(G,T))$ implies that any $f \in (\mathrm{Aut}^{\mathrm{sm}}_{G/k})^0(k)$ has a $G(k)$-translate that preserves T and a positive system of roots in $\Phi(G,T)$. We may replace f by that translate. But as a point of

$$(\mathrm{Aut}^{\mathrm{sm}}_{G/k})^0(k_s) = (G(k_s) \rtimes Z_{G,C}(k_s))/C(k_s)$$

clearly such an f must come from $Z_{G,C}(k_s)$, so $f \in Z_{G,C}(k)$. □

Remark 6.2.5. In general $(\mathrm{Aut}^{\mathrm{sm}}_{G/k})^0$ is not perfect (unlike for semisimple G), since the inclusion $C/Z_G \subset Z_{G,C}$ can fail to be an equality. This is seen via the explicit description of $Z_{G,C}$ in various cases: see [CGP, Thm. 1.3.9] for the standard case, Lemma 6.1.3 and Proposition 8.5.4 for some non-standard cases with root system of type B, C, F_4, or G_2 over k_s, and [CGP, Prop. 9.8.15] for minimal type cases whose root system over k_s is BC_n $(n \geqslant 1)$.

Corollary 6.2.6. *Let G be a pseudo-semisimple group over a field k, and $Z \subset G$ a central closed k-subgroup scheme such that $\overline{G} = G/Z$ is pseudo-reductive. The action of $(\mathrm{Aut}^{\mathrm{sm}}_{G/k})^0$ on G is trivial on Z and the induced homomorphism $\alpha : (\mathrm{Aut}^{\mathrm{sm}}_{G/k})^0 \to (\mathrm{Aut}^{\mathrm{sm}}_{\overline{G}/k})^0$ is an isomorphism.*

Proof. Let C be a Cartan k-subgroup of G, so its image $\overline{C} = C/Z$ in $\overline{G} = G/Z$ is also a Cartan k-subgroup. The action of $Z_{G,C}$ on G is trivial on C and hence trivial on the schematic center Z_G of G, and clearly the conjugation action of

G on itself is trivial on Z_G. Thus, the explicit description of $(\mathrm{Aut}^{\mathrm{sm}}_{G/k})^0$ shows that this identity component acts trivially on Z_G.

The natural map α is identified with the natural map

$$(G \rtimes Z_{G,C})/C \to (\overline{G} \rtimes Z_{\overline{G},\overline{C}})/\overline{C}.$$

Since the natural map of coset spaces $G/C \to \overline{G}/\overline{C}$ is visibly an isomorphism and $C \to Z_{G,C}$ has kernel Z_G, to show that α is an isomorphism it remains to observe that the natural map $Z_{G,C} \to Z_{\overline{G},\overline{C}}$ is an isomorphism due to Lemma 6.1.2(ii). □

6.3 Tits-style classification

If G is a split connected reductive k-group then $\mathrm{Aut}^0_{G/k} = G/Z_G$ and the étale component group of $\mathrm{Aut}_{G/k}$ is a constant k-group naturally identified with the group of automorphisms preserving a choice of pinning (or equivalently, the group of automorphisms of the based root datum). This description, along with the existence of split forms and the Borel–Tits structure theory via relative root systems, underlies the proof of the classification of connected semisimple groups [Tits, 2.7.1] in terms of the Tits index and anisotropic kernel.

For *every* imperfect field k, much of the familiar structure in the proof of the classification for the semisimple case breaks down with pseudo-semisimple k-groups G: (i) there is generally no useful description of the finite étale component group E_G of $\mathrm{Aut}^{\mathrm{sm}}_{G/k}$ (see Remark 6.3.6), (ii) the identity component $(\mathrm{Aut}^{\mathrm{sm}}_{G/k})^0$ is generally much larger than G/Z_G (Remark 6.2.5), (iii) if the Galois theory of k is sufficiently rich then G may not admit a pseudo-split k_s/k-form (see §C.1), and (iv) in characteristic 2 there is generally no good notion of pinning in the pseudo-split case (due to constructions as in Definition 3.1.2). Despite these deficiencies, remarkably there is a version of the Tits classification theorem in the general pseudo-semisimple case! We shall prove this as an application of relative root systems for pseudo-reductive groups (developed in [CGP, C.2.13–C.2.15]) and the description of $(\mathrm{Aut}^{\mathrm{sm}}_{G/k})^0$ in Proposition 6.2.4.

We are going to use Dynkin diagrams to express the classification theorem, but such diagrams involve a choice of maximal k-torus T in G (or even in $Z_G(S)$ for a maximal split k-torus S in G), and there is no k-rational conjugacy for such choices in general. Thus, we seek a way to work with diagrams that *avoids reference* to the choice of T and is *functorial* with respect to all k-isomorphisms among such k-groups G. Moreover, to ensure that conjugation in $\mathrm{Gal}(k_s/k)$ will not affect the meaning of our classification result, we have to keep track of canonicity with respect to change in k_s/k. The notion of *Dynkin scheme* (over k) in [SGA3, XXIV, 3.2–3.3] is very well-suited to this purpose, but to keep the discussion in a more concrete form we will instead use the equivalent notion of *canonical diagram* (with Galois action) due to Kottwitz.

6.3.1. We first describe the canonical diagram for *pseudo-split* pseudo-reductive k-groups G. Let T be a split maximal k-torus in G, and let Φ^+ be a positive system of roots in $\Phi(G,T)$. If (T', Φ'^+) is another such pair for G then it is carried to (T, Φ^+) via conjugation by some $g \in G(k)$, thereby defining an isomorphism $\mathrm{Dyn}(T', \Phi'^+) \simeq \mathrm{Dyn}(T, \Phi^+)$ between Dynkin diagrams. This isomorphism is independent of g because any $g \in G(k)$ preserving (T, Φ^+) lies in $Z_G(T)(k)$. Indeed, such g lies in $N_G(T)(k)$ with image in $W(G,T)(k) = W(\Phi(G,T))$ that preserves Φ^+ and hence is trivial, so $g \in Z_G(T)(k)$ as desired.

Likewise, by Proposition 6.2.4, if G is pseudo-semisimple then the stabilizer of (T, Φ^+) in $(\mathrm{Aut}^{\mathrm{sm}}_{G/k})^0(k)$ is $Z_{G,C}(k)$ for $C = Z_G(T)$, so for such G every $f \in (\mathrm{Aut}^{\mathrm{sm}}_{G/k})^0(k)$ carrying (T', Φ'^+) to (T, Φ^+) induces the *same* isomorphism $\mathrm{Dyn}(T', \Phi'^+) \simeq \mathrm{Dyn}(T, \Phi^+)$. Hence, if G is pseudo-reductive we have *canonical* isomorphisms $\mathrm{Dyn}(T', \Phi'^+) \simeq \mathrm{Dyn}(T, \Phi^+)$ using any $g \in G(k)$ carrying (T', Φ'^+) to (T, Φ^+), and if $G = \mathscr{D}(G)$ then the isomorphisms can even be determined using $(\mathrm{Aut}^{\mathrm{sm}}_{G/k})^0(k)$. These canonical isomorphisms are associative (in an evident sense) relative to any third such pair (T'', Φ''^+) with split T''.

We may now define the *canonical diagram* $\mathrm{Dyn}(G)$ attached to any pseudo-split pseudo-reductive G to be the inverse limit of the graphs $\mathrm{Dyn}(T, \Phi^+)$ for split maximal k-tori T and positive systems of roots $\Phi^+ \subset \Phi(G,T)$, using as transition maps the canonical isomorphisms induced from the action of $G(k)$ (or even $(\mathrm{Aut}^{\mathrm{sm}}_{G/k})^0(k)$ when G is pseudo-semisimple). Equivalently, it is the set of compatible systems of vertices and directed edges with multiplicity in such diagrams, with compatibility defined via the canonical diagram isomorphisms $\mathrm{Dyn}(T', \Phi'^+) \simeq \mathrm{Dyn}(T, \Phi^+)$ arising from $G(k)$ (or even from $(\mathrm{Aut}^{\mathrm{sm}}_{G/k})^0(k)$ when G is pseudo-semisimple).

The formation of $\mathrm{Dyn}(G)$ is clearly functorial with respect to isomorphisms in such G and is compatible with separable extension on k. Thus, if we *drop the pseudo-split hypothesis* and k'/k is a Galois extension such that $G_{k'}$ is pseudo-split then functoriality with respect to both scalar extension along a k-automorphism $\gamma : k' \simeq k'$ and the canonical k'-isomorphism $c_\gamma : {}^\gamma(G_{k'}) \simeq G_{k'}$ defines a diagram automorphism

$$[\gamma] : \mathrm{Dyn}(G_{k'}) \simeq \mathrm{Dyn}({}^\gamma(G_{k'})) \overset{\mathrm{Dyn}(c_\gamma)}{\simeq} \mathrm{Dyn}(G_{k'})$$

that is easily seen to define a continuous left action of $\mathrm{Gal}(k'/k)$ on $\mathrm{Dyn}(G_{k'})$. This Galois action is functorial in such k'/k, so it defines a canonical continuous left action of $\mathrm{Gal}(k_s/k)$ on $\mathrm{Dyn}(G_{k_s})$ that is functorial in the choice of k_s/k.

Definition 6.3.2. The *canonical diagram* $\mathrm{Dyn}(G)$ is $\mathrm{Dyn}(G_{k_s})$ equipped with the action of $\mathrm{Gal}(k_s/k)$ defined above.

It is clear from the construction that naturally $\mathrm{Dyn}(\mathscr{D}(G)) = \mathrm{Dyn}(G)$ and the formation of $\mathrm{Dyn}(G)$ equipped with its Galois action is functorial with respect to both k-isomorphisms in G and separable extension on k (equipped with a compatible extension between separable closures defining the Galois groups under consideration). In particular, for pseudo-semisimple G the natural action of $\mathrm{Aut}^{\mathrm{sm}}_{G/k}(k_s) = \mathrm{Aut}_{G/k}(k_s)$ on $\mathrm{Dyn}(G_{k_s})$ is a $\mathrm{Gal}(k_s/k)$-equivariant action on the canonical diagram $\mathrm{Dyn}(G)$ under which $(\mathrm{Aut}^{\mathrm{sm}}_{G/k})^0(k_s)$ acts trivially. For the finite étale component group

$$E_G := \pi_0(\mathrm{Aut}^{\mathrm{sm}}_{G/k}),$$

the finite group $E_G(k_s)$ thereby acts $\mathrm{Gal}(k_s/k)$-equivariantly on $\mathrm{Dyn}(G)$.

Here are three ways to interpret the $\mathrm{Gal}(k_s/k)$-action on the set of vertices of the canonical diagram. Firstly, it agrees with the action arising from the natural identification of the set of vertices of $\mathrm{Dyn}(G_{k_s})$ with the set of $G(k_s)$-conjugacy classes of maximal proper pseudo-parabolic k_s-subgroups of G_{k_s}. Secondly, it agrees with the action on the set of k_s-points of the Dynkin scheme defined similarly to the reductive case as in [SGA3, XXIV, 3.2–3.3]. To give a third interpretation (which will be used for calculations in some proofs below), observe that for a maximal k-torus T of G the natural action of $\mathrm{Gal}(k_s/k)$ on $\mathrm{X}(T_{k_s})$ permutes the set of positive systems of roots Φ^+ of $\Phi(G_{k_s}, T_{k_s})$. Hence, since $W(G_{k_s}, T_{k_s}) = W(\Phi)$ acts simply transitively on the set of such Φ^+'s, upon choosing Φ^+ we see that for every $\gamma \in \mathrm{Gal}(k_s/k)$ there is a unique $w_\gamma \in W(G_{k_s}, T_{k_s})$ such that $w_\gamma(\gamma(\Phi^+)) = \Phi^+$, or equivalently $w_\gamma(\gamma(\Delta)) = \Delta$ where Δ is the basis of Φ^+. This leads to the following notion:

Definition 6.3.3. For a basis Δ of Φ, the $*$-*action* of $\mathrm{Gal}(k_s/k)$ on the Dynkin diagram associated to Δ is defined by $\gamma * a = w_\gamma(\gamma(a))$ for $a \in \Delta$.

Under the natural identification of $\mathrm{Dyn}(G)$ with the diagram Δ associated to the basis of Φ^+, the $\mathrm{Gal}(k_s/k)$-action on $\mathrm{Dyn}(G)$ coincides with the $*$-action of $\mathrm{Gal}(k_s/k)$ on Δ. In particular, $(\gamma, a) \mapsto \gamma * a$ is an action of $\mathrm{Gal}(k_s/k)$ through diagram automorphisms (i.e., it respects the pairing between roots and coroots), as is also easy to check directly. The canonicity of the Galois action on

$\mathrm{Dyn}(G)$ expresses the fact (which is easily verified directly as well) that if Δ' is another basis of $\Phi(G_{k_s}, T_{k_s})$ and $*'$ denotes the associated $\mathrm{Gal}(k_s/k)$-action on Δ' then for the unique $w \in W(G_{k_s}, T_{k_s})$ such that $\Delta' = w(\Delta)$ the action of $w'_\gamma := w w_\gamma \gamma(w)^{-1}$ carries $\gamma(\Delta')$ to Δ', so $\gamma *' (w(a)) = w(\gamma * a)$; i.e., the isomorphism defined by w between the diagrams associated to Δ and Δ' is $\mathrm{Gal}(k_s/k)$-equivariant when using the $*$-action on Δ and the $*'$-action on Δ'.

Proposition 6.3.4. *Let G be a pseudo-semisimple k-group.*

(i) *The action map $E_G(k_s) \to \mathrm{Aut}(\mathrm{Dyn}(G))$ has trivial kernel,*
(ii) $\mathrm{H}^1(k, (\mathrm{Aut}^{\mathrm{sm}}_{G/k})^0)$ *is the set of k-isomorphism classes of pairs $(\mathscr{G}, \varphi)$ where $\mathscr{G}$ is pseudo-semisimple over k and $\varphi : \mathrm{Dyn}(\mathscr{G}) \simeq \mathrm{Dyn}(G)$ is a Galois-equivariant isomorphism between the canonical diagrams such that $\varphi = \mathrm{Dyn}(f)$ for some k_s-isomorphism $f : \mathscr{G}_{k_s} \simeq G_{k_s}$,*
(iii) *the natural map $\mathrm{H}^1(k, (\mathrm{Aut}^{\mathrm{sm}}_{G/k})^0) \to \mathrm{H}^1(k, \mathrm{Aut}^{\mathrm{sm}}_{G/k})$ is the forgetful map sending each pair $(\mathscr{G}, \varphi)$ to the class of $\mathscr{G}$ as a k_s/k-form of G.*

Proof. The meaning of (i) is that the kernel of the action of $\mathrm{Aut}^{\mathrm{sm}}_{G/k}(k_s)$ on $\mathrm{Dyn}(G_{k_s})$ is $(\mathrm{Aut}^{\mathrm{sm}}_{G/k})^0(k_s)$, and this clearly implies the rest. To prove (i) we may assume $k = k_s$. Consider a k-automorphism f of G with trivial action on $\mathrm{Dyn}(G)$; f can be modified by an inner automorphism arising from an element of $G(k)$ in order to preserve a maximal k-torus T and a positive system of roots Φ^+ in $\Phi(G,T)$. The action of f on $\mathrm{Dyn}(G)$ is determined by the natural effect of f on the Dynkin diagram associated to (G, T, Φ^+), so we conclude that f acts trivially on the basis of Φ^+. As $\mathrm{X}(T)_{\mathbf{Q}}$ is generated by that basis (since G is pseudo-semisimple), f acts trivially on T. It follows that f acts trivially on the commutative pseudo-reductive $C := Z_G(T)$ [CGP, Prop. 1.2.2], so f belongs to $Z_{G,C}(k) \subset (\mathrm{Aut}^{\mathrm{sm}}_{G/k})^0(k)$ (see Proposition 6.1.4). □

Remark 6.3.5. Explicitly, (ii) says that $\mathrm{H}^1(k, (\mathrm{Aut}^{\mathrm{sm}}_{G/k})^0)$ classifies isomorphism classes of pairs $(\mathscr{G}, \varphi)$ consisting of a pseudo-semisimple k-group $\mathscr{G}$ and a $*$-compatible isomorphism φ from the canonical diagram of $\mathscr{G}_{k_s}$ to that of G_{k_s} such that φ is induced by a k_s-isomorphism $\mathscr{G}_{k_s} \simeq G_{k_s}$.

Remark 6.3.6. If T is a maximal k-torus of G and Φ^+ is a positive system of roots in $\Phi(G_{k_s}, T_{k_s})$ then by Proposition 6.3.4(i) and [CGP, Thm. 2.3.10] the group $E_G(k_s)$ is naturally a subgroup of the automorphism group of the based root datum (R, Δ) attached to $(G^{\mathrm{ss}}_{\overline{k}}, T_{\overline{k}}, \Phi^+_{\mathrm{nm}})$, where Φ^+_{nm} is the set of non-multipliable roots in Φ^+. This inclusion is well-known to be an equality in the reductive case, and in Proposition 6.3.10 we will show that it is an equality in the absolutely pseudo-simple case except for G arising from a special construction of type D_{2n} ($n \geqslant 2$) over imperfect fields of characteristic 2. Allowing G to be pseudo-semisimple but possibly not absolutely pseudo-simple, over *every* imperfect field there exist such G for which $E_G(k_s) \hookrightarrow \mathrm{Aut}(R, \Delta)$ fails to be an equality, due to phenomena illustrated in Examples 6.3.8 and C.1.6.

For pseudo-semisimple G, a characterization of $E_G(k_s)$ inside the automorphism group of the based root datum can be obtained by using Theorem 6.1.1. However, such a result is cumbersome to state (and use) due to the presence of "moduli" (finite extensions of k_s, and in characteristic 2 some homothety classes of subspaces of such extensions), so we do not record it here.

Proposition 6.3.7. *If G is a pseudo-split pseudo-semisimple k-group then E_G is constant and* $\mathrm{Aut}_k(G) \to E_G(k)$ *is surjective (equivalently,* $\mathrm{Aut}_k(G)$ *meets every connected component of* $(\mathrm{Aut}^{\mathrm{sm}}_{G/k})_{k_s}$).

Proof. It suffices to show that $\mathrm{Aut}_k(G) \to E_G(k_s)$ is surjective. Let S be a split maximal k-torus in G, Φ^+ a positive system of roots in $\Phi(G, S)$, and Δ the basis of Φ^+. Since all maximal split k-tori in G are $G(k)$-conjugate and $N_G(S)(k) \to W(\Phi(G, S))$ is surjective, for every k-automorphism f of G there exists $g \in G(k)$ such that the composition of f with conjugation by g preserves S and the induced action on $\Phi(G, S)$ preserves Φ^+. The same likewise holds over k_s, so by Proposition 6.3.4(i) it suffices to show that for any k_s-automorphism θ of $(G_{k_s}, S_{k_s}, \Phi^+)$ there exists a k-automorphism φ of (G, S, Φ^+) such that φ_{k_s} and θ have the same effect on Δ. To build such a φ we shall use Theorem 6.1.1.

The restriction θ_0 of θ to $Z_{G_{k_s}}(S_{k_s}) = Z_G(S)_{k_s}$ is defined over k. Indeed, its restriction to S_{k_s} is defined over k (as S is a split k-torus), so the Galois-equivariance of θ_0 follows from the fact (part of [CGP, Prop. 1.2.2]) that any isomorphism between commutative pseudo-reductive groups is uniquely determined by its restriction between the unique maximal tori.

Let f be the k-automorphism of $Z_G(S)$ descending θ_0, and let f_S be its restriction to S. Since θ preserves Φ^+, clearly f_S carries Δ onto itself. For any $a \in \Delta$ and $a' := \mathrm{X}(f_S)(a) \in \Delta$, the effect of θ on Δ is $a \mapsto a'$ since $\theta|_{S_{k_s}} = (f_S)_{k_s}$. Hence, $f_S \circ a^\vee = a'^\vee$ for all $a \in \Delta$. By the Isomorphism Theorem (i.e., Theorem 6.1.1), to build a k-automorphism of (G, S, Φ^+) restricting to $a \mapsto a'$ on Δ and restricting to f_S on S, it suffices to construct k-isomorphisms $f_a : G_a \simeq G_{a'}$ for all $a \in \Delta$ such that each f_a is equivariant via f_S for the respective actions of S on G_a and $G_{a'}$. (As is shown in the proof of Theorem 6.1.1, any such f_a must agree with f_S on $G_a \cap S = a^\vee(\mathrm{GL}_1)$.)

Since θ restricts to a k_s-isomorphism $\theta_a : (G_a)_{k_s} \simeq (G_{a'})_{k_s}$ with the desired S_{k_s}-equivariance properties, the obstruction to building f_a is given by a degree-

1 Galois cohomology class for k_s/k valued in the group of automorphisms of G_a that restrict to the *identity map* on $a^\vee(\mathrm{GL}_1)$, or equivalently on the Cartan subgroup $C_a := Z_{G_a}(a^\vee(\mathrm{GL}_1))$. Hence, by Hilbert's Theorem 90 it suffices to show that $Z_{G_a,C_a} \simeq \mathrm{R}_{F_a/k}(\mathrm{GL}_1)$ for some finite extension field F_a/k. If G_a has root system A_1 then we may use Lemma 6.1.3, and if G_a has root system BC_1 (so k is imperfect of characteristic 2) then we may use [CGP, (9.8.7)]. □

We now provide some causes for the phenomenon in Remark 6.3.6. For pseudo-semisimple G, consider kernels of projections from $G_{\overline{k}}$ onto the simple factors of its maximal adjoint semisimple quotient $G^{\mathrm{ad}}_{\overline{k}}$. If the minimal fields of definition over k_s for two such kernels are not k_s-isomorphic but the corresponding irreducible components of the root system of G_{k_s} are isomorphic then there is no automorphism of G_{k_s} realizing a swap of those two irreducible components. The combinatorial based root datum does not encode this field-theoretic information. For example:

Example 6.3.8. Let H be a split connected semisimple group that is absolutely simple and simply connected over an imperfect field k of characteristic p. Define $G = \mathrm{R}_{k'/k}(H_{k'}) \times H$ where k'/k is a nontrivial purely inseparable finite extension. There are exactly two pseudo-simple normal k-subgroups of G, namely the factors $\mathrm{R}_{k'/k}(H_{k'})$ and H in the definition, and these are preserved under any k_s-automorphism since they cannot be swapped. The based root datum of G is the disjoint union of two copies of the one for H, so an automorphism swapping these components cannot arise from $\mathrm{Aut}^{\mathrm{sm}}_{G/k}(k_s) = \mathrm{Aut}_{k_s}(G_{k_s})$. The interested reader can make analogous pseudo-simple (but not absolutely pseudo-simple) examples over many k satisfying $[k : k^p] \geqslant p^2$ via the idea in Example C.1.2.

To analyze the finite étale component group E_G of $\mathrm{Aut}^{\mathrm{sm}}_{G/k}$, now it is natural to restrict attention to absolutely pseudo-simple G. Over many imperfect k with $\mathrm{char}(k) = 2$ there are such G for which E_G is smaller than expected: for G as in Proposition C.1.4 the based root datum is that of the adjoint central quotient of Spin_{4n} with $n \geqslant 2$ (so its automorphism group is nontrivial) but it is easy to see that E_G is trivial. This construction can be generalized:

Example 6.3.9. Let k be imperfect of characteristic 2, $\mathscr{G}$ the split k-group Spin_{4n} with $n \geqslant 2$, and $\mathscr{Z} := Z_{\mathscr{G}} \simeq \mu_2 \times \mu_2$. Let K/k be a nontrivial purely inseparable finite extension, and let Z be a closed k-subgroup of $\mathrm{R}_{K/k}(\mathscr{Z}_K)$ containing $\mathscr{Z}$ such that $\mathrm{R}_{K/k}(\mathscr{Z}_K)/Z$ has no nontrivial smooth connected k-subgroup and Z is *not* stable under the natural action on $\mathrm{R}_{K/k}(\mathscr{Z}_K)$ by the group $\Gamma := \pi_0(\mathrm{Aut}_{\mathscr{G}/k})$ of diagram automorphisms of $\mathscr{G}$ (i.e., $\Gamma = \mathbf{Z}/(2)$ if

$n > 2$, $\Gamma = S_3$ if $n = 2$); see (the proof of) Proposition C.1.4 and Remark C.1.5 for the existence of such Z over suitable k.

The k-group $G = \mathrm{R}_{K/k}(\mathscr{G}_K)/Z$ is pseudo-reductive (and therefore absolutely pseudo-simple) due to the same reasoning (based on [CGP, Lemma 9.4.1]) as in the proof of Proposition C.1.4, and $G^{\mathrm{ss}}_{\overline{k}}$ is the adjoint central quotient of $\mathscr{G}_{\overline{k}}$ since $\mathscr{Z} \subset Z$. Thus, the automorphism group of the based root datum of G_{k_s} is Γ. However, $E_G(k_s)$ is a proper subgroup of Γ since Z is not stable under the action of Γ on $\mathrm{R}_{K/k}(\mathscr{Z}_K)$. It is clear that G is not of minimal type.

Example 6.3.9 accounts for all absolutely pseudo-simple groups G for which the k_s-points of E_G do not exhaust the group of diagram automorphisms:

Proposition 6.3.10. *Let G be an absolutely pseudo-semisimple group over a field k, T a maximal k-torus of G, $R = (X, \Phi, X^\vee, \Phi^\vee)$ the (semisimple) root datum for (G_{k_s}, T_{k_s}), and Δ a basis of the irreducible Φ. Let $E_G = \pi_0(\mathrm{Aut}^{\mathrm{sm}}_{G/k})$.*

The natural inclusion $E_G(k_s) \subset \mathrm{Aut}(R, \Delta)$ is an equality except precisely when k is imperfect with $\mathrm{char}(k) = 2$ *and G_{k_s} arises in Example* 6.3.9 *over k_s.*

Proof. We may and do assume k is separably closed. If there are no nontrivial diagram automorphisms then there is nothing to do, so the only cases we need to consider are those with diagram A_n $(n \geqslant 2)$, D_n $(n \geqslant 4)$, or E_6. In particular, by Theorem 3.4.2, G is standard!

Now $G = \mathrm{R}_{K/k}(\mathscr{G}_K)/Z$ for a purely inseparable finite extension field K/k, a (necessarily split) connected absolutely simple k-group $\mathscr{G}$ that is (absolutely) simple and simply connected, and a closed k-subgroup $Z \subset \mathrm{R}_{K/k}(\mathscr{Z}_K)$ for $\mathscr{Z} := Z_{\mathscr{G}}$ such that $\mathrm{R}_{K/k}(\mathscr{Z}_K)/Z$ does not contain a nontrivial smooth connected unipotent k-subgroup. (The final condition encodes that the central quotient G is pseudo-reductive, by Corollary 4.1.4.) We may assume $K \neq k$ (so k is imperfect), as otherwise G is in the well-known absolutely simple case.

Since $1 \to Z \to \mathrm{R}_{K/k}(\mathscr{G}_K) \to G \to 1$ is a smooth k-tame central extension, it is the universal smooth k-tame central extension of G since $\mathscr{G}$ is simply connected. Hence, every k-automorphism of G lifts (uniquely) to an automorphism of $\mathrm{R}_{K/k}(\mathscr{G}_K)$ that preserves Z, and conversely every automorphism of $\mathrm{R}_{K/k}(\mathscr{G}_K)$ preserving Z descends to an automorphism of G. By [CGP, Thm. 1.6.2(2)], every k-automorphism of $\mathrm{R}_{K/k}(\mathscr{G}_K)$ has the form $\mathrm{R}_{K/k}(\varphi)$ for a unique K-automorphism φ of $\mathscr{G}_K$. Away from the D_{2n}-cases (with $n \geqslant 2$), the action of diagram automorphisms on $\mathscr{Z}$ is through the identity or inversion, so Z is preserved by $\mathrm{R}_{K/k}(\varphi)$ for any φ and hence $\mathrm{Aut}^{\mathrm{sm}}_{G/k} = \mathrm{R}_{K/k}(\mathrm{Aut}_{\mathscr{G}_K/K})$. This clearly has the expected component group.

Now suppose the diagram is D_{2n} with $n \geqslant 2$. If $\mathrm{char}(k) \neq 2$ then $\mathscr{Z}$ is étale, so $Z = \mathrm{R}_{K/k}(\mathscr{Z}'_K)$ for a k-subgroup $\mathscr{Z}' \subset \mathscr{Z}$ [CGP, Prop. A.5.13] and $G = \mathrm{R}_{K/k}((\mathscr{G}/\mathscr{Z}')_K)$ with $G^{\mathrm{red}}_{\overline{k}} = (\mathscr{G}/\mathscr{Z}')_{\overline{k}}$. Thus, in such cases $\mathrm{Aut}^{\mathrm{sm}}_{G/k} = \mathrm{R}_{K/k}(\mathrm{Aut}_{(\mathscr{G}/\mathscr{Z}')_K/K})$, so its finite component group is the same as that of the k-group $\mathrm{Aut}_{(\mathscr{G}/\mathscr{Z}')/k}$; hence, this component group is the automorphism group of the common based root datum of G and $\mathscr{G}/\mathscr{Z}'$.

Finally, assume $\mathrm{char}(k) = 2$ (and the diagram is D_{2n} with $n \geqslant 2$). If Z is stable under the action by $\pi_0(\mathrm{Aut}_{\mathscr{G}/k}) = \pi_0(\mathrm{R}_{K/k}(\mathrm{Aut}_{\mathscr{G}_K/K}))$ on $\mathrm{R}_{K/k}(\mathscr{Z}_K)$ then the same reasoning as above may be applied, so we may assume that Z is not stable under this action. If $\mathscr{Z} \subset Z$ then we are exactly in the cases from Example 6.3.9, so we may assume that $Z \cap \mathscr{Z}$ is a proper k-subgroup of $\mathscr{Z} = \mu_2 \times \mu_2$. Recall that over a *field*, there are no nontrivial homomorphisms between unipotent group schemes and multiplicative type group schemes [SGA3, XVII, Prop. 2.4]. We may assume Z is nontrivial, so Z is not unipotent (as $\mathrm{R}_{K/k}(\mathscr{Z}_K)$ does not contain a nontrivial unipotent k-subgroup). Thus, $Z \cap \mathscr{Z} \neq 1$ since the k-subgroup scheme $\mathrm{R}_{K/k}(\mathscr{Z}_K)/\mathscr{Z} \subset (\mathrm{R}_{K/k}(\mathrm{GL}_1)/\mathrm{GL}_1)^2$ is unipotent (as the k-group $\mathrm{R}_{K/k}(\mathrm{GL}_1)/\mathrm{GL}_1$ is unipotent). Hence, the proper k-subgroup $\mu := Z \cap \mathscr{Z} \subset \mathscr{Z} = \mu_2 \times \mu_2$ is one of the three copies of μ_2. Since Z/μ inherits unipotence from $\mathrm{R}_{K/k}(\mathscr{Z}_K)/\mathscr{Z}$, the natural map $Z_K \to \mathscr{Z}_K$ corresponding to the inclusion $Z \hookrightarrow \mathrm{R}_{K/k}(\mathscr{Z}_K)$ induces a map $(Z/\mu)_K \to (\mathscr{Z}/\mu)_K$ that must be trivial (as Z/μ is unipotent and $\mathscr{Z}/\mu$ is of multiplicative type), so $\mu \subset Z \subset \mathrm{R}_{K/k}(\mu_K)$ and $G^{\mathrm{red}}_{\overline{k}} = (\mathscr{G}/\mu)_{\overline{k}}$.

We may assume that the common based root datum (R, Δ) of G and $\mathscr{G}/\mu$ has a nontrivial automorphism (or else there is nothing to do), so the quotient $\mathscr{G}/\mu$ of $\mathscr{G} = \mathrm{Spin}_{4n}$ is isomorphic to SO_{4n} as a k-group (even if $n = 2$) and the automorphism group of the based root datum has order 2 with trivial action on μ (and so also on Z). Thus, the evident map $\mathrm{Aut}^{\mathrm{sm}}_{G/k} \to \mathrm{R}_{K/k}(\mathrm{Aut}_{(\mathscr{G}/\mu)_K/K})$ is an isomorphism, so $\pi_0(\mathrm{Aut}^{\mathrm{sm}}_{G/k}) = \pi_0(\mathrm{Aut}_{(\mathscr{G}/\mu)/k}) = \mathrm{Aut}(R, \Delta)$ as desired. □

Now we prepare to define the anisotropic kernel of a pseudo-reductive group over a field k. Recall (see [CGP, C.2.3–C.2.5]) that if G is a smooth connected affine k-group then the minimal pseudo-parabolic k-subgroups P and the maximal k-split tori S each constitute a single $G(k)$-conjugacy class (any pseudo-parabolic k-subgroup contains such an S), and if G is pseudo-reductive and $S \subset P$ then $P = Z_G(S) \ltimes U$ for the k-split $U = \mathscr{R}_{u,k}(P)$.

Fix a minimal pseudo-parabolic k-subgroup P of G and a maximal k-split torus $S \subset P$. Let $M = Z_G(S)$. Every maximal k-torus T of M contains S and is clearly maximal in G. Assume G is pseudo-reductive, so M is pseudo-reductive. By [CGP, Lemma 1.2.5], T is an almost direct product of the maximal central k-torus of M and the maximal k-torus $T' := T \cap \mathscr{D}(M)$ of $\mathscr{D}(M)$, so the pseudo-semisimple derived group $\mathscr{D}(M)$ is k-anisotropic (i.e., it does not contain a nontrivial k-split torus); we call $\mathscr{D}(M)$ the *anisotropic kernel* of G. For our purposes this will only matter through its maximal central quotient (which coincides with $\mathscr{D}(M/Z_M)$ since the triviality of the center of M/Z_M via Proposition 4.1.3 ensures that $\mathscr{D}(M/Z_M)$ has trivial center, due to [CGP, Lemma 1.2.5(ii), Prop. 1.2.6]).

There is an evident Galois-equivariant diagram inclusion

$$\mathrm{Dyn}(\mathscr{D}(M)) = \mathrm{Dyn}(M) \subset \mathrm{Dyn}(G).$$

In more explicit terms, if T is a maximal k-torus of M and $T' := T \cap \mathscr{D}(M)$ then for a positive system of roots $\Phi^+ \subset \Phi(G_{k_s}, T_{k_s})$ contained in $\Phi(P_{k_s}, T_{k_s})$ and the positive system of roots $\Psi^+ = \Psi \cap \Phi^+$ in $\Psi := \Phi(M_{k_s}, T_{k_s})$, the diagram of $(\mathscr{D}(M)_{k_s}, T'_{k_s}, \Psi^+)$ is naturally a subdiagram of that of $(G_{k_s}, T_{k_s}, \Phi^+)$ compatibly with the $*$-action of $\mathrm{Gal}(k_s/k)$ on each (seen classically by the fact that $N_M(T)(k_s)$ acts transitively on the set of choices of Φ^+ that lie inside $\Phi(P_{k_s}, T_{k_s})$, as $P = M \ltimes U$). The vertices of $\mathrm{Dyn}(\mathscr{D}(M)) = \mathrm{Dyn}(M)$ are thereby identified with the *non-distinguished* vertices of $\mathrm{Dyn}(G)$ (i.e., those corresponding to the simple roots whose restriction to S_{k_s} is trivial).

The use of Dynkin diagrams to interpret $\mathrm{H}^1(k, (\mathrm{Aut}^{\mathrm{sm}}_{G/k})^0)$ in Proposition 6.3.4(ii) and Remark 6.3.5 leads to a pseudo-semisimple generalization of the Tits classification of connected semisimple groups:

Theorem 6.3.11 (Relative Isomorphism Theorem). *Let (G, S) be as above with G pseudo-semisimple, and let $M = Z_G(S)$. Naturally identify* $\mathrm{Dyn}(\mathscr{D}(M)) = \mathrm{Dyn}(\mathscr{D}(M/Z_M))$ *with a subdiagram of* $\mathrm{Dyn}(G)$.

If $(\mathscr{G}, \mathscr{S})$ is another such pair and $\mathscr{M} := Z_{\mathscr{G}}(\mathscr{S})$ then $\mathscr{G} \simeq G$ if and only if there exists a Galois-equivariant diagram isomorphism $\theta : \mathrm{Dyn}(\mathscr{G}) \simeq \mathrm{Dyn}(G)$ restricting to an isomorphism $\theta_0 : \mathrm{Dyn}(\mathscr{D}(\mathscr{M})) \simeq \mathrm{Dyn}(\mathscr{D}(M))$ such that θ arises from a k_s-isomorphism $f : \mathscr{G}_{k_s} \simeq G_{k_s}$ and θ_0 arises from a k-isomorphism $f_0 : \mathscr{D}(\mathscr{M}/Z_{\mathscr{M}}) \simeq \mathscr{D}(M/Z_M)$.

In particular, G is determined up to k-isomorphism by the isomorphism class of the triple $(G_{k_s}, \mathscr{D}(M/Z_M), \iota)$, where ι is the $\mathrm{Gal}(k_s/k)$*-equivariant inclusion of* $\mathrm{Dyn}(\mathscr{D}(M)) = \mathrm{Dyn}(\mathscr{D}(M/Z_M))$ *as a subdiagram of* $\mathrm{Dyn}(G)$.

In the final assertion of the theorem, an *isomorphism* between two such triples is a pair (f, f_0) as above (so $\mathrm{Dyn}(f)$ is $\mathrm{Gal}(k_s/k)$-equivariant and the isomorphisms $\mathrm{Dyn}(f)$ and $\mathrm{Dyn}(f_0)$ intertwine the given diagram inclusions). If G is semisimple then θ uniquely extends to an isomorphism between "indexed root data" (see [Spr, 16.2.1]), so the semisimple case of Theorem 6.3.11 is exactly the well-known theorem of Tits.

The necessity in Theorem 6.3.11 is obvious, and to prove sufficiency we assume that θ, f, f_0 are given. Let $P \subset G$ and $\mathscr{P} \subset \mathscr{G}$ be minimal pseudo-parabolic k-subgroups containing S and $\mathscr{S}$ respectively. Let T be a maximal k-torus of M, so T contains S and is maximal in G. The relationship between $\Phi := \Phi(G_{k_s}, T_{k_s})$ and $\Psi := \Phi(M_{k_s}, T_{k_s}) = \Phi(\mathscr{D}(M)_{k_s}, T'_{k_s})$ for $T' := T \cap \mathscr{D}(M)$ works out as follows. Choose a positive system of roots Φ^+ of Φ contained in the parabolic subset $\Phi(P_{k_s}, T_{k_s}) \subset \Phi$ and define $\Psi^+ = \Psi \cap \Phi^+$. Let Δ be the basis of Φ^+, so $\Delta_0 := \Psi \cap \Delta$ is the basis of Ψ^+. The root system Ψ consists of the elements of Φ that have trivial restriction to S_{k_s}. The restriction map $\Delta - \Delta_0 \to \mathrm{X}(S_{k_s}) = \mathrm{X}(S)$ (whose fibers are orbits under the $*$-action of $\mathrm{Gal}(k_s/k)$) has image ${}_k\Delta$ that is a basis of the relative root system ${}_k\Phi := \Phi(G, S)$ (as shown in the proof of [CGP, Thm. C.2.15]). For $c \in \Psi$, the c-root groups of (M_{k_s}, T_{k_s}) and (G_{k_s}, T_{k_s}) *coincide*. In particular, for each $c \in \Psi$ the subgroup $(G_{k_s})_c$ generated by the $\pm c$-root groups of G_{k_s} coincides with $(\mathscr{D}(M)_{k_s})_c$.

The set of pseudo-parabolic k_s-subgroups of G_{k_s} containing T_{k_s} is in natural bijective correspondence with the set of parabolic subsets of $\Phi = \Phi(G_{k_s}, T_{k_s})$ via $Q \mapsto \Phi(Q, T_{k_s})$ [CGP, Prop. 3.5.1(2)], so the set of pseudo-parabolic k_s-subgroups containing T_{k_s} and corresponding to parabolic subsets containing Δ is in natural inclusion-preserving bijection with the set of subsets of Δ. The restriction map $\mathrm{X}(T_{k_s}) \to \mathrm{X}(S_{k_s}) = \mathrm{X}(S)$ carries $\Delta - \Delta_0$ onto ${}_k\Delta$, so minimality of P implies that $\Phi(P, S)$ is the positive system of roots in ${}_k\Phi$ with basis ${}_k\Delta$. Thus, P_{k_s} corresponds to $\Delta_0 \subset \Delta$, and similarly for $(\mathscr{G}, \mathscr{P})$. Hence, we may compose f with conjugation by an element of $G(k_s)$ so that $f(\mathscr{P}_{k_s}) = P_{k_s}$; note that composing with such conjugation has no effect on $\mathrm{Dyn}(f)$.

To simplify notation, let $A = (\mathrm{Aut}^{\mathrm{sm}}_{G/k})^0$. The k-group $\mathscr{G}$ is isomorphic to the twist of G by the continuous 1-cocycle $c : \mathrm{Gal}(k_s/k) \to A(k_s)$ defined by

$$c(\gamma) = f \circ ({}^{\gamma}f)^{-1}$$

(where ${}^{\gamma}f$ denotes the γ-twist of f); this is valued in $A(k_s)$ due to Proposition 6.3.4(i) since $\mathrm{Dyn}(f) = \theta$ is $\mathrm{Gal}(k_s/k)$-equivariant. By Proposition 6.2.4,

$$A(k_s) = (G(k_s) \rtimes Z_{G,C}(k_s))/C(k_s)$$

where $C := Z_G(T)$. Since $f(\mathscr{P}_{k_s}) = P_{k_s}$, c is valued in the stabilizer of P_{k_s} in $A(k_s)$. The triviality of the $Z_{G,C}$-action on T implies that $Z_{G,C}$ preserves P, so the stabilizer of P in A is $(P \rtimes Z_{G,C})/C$ since $N_G(P) = P$ [CGP, Prop. 3.5.7].

The $Z_{G,C}$-action on G preserves S, so it preserves $M = Z_G(S)$. Hence, it makes sense to consider the inclusion of k-groups

$$j : (M \rtimes Z_{G,C})/C \hookrightarrow (P \rtimes Z_{G,C})/C.$$

The retraction $P \twoheadrightarrow P/U \simeq M$ onto M (for $U := \mathscr{R}_{u,k}(P)$) is clearly $Z_{G,C}$-equivariant and so defines a retraction

$$r : (P \rtimes Z_{G,C})/C \twoheadrightarrow (M \rtimes Z_{G,C})/C$$

whose kernel is identified with the k-split smooth connected unipotent k-group U. Every k_s/k-form of U is k-split [CGP, Thm. B.3.4], so the twisting method in Galois cohomology implies that $\mathrm{H}^1(r)$ is a bijection. Hence, $\mathrm{H}^1(j)$ is bijective, so by replacing c with a cohomologous 1-cocycle we may and do assume that c is valued in $(M(k_s) \rtimes Z_{G,C}(k_s))/C(k_s)$.

The $Z_{G,C}$-action on M preserves $\mathscr{D}(M)$, so we get a natural map

$$q : (M \rtimes Z_{G,C})/C \to (\mathrm{Aut}^{\mathrm{sm}}_{\mathscr{D}(M)/k})^0.$$

Corollary 6.2.6 identifies the target of q with $(\mathrm{Aut}^{\mathrm{sm}}_{\mathscr{D}(M/Z_M)/k})^0$, and in this way via Proposition 6.3.4(ii) the map $\mathrm{H}^1(q)$ carries the cohomology class $[c]$ to the class of the pair $(\mathscr{D}(\mathscr{M}/Z_{\mathscr{M}}), \theta_0)$. The existence of the k-isomorphism f_0 therefore implies that $\mathrm{H}^1(q)([c]) = 1$. To exploit this triviality, we shall use:

Proposition 6.3.12. *The map q is surjective with kernel of the form $\mathrm{R}_{F/k}(\mathrm{GL}_1)$ for a nonzero finite reduced k-algebra F. In particular, q is a smooth surjection.*

Granting the proposition, let us conclude the proof of Theorem 6.3.11. Since q is a smooth surjection and $\mathrm{H}^1(q)([c]) = 1$, $[c]$ comes from $\mathrm{H}^1(k, \ker q)$. But the description of $\ker q$ as a Weil restriction implies that $\mathrm{H}^1(k, \ker q) = 1$ by Shapiro's Lemma and Hilbert's Theorem 90, so $[c] = 1$. Thus, by the construction of c, the class of $(\mathscr{G}, \theta)$ in $\mathrm{H}^1(k, (\mathrm{Aut}^{\mathrm{sm}}_{G/k})^0)$ is trivial, so $\mathscr{G} \simeq G$.

To prove Proposition 6.3.12, we first describe q in more concrete terms. Let T' be the maximal k-torus $T \cap \mathscr{D}(M)$ of $\mathscr{D}(M)$ and let Z be the maximal central k-torus of M, so $T = Z \cdot T'$ by [CGP, Lemma 1.2.5(ii)]. Thus, the intersection $C' := C \cap \mathscr{D}(M) = Z_{\mathscr{D}(M)}(T')$ is a Cartan k-subgroup of $\mathscr{D}(M)$. Since $M = C \cdot \mathscr{D}(M)$, we have $(M \rtimes Z_{G,C})/C = (\mathscr{D}(M) \rtimes Z_{G,C})/C'$. Also, by Proposition 6.2.4 (applied to $\mathscr{D}(M)$) we have

$$(\mathrm{Aut}^{\mathrm{sm}}_{\mathscr{D}(M)/k})^0 = (\mathscr{D}(M) \rtimes Z_{\mathscr{D}(M),C'})/C',$$

so q is identified with the canonical map

$$(\mathscr{D}(M) \rtimes Z_{G,C})/C' \to (\mathscr{D}(M) \rtimes Z_{\mathscr{D}(M),C'})/C'$$

induced by the natural restriction map

$$\rho : Z_{G,C} \to Z_{\mathscr{D}(M),C'} = Z_{M,C}$$

(arising from the property that $Z_{G,C}$ preserves M and $\mathscr{D}(M)$ inside G; the equality is Lemma 6.1.2(i)).

The surjectivity of q is equivalent to that of ρ, and $\ker q \simeq \ker \rho$. Hence, to prove Proposition 6.3.12 it is equivalent to prove the surjectivity of ρ and to determine its kernel. Let E/k be a finite Galois extension that splits T. The surjectivity of ρ is verified by computing over E via applying Proposition 6.1.4 to the pairs (G, C) and (M, C): since $\Phi(M_E, T_E)$ has Δ_0 as a basis and $(M_E)_a = (G_E)_a$ for each $a \in \Phi(M_E, T_E)$, clearly ρ_E is the natural projection

$$\prod_{a \in \Delta} Z_{(G_E)_a,(C_E)_a} \to \prod_{a \in \Delta_0} Z_{(G_E)_a,(C_E)_a},$$

so ρ is surjective and $\ker \rho$ is commutative and pseudo-reductive of minimal type (Proposition 6.1.4). It remains to show $\ker \rho \simeq \mathrm{R}_{F/k}(\mathrm{GL}_1)$ for some F.

To analyze $\ker \rho$ as a k-group we will first reduce to the case where G is absolutely pseudo-simple. Consider $\overline{G} = G/Z_G$; this is pseudo-semisimple with trivial center by Proposition 4.1.3, and the image $\overline{S}$ of S is a maximal k-split torus (due to Proposition 5.3.1(i)). Thus, the central quotient $\overline{M} = M/Z_G$ of M coincides with $Z_{\overline{G}}(\overline{S})$. By Lemma 6.1.2(ii), the formation of ρ is unaffected by passage to $\overline{G}$, $\overline{C}$, $\overline{S}$. Hence, we may (and will) assume that Z_G is trivial.

Lemma 6.3.13. *There exists a nonzero finite étale k-algebra k' and a smooth affine k'-group G' with trivial center and absolutely pseudo-simple fibers over the factor fields of k' such that $G \simeq \mathrm{R}_{k'/k}(G')$. If $(k''/k, G'')$ is a second such pair then any k-isomorphism $\mathrm{R}_{k'/k}(G') \simeq \mathrm{R}_{k''/k}(G'')$ arises from a unique pair (φ, α) consisting of a k-algebra isomorphism $\alpha : k' \simeq k''$ and a group isomorphism $\varphi : G' \simeq G''$ over α.*

Proof. By Galois descent and the uniqueness assertions we may assume $k = k_s$. If $\{G'_i\}_{i \in I}$ is the set of pseudo-simple normal k-subgroups of G then they pairwise commute and multiplication $m : \prod G'_i \to G$ is a central quotient map [CGP, Prop. 3.1.8], so the triviality of Z_G forces all $Z_{G'_i}$ to be trivial and hence m is an isomorphism. This provides existence (using $k' = k^I$ and $G' = \coprod G'_i$), and uniqueness follows from the fact that the product $\prod G_i'$ uniquely characterizes its k-subgroups G_i' as its pseudo-simple normal k-subgroups. □

Let k' and G' be as in the preceding lemma. We identify G with $\mathrm{R}_{k'/k}(G')$. By [CGP, Prop. A.5.15], the Cartan k-subgroup C has the form $\mathrm{R}_{k'/k}(C')$ for a unique Cartan k'-subgroup $C' \subset G'$ and there is a unique maximal k'-split torus $S' \subset G'$ for which S is the maximal k-split torus in $\mathrm{R}_{k'/k}(S')$, with $M' := Z_{G'}(S')$ satisfying $\mathrm{R}_{k'/k}(M') = Z_G(\mathrm{R}_{k'/k}(S')) \subset Z_G(S) = M$. To prove the reverse inclusion $\mathrm{R}_{k'/k}(M') \supset M$ it suffices to treat the case where k' is a field, and in that case it suffices to show that the natural map $M_{k'} \to G'$ lands inside M'. But the natural map $S_{k'} \to S'$ induced by $\phi : G_{k'} \to G'$ is easily seen to be an isomorphism, so ϕ carries $M_{k'} = Z_{G_{k'}}(S_{k'})$ into $Z_{G'}(S') = M'$ as desired.

Proposition 6.1.7 applied to G' and M' identifies ρ with $\mathrm{R}_{k'/k}(\rho')$, where ρ' is the analogous map over k' associated to (G', S', C') (working over each factor field of k' separately). This permits us to replace (G, S, C) with the fibers of (G', S', C') over the factor fields of k', so now we have arranged that G is absolutely pseudo-simple.

Consider the identification

$$(\ker \rho)_E = \prod_{a \in \Delta - \Delta_0} Z_{(G_E)_a,(C_E)_a}. \tag{6.3.13}$$

Note that all elements of $\Delta - \Delta_0$ in a common orbit under the $*$-action have the same length in the irreducible (possibly non-reduced) root system $\Phi(G_E, T_E)$.

To use (6.3.13), we need to describe the factors in this direct product:

Lemma 6.3.14. *Let $\mathscr{H}$ be a pseudo-split absolutely pseudo-simple E-group of rank 1, and define $\mathscr{C} := Z_{\mathscr{H}}(\mathscr{T})$ for a split maximal E-torus $\mathscr{T} \subset \mathscr{H}$. For a basis $\{a\}$ of $\Phi(\mathscr{H}, \mathscr{T})$ there is a unique purely inseparable finite extension F/E and unique E-isomorphism*

$$Z_{\mathscr{H},\mathscr{C}} \simeq \mathrm{R}_{F/E}(\mathrm{GL}_1) \tag{6.3.14}$$

carrying the natural map $\mathscr{T} \to Z_{\mathscr{H},\mathscr{C}}$ over to the composition of $a : \mathscr{T} \to \mathrm{GL}_1$ with the natural inclusion $\mathrm{GL}_1 \hookrightarrow \mathrm{R}_{F/E}(\mathrm{GL}_1)$.

This isomorphism is functorial with respect to isomorphisms in the triple $(\mathscr{H}, \mathscr{T}, \{a\})$ and compatible with separable extension on E.

Note that F/E is independent of the choice of $\mathscr{T}$ (by working over E_s).

Proof. By Lemma 6.1.2(ii), we may replace $\mathscr{H}$ with its maximal quotient of minimal type, and then pass to the universal smooth E-tame central extension so that $\mathscr{H}_{\overline{k}}^{\mathrm{ss}}$ is simply connected. The uniqueness assertions are clear, assuming existence. If the root system is A_1 then we may conclude by Lemma 6.1.3, and for the root system BC_1 we may use the rank-1 case of [CGP, Prop. 9.8.15]. □

6.3.15. Completion of the proof of Proposition 6.3.12. Let $\pi : \Delta - \Delta_0 \twoheadrightarrow {}_k\Delta$ be the natural map induced by $\mathrm{X}(T_E) \twoheadrightarrow \mathrm{X}(S_E) = \mathrm{X}(S)$, so the fibers of π are the $*$-orbits. Applying Lemma 6.3.14 to the E-groups $(G_E)_a$ for $a \in \Phi = \Phi(G_E, T_E)$ defines purely inseparable finite extensions F'_a/E for all $a \in \Phi$. Any two elements of the irreducible (possibly non-reduced) root system Φ with the same length are conjugate under $N_G(T)(E)$, so F'_a only depends on a through its length. But clearly if $\gamma \in \mathrm{Gal}(E/k)$ then canonically ${}^{\gamma}(F'_a) \simeq F'_{\gamma(a)} = F'_a$ as E-algebras (induced by ${}^{\gamma}((G_E)_a) \simeq (G_E)_{\gamma(a)}$ carrying ${}^{\gamma}((C_E)_a)$ onto $(C_E)_{\gamma(a)}$), thereby defining an E/k-descent datum on F'_a. Let F_a/k be the associated Galois descent,

For each $\alpha \in {}_k\Delta$ and a in the $*$-orbit $\pi^{-1}(\alpha)$ inside $\Delta - \Delta_0$, clearly the purely inseparable finite extension F_a/k only depends on α. Denote this extension as F_α, so (6.3.13) gives a natural isomorphism

$$(\ker \rho)_E \simeq \prod_{\alpha \in {}_k\Delta} \mathrm{R}_{F_\alpha/k}(\mathrm{GL}_1^{\pi^{-1}(\alpha)})_E. \tag{6.3.15}$$

It suffices to show that the natural $\mathrm{Gal}(k_s/k)$-action on $(\ker \rho)(k_s)$ (or alternatively the E/k-descent datum on $(\ker \rho)_E$) corresponds to the permutation action on the right side through the $*$-action on each index set $\pi^{-1}(\alpha)$. Indeed, once this is established, if L_α/k is the finite separable subextension of E/k corresponding to the $\mathrm{Gal}(E/k)$-stabilizer of a chosen element of $\pi^{-1}(\alpha)$ then

$$\ker \rho \simeq \prod_{\alpha \in {}_k\Delta} \mathrm{R}_{F_\alpha/k}(\mathrm{R}_{L_\alpha F_\alpha/F_\alpha}(\mathrm{GL}_1)) = \mathrm{R}_{F/k}(\mathrm{GL}_1)$$

with $F := \prod_{\alpha \in {}_k\Delta} L_\alpha F_\alpha$. (It is for absolutely pseudo-simple G that each factor field of F is a compositum of a separable and a purely inseparable extension.)

Clearly $S \subset \ker \rho$ (as $M = Z_G(S)$), and for $a \in \Delta - \Delta_0 = \coprod_{\alpha \in {}_k\Delta} \pi^{-1}(\alpha)$ the 1-dimensional maximal E-torus inside the a-factor of (6.3.15) is by definition identified with GL_1 via the nontrivial character $a : S_E \twoheadrightarrow \mathrm{GL}_1$. Since an isomorphism between commutative pseudo-reductive groups is uniquely determined by its restriction between maximal tori, for any $a \in \Delta - \Delta_0$ and $\gamma \in \mathrm{Gal}(E/k)$ it suffices to show:

(i) $\gamma(a)|_{S_E} = (\gamma * a)|_{S_E}$,

(ii) $Z_{(G_E)_{\gamma(a)},(C_E)_{\gamma(a)}} = Z_{(G_E)_{\gamma*a},(C_E)_{\gamma*a}}$ inside Z_{G_E,C_E}, where the left side is embedded using the basis $\gamma(\Delta)$ of Φ and the right side is embedded using the basis $\gamma * \Delta = \Delta$ of Φ.

By definition, $\gamma * a = w_\gamma(\gamma(a)) = \gamma(a) \circ \mathrm{Int}\, n_\gamma^{-1}$ for $n_\gamma \in N_G(T)(E)$ representing the unique $w_\gamma \in W(\Phi)$ carrying $\gamma(\Phi^+)$ onto Φ^+. But Φ^+ was chosen inside the $\mathrm{Gal}(E/k)$-stable parabolic subset $\Phi(P_E, T_E) = \Phi(P_{k_s}, T_{k_s}) \subset \Phi$, so $\gamma(\Phi^+) \subset \Phi(P_E, T_E)$. Thus, Φ^+ and $\gamma(\Phi^+)$ correspond to two minimal pseudo-parabolic E-subgroups of G_E that are contained in $P_E = M_E \ltimes U_E$ and contain the split maximal E-torus T_E. By [CGP, Lemma 3.5.5], these correspond to minimal pseudo-parabolic E-subgroups of $P_E/U_E = M_E$ containing T_E, so $W(M_E, T_E)$ carries one to the other. Hence, we can choose $n_\gamma \in N_M(T)(E)$, so n_γ centralizes S_E and thus (i) holds. Also, $Z_{(G_E)_a,(C_E)_a}$ viewed inside Z_{G_E,C_E} via Δ acts trivially on M_E by definition of ρ since $a \in \Delta - \Delta_0$, so likewise its γ-twist $Z_{(G_E)_{\gamma(a)},(C_E)_{\gamma(a)}}$ inside $Z_{G_E,C_E} = (Z_{G,C})_E$ (embedded via $\gamma(\Delta)$!) acts trivially on M_E too. Since $n_\gamma \in M(E)$, we obtain (ii) as claimed, completing the proof of Proposition 6.3.12 (and so also of Theorem 6.3.11).

Remark 6.3.16. The k-isomorphism class of a pseudo-semisimple k-group G is determined by a 4-tuple $(\mathsf{G}, \tau, M_0, \iota)$ where G is a pseudo-semisimple k_s-group (a root datum in the semisimple case), τ is a continuous action of $\mathrm{Gal}(k_s/k)$ on $\mathrm{Dyn}(\mathsf{G})$, M_0 is a k-anisotropic pseudo-semisimple k-group with trivial center, and ι is a Galois-equivariant diagram inclusion $\mathrm{Dyn}(M_0) \hookrightarrow \mathrm{Dyn}(\mathsf{G})$.

The discussion preceding Lemma C.2.2 shows that G is quasi-split (i.e., its minimal pseudo-parabolic k-subgroups are solvable) precisely when $M_0 = 1$, so G is pseudo-split precisely when $M_0 = 1$ and τ is the trivial action. The special case that a pseudo-split pseudo-semisimple k-group G is determined by G_{k_s} is immediate from Theorem 6.3.11 and is generalized to pseudo-reductive G in Proposition C.1.1. (See Proposition C.2.8 for variant in the quasi-split case.)

Theorem 6.1.1 determines G in terms of: $\mathsf{C} = Z_{\mathsf{G}}(\mathsf{T})$ for a maximal k_s-torus T, the root system $\Phi = \Phi(\mathsf{G}, \mathsf{T})$, and the T-equivariant rank-1 pseudo-simple subgroups associated to roots in a basis of Φ. That does not address the existence problem over k_s when such data are given. For irreducible Φ, the pseudo-split possibilites *of minimal type* over k are classified alongside an existence result: see Theorem 3.4.1 for reduced Φ and [CGP, Thm. 9.8.6] for non-reduced Φ.

What about the *general existence problem over k*? Beyond the absolutely pseudo-simple case this is nontrivial since for many k (occurring in any positive characteristic) Examples C.1.2 and C.1.6 provide standard pseudo-semisimple (but not absolutely pseudo-simple) k-groups with no pseudo-split k_s/k-form. The absolutely pseudo-simple case is addressed in Remark C.2.13.

7

Constructions with regular degenerate quadratic forms

For an imperfect field k with char$(k) = p$, the non-standard pseudo-split absolutely pseudo-simple k-groups G of minimal type with a reduced root system Φ of rank $n \geqslant 2$ are classified in Theorem 3.4.1(ii),(iii) in terms of field-theoretic and linear-algebraic data. These only exist when Φ has an edge of multiplicity p (so $p \in \{2,3\}$).

For type G_2 in characteristic 3 and type F_4 in characteristic 2 this involves just field-theoretic data: a nontrivial purely inseparable finite extension K/k and a proper subfield $K_> \subset K$ containing kK^p. The non-standard G_2-cases and non-standard F_4-cases as handled in [CGP, Ch. 7–8] amount to the basic exotic construction reviewed in §2.2, and for the convenience of the reader we reproved what we need about G_2 and F_4 in §3.4.

Non-standard G of types B_n and C_n are more subtle: they are classified by a subfield $K_> \subset K$ containing kK^2 and a pair of nonzero kK^2-subspaces $V, V_> \subset K$ satisfying certain conditions (with $V = K$ for type C_n with $n \geqslant 3$, and $V_> = K_>$ for type B_n with $n \geqslant 3$). For any such G, by Proposition 4.1.3 the quotient G/Z_G is absolutely pseudo-simple with trivial center, its root system over k_s is clearly Φ, and it is non-standard because any pseudo-reductive central extension of a standard pseudo-reductive group is standard (Lemma 3.2.8). The aim of this chapter is to use *degenerate* quadratic forms and quadrics in Severi–Brauer varieties to give a geometric description of all non-standard absolutely pseudo-simple k-groups G of minimal type with root system B_n over k_s such that $Z_G = 1$ and the Cartan k-subgroups of G are *tori*. (Even in the pseudo-split case this will provide more information than Theorem 3.4.1(iii).) Our inspiration is the fact that the adjoint connected semisimple groups of type B_n over a field F are precisely $\mathrm{SO}(q)$ for non-degenerate quadratic spaces (V, q)

over F of dimension $2n+1$, with (V,q) determined up to conformal isometry by the isomorphism class of $\mathrm{SO}(q)$.

In Chapter 8 we will use universal smooth k-tame central extensions of these *adjoint* type-B constructions, as well as fiber products in the spirit of the basic exotic construction in §2.2, to build non-standard G of minimal type such that $G_{\overline{k}}^{\mathrm{ss}}$ is simply connected of type B or C. (The link to type C rests on the unipotent isogeny $\mathrm{SO}_{2n+1} \to \mathrm{Sp}_{2n}$ from type B to type C in characteristic 2.)

7.1 Regular degenerate quadratic forms

Let k be an imperfect field with $\mathrm{char}(k) = 2$, and let (V,q) be a quadratic space over k such that $d := \dim(V)$ is finite and $q \neq 0$. Non-degeneracy of q is *defined* to mean k-smoothness of the quadric hypersurface $H_q := (q=0) \subset \mathbf{P}(V^*)$. Let $B_q(v,w) = q(v+w) - q(v) - q(w)$ be the associated symmetric bilinear form (also alternating since $\mathrm{char}(k) = 2$), and let $V^{\perp}$ be the defect space, so (as we saw in Example 1.3.1) $\dim(V/V^{\perp}) = 2n$ is even.

Assume (V,q) is *regular* in the sense that the 2-linear map $q : V^{\perp} \to k$ is injective. Regularity of q has geometric meaning: it amounts to regularity (or equivalently, k-smoothness) of the projective quadric H_q at its k-points. Indeed, such regularity is equivalent to that of the affine quadric $\{q=0\}$ at its nonzero k-points $v_0 \in V$ and the tangent space at such v_0 has dimension $d-1$ precisely when the degree-2 polynomial $q(v+v_0) - q(v_0) = B_q(v,v_0) + q(v)$ on V has nonzero linear part. This linear part is $B_q(v,v_0)$, so it vanishes if and only if $v_0 \in V^{\perp}$. Hence, regularity at all k-points is precisely the condition that $\ker(q|_{V^{\perp}}) = \{0\}$, as asserted.

Lemma 7.1.1. *If k'/k is a separable extension field then $(V_{k'}, q_{k'})$ is regular.*

Proof. The formation of $V^{\perp}$ commutes with any extension on k. Fix a k-basis $\{e_1,\dots,e_{d-2n}\}$ of $V^{\perp}$. In the coordinates determined by this basis, the 2-linear $q|_{V^{\perp}}$ is given by $c_1x_1^2 + \cdots + c_{d-2n}x_{d-2n}^2$ with $c_i \in k$. The regularity hypothesis says that $\{c_i\}$ is k^2-linearly independent inside k, and we just have to check that $\{c_i\}$ is k'^2-linearly independent inside k'. Since k'^2 is separable over k^2, and k is purely inseparable algebraic over k^2, the tensor product $k'^2 \otimes_{k^2} k$ is a field. Thus, the natural map of k-algebras $k'^2 \otimes_{k^2} k \to k'$ is injective, so we are done. □

Assume $0 < \dim V^{\perp} < \dim V$, which is to say B_q is degenerate but nonzero, so $d \geqslant 2n+1 \geqslant 3$. By regularity $q|_{V^{\perp}} \neq 0$, so q is non-degenerate if and only if $\dim V^{\perp} = 1$ (equivalently, $d = 2n+1$). In the non-degenerate case the group scheme $\mathrm{O}(q)$ is identified with $\mathrm{SO}(q) \times \mu_2$ where $\mathrm{SO}(q)$ is an adjoint connected semisimple group that is absolutely simple of type B_n, so in such cases $\mathrm{SO}(q)$ is the maximal smooth closed subgroup of $\mathrm{O}(q)$. Motivated by the non-degenerate case, in general we define $\mathrm{SO}(q)$ to be the maximal smooth closed k-subgroup of the k-group scheme $\mathrm{O}(q)$; this is the k_s/k-descent of the Zariski closure of $\mathrm{O}(q)(k_s)$ in $\mathrm{O}(q)_{k_s}$.

Proposition 7.1.2. *The* $\mathrm{SO}(q)$*-action on* V *is trivial on* $V^{\perp}$*. The* k*-group* $\mathrm{SO}(q)$ *is absolutely pseudo-simple of type* B_n *with trivial center and its Cartan* k*-subgroups are tori. For a linear complement* V' *to* $V^{\perp}$ *in* V*, the quadratic space* $(V', q|_{V'})$ *is non-degenerate of dimension* $2n$ *and if* $q|_{V'}$ *is split (i.e., hyperbolic) then* $\mathrm{SO}(q)$ *is pseudo-split. Moreover,* $\mathrm{SO}(q)$ *admits a pseudo-split* k_s/k*-form and every maximal* k*-torus* T *of* $\mathrm{SO}(q)$ *is contained in a Levi* k*-subgroup.*

As in the semisimple adjoint type-B case, Proposition 7.2.2 will show that if (V', q') is a second such regular quadratic space then all k-isomorphisms $\mathrm{SO}(q) \simeq \mathrm{SO}(q')$ arise from conformal isometries $(V, q) \simeq (V', q')$. (See §7.2 for the notion of conformal isometry.) Also, Proposition 7.1.6 will relate the k-rank of $\mathrm{SO}(q)$ and the dimension of maximal hyperbolic subspaces of V.

Proof. It is clear that $V^{\perp}$ is stable under the action of $\mathrm{O}(q)$ on V. For $g \in \mathrm{SO}(q)(k_s) = \mathrm{O}(q)(k_s)$ and $v \in V^{\perp}_{k_s}$,

$$q(g(v) - v) = q(g(v)) - q(v) = q(v) - q(v) = 0.$$

But $g(v) - v \in V^{\perp}_{k_s}$ and $q : V^{\perp}_{k_s} \to k_s$ is injective, so $g(v) - v = 0$. The Zariski-density of $\mathrm{SO}(q)(k_s)$ in $\mathrm{SO}(q)_{k_s}$ then implies that $\mathrm{SO}(q)$ acts trivially on $V^{\perp}$.

For any linear complement V' to $V^{\perp}$ in V, the natural map $V' \to V/V^{\perp}$ is an isomorphism identifying B_q on V' with the non-degenerate symplectic form $\overline{B}_q$ on $V/V^{\perp}$ induced by B_q, so $\dim V' = 2n$ and $q|_{V'}$ is non-degenerate. For a line $\ell \subset V^{\perp}$ and $W := V' \oplus \ell$, $q|_W$ is non-degenerate and $\mathrm{SO}(q|_W)$ can be realized as a type-B_n adjoint connected semisimple k-subgroup of $\mathrm{SO}(q)$ acting trivially on $V^{\perp}$. Thus, any maximal torus of $\mathrm{SO}(q|_W)$ has dimension n.

Letting T be a maximal k-torus of $\mathrm{SO}(q)$, by taking V' to be a T-equivariant complement to $V^{\perp}$ in V we see that the k-subgroup $\mathrm{SO}(q|_W) \subset \mathrm{SO}(q)$ contains T as a maximal torus (since T is a maximal torus of $\mathrm{SO}(q)$). Hence, no nonzero vector of V' is fixed under T, and V' is the unique T-equivariant linear complement to $V^{\perp}$ in V. It also follows that $\dim T = n$. Note that if $q|_{V'}$ is split then $\mathrm{SO}(q|_W)$ is k-split; i.e., $\mathrm{SO}(q|_W)$ contains a k-split maximal k-torus. Any such torus is then a maximal k-torus of $\mathrm{SO}(q)$, so in such cases $\mathrm{SO}(q)$ contains a

k-split maximal k-torus. To show in general (allowing non-split T) that $\mathrm{SO}(q)$ is absolutely pseudo-simple with trivial center and admits $\mathrm{SO}(q|_W)$ as a Levi k-subgroup, it suffices to work over k_s.

We will show at the end of this proof that $\mathrm{SO}(q)$ admits a pseudo-split k_s/k-form, so for now we may and do assume $k = k_s$. Hence, $\mathrm{SO}(q)$ is the Zariski closure in $\mathrm{O}(q)$ of the group $\mathrm{O}(q)(k)$ of k-points. The action of $\mathrm{O}(q)$ on V preserves $V^{\perp}$ so it defines a k-homomorphism

$$\pi : \mathrm{O}(q) \to \mathrm{Sp}(\overline{B}_q) \simeq \mathrm{Sp}_{2n}.$$

Fix a maximal k-torus T of $\mathrm{SO}(q)$ and let V' be the T-equivariant complement to $V^{\perp}$ in V. As above, let $W = V' \oplus \ell$ and naturally identify $L := \mathrm{SO}(q|_W)$ with a k-subgroup of $\mathrm{SO}(q)$ containing the maximal k-torus T. The restriction of π to $\mathrm{SO}(q|_W)$ is the natural exceptional isogeny from a group of adjoint type B_n onto a simply connected group of type C_n in characteristic 2, so π carries $\mathrm{SO}(q)$ *onto* the semisimple group $\mathrm{Sp}(\overline{B}_q)$. Hence π must kill the smooth connected unipotent normal k-subgroup $\mathscr{R}_{u,k}(\mathrm{SO}(q)^0)$. Therefore, to show that this k-unipotent radical is trivial it would suffice to check that $(\ker \pi)(k)$ is trivial. If $g \in (\ker \pi)(k)$ then $g(v) - v \in V^{\perp}$ for all $v \in V$, so

$$q(v) = q(g(v)) = q((g(v) - v) + v) = q(g(v) - v) + q(v)$$

and thus $g(v) - v \in \ker(q|_{V^{\perp}}) = \{0\}$ for all $v \in V$. This says $g = 1$, as desired.

We have shown that $\mathrm{SO}(q)^0$ is pseudo-reductive, so

$$H := \mathscr{D}(\mathrm{SO}(q)^0)$$

is pseudo-semisimple. The isogeny $\pi|_L : L \to \mathrm{Sp}(\overline{B}_q) =: \overline{L}$ with unipotent kernel maps T isomorphically onto a maximal torus $\overline{T}$ of $\overline{L}$. But clearly $Z_{\overline{L}}(\overline{T}) = \overline{T}$, so $Z_{\mathrm{SO}(q)}(T)$ is a smooth k-subgroup of $\mathrm{SO}(q)$ whose image in $\overline{L}$ is the isomorphic copy $\overline{T}$ of T. Hence, $Z_{\mathrm{SO}(q)}(T)(k) = T(k)$ since $(\ker \pi)(k) = 1$, so $Z_{\mathrm{SO}(q)}(T) = T$. In particular, all Cartan k-subgroups of $\mathrm{SO}(q)^0$ and H are tori, and the schematic center $Z_{\mathrm{SO}(q)}$ is contained in T. Thus, $Z_{\mathrm{SO}(q)} = 1$ since the connected semisimple subgroup $\mathrm{SO}(q|_W) \simeq \mathrm{SO}_{2n+1}$ has trivial center.

Let $f : \mathrm{SO}(q)^0 \twoheadrightarrow \overline{L}$ be the restriction of π to $\mathrm{SO}(q)^0$. Since $f|_T : T \to \overline{T}$ is an isomorphism, $(\ker f_{\overline{k}})^0_{\mathrm{red}}$ is unipotent. Thus, by [CGP, Rem. 2.3.6], the sets of roots $\Phi := \Phi(\mathrm{SO}(q)^0, T)$ and $\Phi(\overline{L}, \overline{T})$ coincide up to positive rational multipliers on the roots, so the Weyl groups coincide and if $\lambda \in \mathrm{X}_*(T)$ is "regular" in the sense that $\langle a, \lambda \rangle \neq 0$ for all $a \in \Phi(\mathrm{SO}(q)^0, T)$ then the cocharacter $\overline{\lambda} := f \circ \lambda \in \mathrm{X}_*(\overline{T})$ is likewise regular for $(\overline{L}, \overline{T})$. For such λ, it follows that $B := P_{\mathrm{SO}(q)^0}(\lambda)$ is a minimal pseudo-parabolic k-subgroup of $\mathrm{SO}(q)^0$ that meets H in a minimal pseudo-parabolic k-subgroup $P_H(\lambda)$, and the image of each under f is a common Borel k-subgroup $\overline{B}$ of $\overline{L}$.

Since H is visibly normal in $\mathrm{SO}(q)$, for any $g \in \mathrm{SO}(q)(k)$ we see that gTg^{-1} is also a maximal k-torus in H and hence is an $H(k)$-conjugate of T. To show that $g \in H(k)$ (thereby proving that $\mathrm{SO}(q)$ is (connected and) pseudo-semisimple), we may replace g with a left $H(k)$-translate so that g normalizes T. The equality of Weyl groups $W(H,T) = W(\mathrm{SO}(q)^0, T) = W(\Phi)$ further allows us to replace g with another $H(k)$-translate that also normalizes B. Hence, $f(g)$-conjugation on $\overline{L}$ preserves $(\overline{B}, \overline{T})$, so $f(g) \in \overline{T}(k)$. But $f : T \to \overline{T}$ is an isomorphism, so by replacing g with a suitable left $T(k)$-translate we can arrange that $g \in (\ker f)(k) \subset (\ker \pi)(k) = 1$. This completes the proof that $\mathrm{SO}(q)$ is pseudo-semisimple.

The comparison of root systems and Weyl groups for $(\mathrm{SO}(q), T)$ and $(\overline{L}, \overline{T})$ via f implies that $\Phi := \Phi(\mathrm{SO}(q), T)$ is irreducible (as its Weyl group acts irreducibly on $\mathrm{X}(T)_{\mathbf{Q}}$), so $\mathrm{SO}(q)$ is absolutely pseudo-simple with rank n. By the classification of irreducible root systems, the containment $\Phi(L,T) \subset \Phi$ then forces Φ to be either B_n or BC_n, and in the former case L is a Levi k-subgroup of $\mathrm{SO}(q)$ by (the proof of) [CGP, Thm. 3.4.6]. To rule out the possibility that Φ is non-reduced, it suffices to show that for a short root $b \in \Phi(L,T)$ the corresponding root group U_b of $\mathrm{SO}(q)$ has Lie algebra supporting only the T-weight b (and not also $2b$).

Note that $2b$ is a long root in $\Phi(\overline{L}, \overline{T})$, so if $2b \in \Phi(\mathrm{SO}(q), T)$ then the nontrivial smooth connected $2b$-root group U_{2b} of $\mathrm{SO}(q)$ must map onto the 1-dimensional $2b$-root group $\overline{U}_{2b} \subset \overline{L}$ under f. By equivariance of $f : U_{2b} \to \overline{U}_{2b}$ with respect to $T \simeq \overline{T}$, this quotient map between root groups corresponds to a linear map between vector spaces, so it has smooth kernel that must be trivial (as $(\ker f)(k)$ is trivial). Hence, the quotient $f : U_b \twoheadrightarrow \overline{U}_{2b}$ splits as a direct product, thereby providing a nontrivial smooth (connected) k-subgroup of $\ker f$, a contradiction. This shows that $\Phi = \Phi(L,T)$.

Finally, we allow k to be arbitrary (of characteristic 2) and will show that $\mathrm{SO}(q)$ admits a pseudo-split k_s/k-form. Letting V' be a linear complement to $V^{\perp}$ in V, (V,q) is the orthogonal sum of the restrictions (V', q') and $(V^{\perp}, q^{\perp})$ of q to V' and $V^{\perp}$ respectively. Since $\dim V' = 2n$ is even, there is a split k_s/k-form $(\mathscr{V}', \mathfrak{q}')$ of (V', q'), so for the orthogonal sum $(\mathscr{V}, \mathfrak{q})$ of $(\mathscr{V}', \mathfrak{q}')$ and $(V^{\perp}, q^{\perp})$ we see that $\mathrm{SO}(\mathfrak{q})$ is a pseudo-split k_s/k-form of $\mathrm{SO}(q)$. □

We continue to assume that $0 < \dim V^{\perp} < \dim V$ and that (V,q) is regular. The k-group $\mathrm{SO}(q)$ is non-reductive whenever q is degenerate, which is to say whenever $\dim V^{\perp} > 1$, due to the following result.

Proposition 7.1.3. *The dimensions of the root groups of* $\mathrm{SO}(q)_{k_s}$ *are* $\dim V^{\perp}$ *for short roots and* 1 *for long roots, with the convention that if* $n = 1$ *then the roots are short, so* $\mathrm{SO}(q)$ *is not reductive whenever* $\dim V^{\perp} > 1$.

The minimal field of definition over k *for the geometric unipotent radical of* $\mathrm{SO}(q)$ *is the subfield* $K \subset k^{1/2}$ *generated over* k *by the square roots* $(q(v)/q(v'))^{1/2}$ *for* $v, v' \in V^{\perp} - \{0\}$. *In particular,* $[K:k] \geqslant \dim V^{\perp}$.

If $\mathrm{SO}(q)$ *is pseudo-split of type* $\mathrm{A}_1 = \mathrm{B}_1$ *then it is isomorphic to* $\mathrm{P}H_{V^{\perp},K/k}$.

Proof. By the regularity of q, for any nonzero $v_0 \in V^{\perp}$ the k-linear map $V^{\perp} \to k^{1/2}$ defined by $v \mapsto (q(v)/q(v_0))^{1/2}$ is injective, so $[K:k] \geqslant \dim V^{\perp}$. Since the formation of K is compatible with separable extension on k, and an equality between purely inseparable extensions of k holds if and only if it does so after a separable extension of k, we may assume that the group $\mathrm{SO}(q)$ is pseudo-split. We fix a split maximal k-torus T in this group and will use notation (such as L, $\overline{L}$, and $\overline{T}$) introduced in the proof of Proposition 7.1.2.

If $n > 1$ then the long roots in $\Phi := \Phi(\mathrm{SO}(q), T)$ are the short roots in $\Phi(\overline{L}, \overline{T})$. By equivariance with respect to the isomorphism $T \simeq \overline{T}$ induced by $f : \mathrm{SO}(q) \twoheadrightarrow \overline{L}$, the associated quotient map between corresponding root groups of $\mathrm{SO}(q)$ and $\overline{L}$ for such roots corresponds to a linear map between vector spaces and hence must be an isomorphism (as $(\ker f)(k)$ is trivial). Thus, if $n > 1$ then the long root groups of $\mathrm{SO}(q)$ are 1-dimensional.

We next show (for any $n \geqslant 1$) that the root group for a short root b has dimension $d - 2n = \dim V^{\perp}$ for all $n \geqslant 1$. Since $k = k_s$, we can choose a k-basis $\{e_1, \ldots, e_d\}$ of V and k-isomophism $\mathrm{GL}_1^n \simeq T$ so that two properties hold:

$$q = x_1x_2 + \cdots + x_{2n-1}x_{2n} + c_{2n+1}x_{2n+1}^2 + \cdots + c_d x_d^2 \tag{7.1.3}$$

with $\{c_i\}$ linearly independent over k^2, and any point $t = (t_1, \ldots, t_n) \in \mathrm{GL}_1^n = T$ acts on the k-vector space $V = k^d$ via

$$t.(x_1, \ldots, x_d) = (t_1x_1, t_1^{-1}x_2, \ldots, t_n x_{2n-1}, t_n^{-1}x_{2n}, x_{2n+1}, \ldots, x_d)$$

with $b(t_1, \ldots, t_n) := t_1^{-1}$. The subtorus $S := (\ker b)^0_{\mathrm{red}}$ consists of those $t \in T$ such that $t_1 = 1$. Clearly the b-root group U_b lies in $\mathscr{D}(Z_{\mathrm{SO}(q)}(S)) =: H$.

For $V_0 = ke_1 \oplus ke_2 \oplus V^{\perp}$ and $q_0 = q|_{V_0}$, the natural k-subgroup $\mathrm{SO}(q_0) \subset \mathrm{SO}(q)$ is centralized by S and so lies inside H. But all points of $Z_{\mathrm{SO}(q)}(S)$ preserve the weight spaces $\{ke_i\}_{2<i\leqslant 2n}$ for the nontrivial S-weights on V, so

H acts trivially on these lines. That is, $H \subset \mathrm{SO}(q_0)$ as well, so $H = \mathrm{SO}(q_0)$. Thus, for the determination of the common dimension of all short root groups we reduce to the case $n = 1$, so $b : T \to \mathrm{GL}_1$ is an isomorphism.

We may choose an isomorphism $\overline{L} = \mathrm{SL}(V/V^{\perp}) = \mathrm{SL}_2$ so that the b-root group is the lower-triangular unipotent subgroup (i.e., $b = t_1^{-1}$). Thus, using the ordered basis of V given by $\{e_1, e_2\}$ followed by an ordered basis of $V^{\perp}$, we see via triviality of the $\mathrm{SO}(q)$-action on $V^{\perp}$ that U_b lies in the part of the lower-triangular unipotent subgroup of $\mathrm{GL}(ke_1 \oplus ke_2 \oplus V^{\perp})$ that acts trivially on $V^{\perp}$. Hence, points $u \in U_b(k)$ act on V via the identity on $V^{\perp}$ and via

$$e_1 \mapsto e_1 - ae_2 + v, \ e_2 \mapsto e_2 + v'$$

for some $a \in k$ and $v, v' \in V^{\perp}$. Since $q(e_1)$ and $q(e_2)$ vanish, orthogonality of u forces $a = q(v)$ and $q(v') = 0$, so $v' = 0$ by regularity. Turning this around, if $\underline{V}^{\perp}$ denotes the k-vector group associated to $V^{\perp}$ then for any $v \in \underline{V}^{\perp}$ the linear automorphism $[v]$ of $\underline{V}$ that fixes e_2 and all points of $\underline{V}^{\perp}$ and carries e_1 to $e_1 - q(v)e_2 + v$ is easily seen to be orthogonal, and for $t \in T = \mathrm{GL}_1$ (equality via b) we have $t.e_1 = t^{-1}e_1$ and $t.e_2 = te_2$.

Triviality of the T-action on $V^{\perp}$ implies that $t[v]t^{-1}$ has the following properties: it fixes e_2, it fixes $\underline{V}^{\perp}$ pointwise, and it carries e_1 to

$$t.([v](te_1)) = t.(t[v](e_1)) = t.(te_1 - q(v)te_2 + tv) = e_1 - q(tv)e_2 + tv.$$

Thus, $t[v]t^{-1} = [tv]$ in $\mathrm{GL}(V)$, so $v \mapsto [v]$ defines a k-group inclusion $\underline{V}^{\perp} \hookrightarrow \mathrm{O}(q)^{\mathrm{sm}} = \mathrm{SO}(q)$ that is equivariant for T-conjugation on $\mathrm{SO}(q)$ and the T-action on $\underline{V}^{\perp}$ through b. By the dynamic definition of root groups this identifies $\underline{V}^{\perp}$ with a k-subgroup of U_b, and the preceding calculations show that this inclusion is an equality on k_s-points and hence is an equality of k-groups (carrying the natural linear structure on $\underline{V}^{\perp}$ over to the unique T-equivariant linear structure on U_b).

Finally, returning to general n, we compute the minimal field of definition over k for the geometric unipotent radical of $\mathrm{SO}(q)$. By Proposition 3.2.5 and the description of long and short root groups above, our problem is reduced to the case $n = 1$. The case $d = 3$ (i.e., non-degenerate q) is well-known, so assume $d \geqslant 4$. The explicit form of q identifies K with $k(\sqrt{c_4/c_3}, \ldots, \sqrt{c_d/c_3})$. We k-linearly embed $V^{\perp}$ into K via $v \mapsto \sqrt{q(v)/q(e_3)}$, so $V^{\perp}$ is thereby a k-subspace of K containing 1 that generates K as a k-algebra. This subspace defines the k-subgroup $\mathrm{P}H_{V^{\perp},K/k} \subset \mathrm{R}_{K/k}(\mathrm{PGL}_2)$ whose diagonal k-torus is a Cartan k-subgroup since $K^2 \subset k$ (see Definition 3.1.2 and Proposition 3.1.4).

For the K-isogeny $\Pi : \mathrm{PGL}_2 \to \mathrm{SL}_2$ it is easy to check via consideration of open cells that $\mathrm{R}_{K/k}(\Pi)$ carries $\mathrm{P}H_{V^\perp,K/k}$ into the natural k-subgroup $\mathrm{SL}_2 \subset \mathrm{R}_{K/k}(\mathrm{SL}_2)$ via a k-homomorphism that is injective on k-points (since $q : V^\perp \to k$ is injective).

Using the identification of each T-root group in $\mathrm{SO}(q)$ with $\underline{V}^\perp$ (computed above for one root, and done likewise for the opposite root with the roles of e_1 and e_2 swapped), we get an evident k-scheme isomorphism ϕ from an open cell Ω of $(\mathrm{SO}(q), T)$ to that of $(\mathrm{P}H_{V^\perp,K/k}, \overline{D})$ over the natural map from each side into $\overline{L} = \mathrm{SL}_2$. The map ϕ is a rational homomorphism because this can be checked on k-points in a dense open subset of $\Omega \times \Omega$ (using that $(\ker \Pi)(K) = 1$), so ϕ uniquely extends to an isomorphism of k-groups. Thus, the minimal field of definition over k for $\mathscr{R}_u(\mathrm{SO}(q)_{\overline{k}}) \subset \mathrm{SO}(q)_{\overline{k}}$ coincides with that of $\mathrm{P}H_{V^\perp,K/k}$. By Proposition 3.1.4 this field is (uniquely) k-isomorphic to $k[V^\perp] = K$. (In the initial setup with any $n \geqslant 1$, this proof shows that if $k = k_s$ then for any short $b \in \Phi(\mathrm{SO}(q), T)$ we have $\mathrm{SO}(q)_b \simeq \mathrm{P}H_{j_{v_0}(V^\perp),K/k}$ where $j_{v_0}(v) = \sqrt{q(v)/q(v_0)}$ for any $v_0 \in V^\perp - \{0\}$.) □

Remark 7.1.4. The description of K in Proposition 7.1.3 shows that the kernel $\mathscr{V} := \ker(q_K|_{V_K^\perp}) \subset V_K^\perp$ is a hyperplane, and the $\mathrm{SO}(q)_K$-action on V_K is trivial on $\mathscr{V}$ (it is even trivial on $V_K^\perp$, by Proposition 7.1.2). Thus, the quadratic form

$$\overline{q}_K := q_K \bmod \mathscr{V} : V_K/\mathscr{V} \to K \tag{7.1.4}$$

is well-defined and non-degenerate on the space $V_K/\mathscr{V}$ of K-dimension $2n+1$, and there is an evident natural map $h : \mathrm{SO}(q)_K \to \mathrm{SO}(\overline{q}_K)$. In the proof of Proposition 7.1.2 we built Levi k-subgroups $L \subset \mathrm{SO}(q)$ of the form $\mathrm{SO}(q|_W)$ where $W = V' \oplus \ell$ for any linear complement V' to $V^\perp$ in V and line ℓ in $V^\perp$. The restriction of h to L_K is an isomorphism since $W_K \to V_K/\mathscr{V}$ is an isomorphism, so h identifies $\mathrm{SO}(q)_K^{\mathrm{ss}}$ with $\mathrm{SO}(\overline{q}_K)$.

Proposition 7.1.5. *Let T be a maximal k-torus of $\mathrm{SO}(q)$. There is a unique T-equivariant linear complement V_0 to $V^\perp$ inside V. For every line $\ell \subset V^\perp$ the restriction of q to the $(2n+1)$-dimensional subspace $W := V_0 \oplus \ell$ is non-degenerate, the k-subgroup scheme L of points $g \in \mathrm{SO}(q)$ such that $g(W) = W$ is a Levi k-subgroup isomorphic to $\mathrm{SO}(q|_W)$ via $g \mapsto g|_W$, and every Levi k-subgroup of $\mathrm{SO}(q)$ containing T arises in this manner for a unique ℓ.*

If $n \geqslant 2$ then there is a unique connected semisimple subgroup of $\mathrm{SO}(q)$ of type D_n containing T, and it is $\mathrm{SO}(q_0)$ for the non-degenerate $q_0 := q|_{V_0}$ (with $\mathrm{SO}(q_0) \hookrightarrow \mathrm{SO}(q)$ defined via isometries of $V = V_0 \oplus V^\perp$ that preserve V_0).

Proof. The uniqueness assertions permit us to assume $k = k_s$, so T is split and we may decompose V as a direct sum of T-weight spaces. Since all maximal k-tori in $G := \mathrm{SO}(q)$ are $\mathrm{SO}(q)(k)$-conjugate (as $k = k_s$), for the rest of the proof it suffices to treat a single T. We fix a k-linear complement V_0 to $V^{\perp}$ in V, so $q_0 := q|_{V_0}$ is non-degenerate of rank $2n$ and (V, q) is the orthogonal sum of q_0 and $q|_{V^{\perp}}$. Since $k = k_s$, we may choose a k-isomorphism $q_0 \simeq x_1x_2 + \cdots + x_{2n-1}x_{2n}$, so we obtain a maximal k-torus $\mathrm{GL}_1^n \subset G$ preserving V_0 via the action

$$(t_1, \ldots, t_n).(x_1, \ldots, x_{2n}) = (t_1x_1, t_1^{-1}x_2, \ldots, t_nx_{2n-1}, t_n^{-1}x_{2n}) \qquad (7.1.5)$$

on V_0 and the trivial action on $V^{\perp}$. We may (and do) take T to be this torus. The T-weights occurring on V_0 are visibly nontrivial and have 1-dimensional weight spaces, so V_0 is the unique T-equivariant linear complement to $V^{\perp}$ in V. The $2n$ characters $\chi_i^{\pm} : (t_1, \ldots, t_n) \mapsto t_i^{\pm 1}$ of T are exactly the short roots in $\Phi(G, T) = \Phi(G_K^{\mathrm{ss}}, T_K) = \Phi(\mathrm{SO}(\overline{q}_K), T_K)$.

Consider a line ℓ in $V^{\perp}$. The injectivity of $q|_{V^{\perp}}$ implies that $q|_{\ell}$ is nonzero, so $W = V_0 \oplus \ell$ equipped with the quadratic form $q|_W$ is the orthogonal sum of $q|_{\ell}$ and q_0. Hence, $(W, q|_W)$ is non-degenerate of dimension $2n + 1$, and in the proof of Proposition 7.1.2 we showed that the inclusion $\mathrm{SO}(q|_W) \hookrightarrow G$ defined by imposing the trivial action on $V^{\perp}$ is a Levi k-subgroup. This is precisely the k-subgroup scheme of G defined by the combined conditions of preservation of W and triviality on $V^{\perp}$ because the orthogonal group scheme $\mathrm{O}(q|_W)$ coincides with $\mathrm{SO}(q|_W) \times \mu_2$ and triviality on $\ell = W \cap V^{\perp}$ characterizes $\mathrm{SO}(q|_W)$ inside $\mathrm{O}(q|_W)$. The subspace $W \subset V$ (and hence the line $\ell = W \cap V^{\perp}$) is uniquely determined by the maximal k-torus $T \subset G$ and the Levi k-subgroup $\mathrm{SO}(q|_W) \subset G$ containing T because of the elementary fact that W is the $\mathrm{SO}(q|_W)$-span of V_0 (see the explicit description of unipotent orthogonal automorphisms arising from short root groups for the case $n = 1$ in the proof of Proposition 7.1.3).

Since $V^{\perp}$ is preserved by the action of G on V, to show that every Levi k-subgroup L of G containing T arises by the above construction for some (necessarily unique) line in $V^{\perp}$ we recall from [CGP, Thm. 3.4.6] that any such L is uniquely determined by its T-root groups inside those of G for roots belonging to a fixed basis Δ of the common root system $\Phi(L, T) = \Phi(G, T)$. The long root groups of (G, T) are 1-dimensional, so they are contained in L. We may and do choose Δ so that its unique short root b is the character χ_1^- of $T = \mathrm{GL}_1^n$. Equip the b-root group $U_b \subset G$ with its unique T-equivariant linear structure. The Levi k-subgroup $L \subset G$ containing T is uniquely determined by its b-root group that in turn corresponds to a line ℓ_0 inside $U_b(k)$. The k-vector group

$U_b \subset \mathrm{GL}(V)$ was identified with $V^{\perp}$ in the proof of Proposition 7.1.3: to each $v \in V^{\perp}$ we associate the orthogonal automorphism of V that pointwise fixes both $V^{\perp}$ and every e_j for $j \neq 1$, and carries e_1 to $e_1 - q(v)e_2 + v$. The b-root group of (L,T) corresponds to those v lying in the line ℓ_0, but the same calculations applied to the non-degenerate $q|_{V_0 \oplus \ell_0}$ show that Levi k-subgroup L_0 of G associated to (T, ℓ_0) via the above construction has the same b-root group inside U_b. Hence, $L = L_0$.

Now assume $n \geqslant 2$. The group $\mathrm{SO}(q_0)$ is naturally identified with a connected semisimple k-subgroup $H \subset \mathrm{SO}(q)$ of type D_n containing T as a maximal k-torus. As any connected semisimple subgroup of G of type D_n containing T is generated by its root groups, which in turn are precisely the 1-dimensional long root groups of $(\mathrm{SO}(q), T)$, the uniqueness of H follows. □

Assume $1 < \dim V^{\perp} < \dim V$. It can happen that the *non-reductive* absolutely pseudo-simple k-group $\mathrm{SO}(q)$ is k-anisotropic (i.e., does not contain GL_1 as a k-subgroup), even when the K-group $\mathrm{SO}(q)_K^{\mathrm{ss}} = \mathrm{SO}(\overline{q}_K)$ (see Remark 7.1.4) is K-split. This is interesting because for k of any characteristic and an absolutely pseudo-simple *standard* pseudo-reductive k-group $\mathscr{G}$ with minimal field of definition K/k for $\mathscr{R}_u(\mathscr{G}_{\overline{k}}) \subset \mathscr{G}_{\overline{k}}$, the k-rank of $\mathscr{G}$ coincides with the K-rank of the connected semisimple $\mathscr{G}_K^{\mathrm{red}}$ (seen by inspection of the standard construction).

To make such (V,q), first observe that anisotropicity of (V,q) implies that the pseudo-reductive group $\mathrm{SO}(q)$ is k-anisotropic. Indeed, more generally:

Proposition 7.1.6. *Let (V,q) be a regular quadratic space over a field k. There exists an orthogonal sum of r hyperbolic planes in (V,q) if and only if the k-group scheme $\mathrm{O}(q)$ contains a split k-torus of dimension r. In particular, q has an isotropic vector if and only if $\mathrm{O}(q)$ contains a nontrivial split k-torus.*

Proof. The case of non-degenerate (V,q) is classical (by consideration of weight vectors). Thus, we may assume (V,q) is degenerate, so by regularity necessarily $\mathrm{char}(k) = 2$. For any pairwise orthogonal hyperbolic planes $H_1, \dots, H_r$ in V, clearly $V = H_1 \perp \cdots \perp H_r \perp V'$ for the orthogonal complement V' of $\perp H_i$, so there is an evident split r-torus inside $\mathrm{O}(q)$.

Now suppose $\mathrm{O}(q)$ contains $T = \mathrm{GL}_1^r$ as a k-subgroup. We seek a pairwise orthogonal collection of r hyperbolic planes inside V, so we may assume $r > 0$. Since T acts on V through a k-subgroup inclusion of T into $\mathrm{GL}(V)$, the weights for T on V generate $\mathrm{X}(T)_{\mathbf{Q}}$. Hence, we can find r linearly independent nontrivial weights $\chi_1, \dots, \chi_r$ for T on V.

If $\chi \in \mathrm{X}(T)$ is nontrivial then any v in the χ-weight space $V(\chi)$ satisfies

$$q(v) = q(t.v) = q(\chi(t)v) = \chi(t)^2 q(v)$$

for all $t \in T$, forcing $q(v) = 0$. Likewise, $V(\chi)$ and $V(\chi')$ are B_q-orthogonal for all nontrivial $\chi' \neq -\chi$ and each is B_q-orthogonal to V^T. Since the 2-linear $q|_{V^\perp}$ has trivial kernel due to regularity, we see that $V^\perp \subset V^T$. For nonzero $v_i \in V(\chi_i)$ (so $v_i \notin V^\perp$) we have $B_q(v_i, \cdot) \neq 0$, so there exists $v_i' \in V(-\chi_i)$ such that $B_q(v_i, v_i') = 1$. Thus, v_i and v_i' span a hyperbolic plane H_i, and the H_i's are pairwise orthogonal. □

We next build examples of non-degenerate anisotropic (V_0, q_0) over some k with $\mathrm{char}(k) = 2$ and $\dim(V_0) = 2n \geqslant 2$ such that $\mathrm{SO}(q_0)$ *splits* over a finite extension K/k for which squaring $\mathfrak{q} : K \to K$ lands inside k and $q_0 \oplus c\mathfrak{q} : V_0 \oplus K \to k$ is k-anisotropic for some $c \in k^\times$. For $(V, q) := (V_0, q_0) \perp (K, c\mathfrak{q})$, the k-group $\mathrm{SO}(q)$ is then anisotropic and the minimal field of definition over k for $\mathscr{R}_u(\mathrm{SO}(q)_{\overline{k}})$ is K/k (as we see via the explicit description in Proposition 7.1.3 of that minimal field extension of k in terms of square roots of ratios of nonzero values of $q|_{V^\perp}$); our examples will have the additional property that for n even, $\mathrm{SO}(\overline{q}_K)$ is K-split. The method is inspired by a construction of Richard Weiss that in turn is modelled on [TW, (14.19)–(14.23)].

Example 7.1.7. Let κ be a field of characteristic 2 and let $k = \kappa(t_1, \dots, t_{n+1})$ be the rational function field over κ in $n+1$ variables with any $n \geqslant 1$. Let F be a quadratic extension of k such that for $1 \leqslant i \leqslant n+1$ the t_i-adic valuation on k has ramification index 1 and residual degree 2 in F. Among such F are the extensions $\kappa'(t_1, \dots, t_{n+1})$ for a quadratic extension κ' of κ (if such κ' exist) or the splitting field of the separable polynomial $x^2 - x - \sum_j t_j \in k[x]$ (which is irreducible over k since $x^2 - x - t$ is irreducible over $L(t)$ for any field L). Define the anisotropic binary quadratic form $\nu = \mathrm{N}_{F/k} : F \to k$ over k.

Let V_0 denote the $2n$-dimensional k-vector space $F^{\oplus n}$, and let $q_i : V_0 \to k$ be the composition of ν with the ith projection $V_0 \to F$. Squaring on $K := \kappa(\sqrt{t_1}, \dots, \sqrt{t_n}, t_{n+1})$ is a quadratic form $\mathfrak{q} : K \to k$ valued in the subfield $k' := \kappa(t_1, \dots, t_n, t_{n+1}^2)$ of k. Consider the quadratic form

$$q = \sum_{1 \leqslant i \leqslant n} t_i q_i + t_{n+1}\mathfrak{q} : V = V_0 \oplus K \to k$$

over k, so if F/k is separable then $\mathrm{SO}(q)$ is absolutely pseudo-simple and $\mathrm{SO}(q)_F$ is pseudo-split (with root system B_n). We shall first prove (for every

$n \geqslant 1$) that q is anisotropic, and then that for suitable separable F/k as above and even $n \geqslant 2$ the connected semisimple K-group $\mathrm{SO}(\overline{q}_K)$ is split.

Since $\mathfrak{q}$ is valued in k', to prove the anisotropicity it suffices to show by induction on n that for any $c_1, \dots, c_n \in F$ and $\xi \in k'$ such that

$$\sum_{i=1}^{n} t_i \nu(c_i) + t_{n+1}\xi = 0, \tag{7.1.7}$$

necessarily $c_1 = \cdots = c_n = 0$ and $\xi = 0$. For all $1 \leqslant i \leqslant n+1$ and all $c \in F^\times$, $\mathrm{ord}_{t_i}(\nu(c))$ is even since the quadratic extension F/k has ramification index 1 and residual degree 2 over the t_i-adic discrete valuation on k. Likewise, $\mathrm{ord}_{t_{n+1}}(\xi)$ is even when $\xi \neq 0$ since $\xi \in k'$.

Define $A_k = \kappa[t_1, \dots, t_{n+1}] \subset k$, so the fraction field of A_k is k. Let A_F denote the module-finite integral closure of A_k in F/k, so A_F has fraction field F and dimension $n+1$. Since A_F is a normal noetherian domain, it coincides with the intersection of its local rings at height-1 primes. By hypothesis the t_i-adic valuation v_i on k has a unique lift w_i on F and t_i is a uniformizer for w_i, so t_i is a unit at all height-1 primes of A_F aside from w_i. Hence, $t_i A_F$ is the prime ideal of elements of A_F with positive w_i-valuation, so $A_F/t_i A_F$ is a *domain* whose fraction field is the residue field at w_i and the natural map $A_k/t_i A_k \to A_F/t_i A_F$ is *injective*.

Consider a solution to (7.1.7). We want to show that all c_i and ξ vanish. Scaling through by the square of a suitable nonzero element of A_k allows us to arrange that $c_i \in A_F$ for all i and $\xi \in \kappa[t_1, \dots, t_n, t_{n+1}^2] =: A_{k'}$. Since the local ring $(A_F)_{w_i}$ at the height-1 prime $t_i A_F$ coincides with the localization of A_F at the height-1 prime $t_i A_k$ for A_k, we see that the induced injective map of discrete valuation rings $(A_k)_{v_i} \to (A_F)_{w_i}$ is *module-finite* and hence free of rank 2. In particular, it makes sense to compute norms relative to this ring extension, so for any $c \in A_F$ the norm $\nu(c) \in A_k$ relative to F/k has mod-$t_i A_k$ reduction that coincides with $\nu_i(c \bmod t_i A_F)$ where ν_i is the norm relative to the induced quadratic extension of residue fields at v_i and w_i. Such compatibility of norms will enable us to carry out induction on n.

If $\xi \neq 0$ then $\sum_{i=1}^n t_i \nu(c_i) \in k^\times$ and its t_{n+1}-adic valuation is equal to $1 + \mathrm{ord}_{t_{n+1}}(\xi) \in 1 + 2\mathbf{Z}$ (see (7.1.7)). Each $t_i \nu(c_i)$ that is nonzero must have even t_{n+1}-adic valuation, so there would have to be at least two nonzero c_i's if $\xi \neq 0$. In particular, this rules out the case $n = 1$, so we may assume $n > 1$ and that the case of $n-1$ is settled. For $1 \leqslant i \leqslant n$, since $A_{k'}/t_i A_{k'} \to A_k/t_i A_k$ is injective by inspection, reducing (7.1.7) modulo t_i permits us to apply induction with $n-1$ to conclude that t_i divides both ξ in $A_{k'}$ and c_j in A_F for every $j \neq i$.

Hence, ξ is divisible by $\tau := \prod_{i=1}^{n} t_i$ in $A_{k'}$ and c_i is divisible by $\tau_i = \tau/t_i$ in A_F for all $1 \leqslant i \leqslant n$ because the *normality* of A_F reduces τ_i-divisibility to t_j-divisibility for each $j \neq i$ separately (as t_j is a unit at all height-1 primes of A_F apart from w_j).

Substituting the resulting identities $\xi = \tau\xi^*$ for $\xi^* \in A_{k'}$ and $c_i = \tau_i c_i^*$ for $c_i^* \in A_F$ into (7.1.7) and canceling a common factor of τ throughout yields

$$\sum_{i=1}^{n} (\tau/t_i)\nu(c_i^*) + t_{n+1}\xi^* = 0.$$

Reducing modulo t_i kills all terms in the sum apart from the ith, so

$$\Big(\prod_{1\leqslant j\leqslant n, j\neq i} t_j\Big)\cdot \nu_i(c_i^* \bmod t_i) = -t_{n+1}(\xi^* \bmod t_i)$$

in the residue field $\mathrm{Frac}(A_k/t_i A_k)$ at v_i (with ν_i the norm into this from the residue field at w_i). If both sides are nonzero then the left side has even t_{n+1}-adic valuation whereas the right side has odd t_{n+1}-adic valuation. This is absurd, so both sides vanish for all i.

We conclude that $c_i = \tau c_i'$ for some $c_i' \in A_F$ and that ξ^* is divisible by $t_1,\ldots,t_n$ in $A_{k'}$, so $\xi = \tau^2\xi'$ for some $\xi' \in A_{k'}$. Substituting these into (7.1.7) and cancelling the common factor of τ^2 throughout brings us to another solution $(c_1',\ldots,c_n',\xi')$ of (7.1.7) with $c_i' \in A_F$ and $\xi' \in A_{k'}$. We can keep iterating the process to conclude that ξ and every c_i are infinitely divisible by $\tau = t_1\cdots t_n$, so $\xi = 0$ and $c_i = 0$ for all i. This completes the proof of k-anisotropicity of q.

Now assume F/k is separable, so $K_F := K\otimes_k F$ is a field and the quadratic form $\overline{q}_K : (K_F)^{\oplus n} \oplus K \to K$ as in (7.1.4) is an orthogonal sum of $t_{n+1}x^2$ and n copies of the non-degenerate rank-2 quadratic space $(K_F, \mathrm{N}_{K_F/K})$. In general, an orthogonal sum of two copies of a non-degenerate rank-2 quadratic space (U,h) over any field L of characteristic 2 is a *split* quadratic space. Indeed, by scaling we can assume $(U,h) = (L\oplus L, x^2+xy+ay^2)$ for some $a \in L$, and if $\{e_1,e_2\}$ and $\{e_1',e_2'\}$ are the associated bases of two orthogonal copies of U then

$$\{e_1+e_1', a(e_1+e_1')+e_2\},\ \{e_2+e_2', e_2+e_2'+e_1'\}$$

are bases of null vectors for orthogonal hyperbolic planes spanning the orthogonal sum $U \perp U$. Thus, if n is *even* then whenever F/k is separable the quadratic form $\overline{q}_K$ is the orthogonal sum of $t_{n+1}x^2$ and n hyperbolic planes, so $\mathrm{SO}(\overline{q}_K)$ is K-split. (If n is *odd* and F/k is separable then typically $\mathrm{SO}(\overline{q}_K)$ has K-rank $n-1$. To be precise, assume in addition that the quadratic residue field extension for F/k at t_{n+1} is separable, or more generally is distinct from that of K/k, as we can certainly arrange. The preceding method and Witt cancellation reduce our task to checking that t_{n+1} viewed in K is not a norm from K_F, as holds since K_F/K has residual degree 2 at t_{n+1}.)

Example 7.1.8. Consider $G = \mathrm{SO}(q)$ for finite-dimensional regular quadratic spaces (V, q) over k such that $0 < \dim V^{\perp} < \dim V$. For any nonzero $v_0 \in V^{\perp}$ we obtain a k-linear injection $j_{v_0} : V^{\perp} \to K$ via $v \mapsto \sqrt{q(v)/q(v_0)}$, and for any nonzero $v_1 \in V^{\perp}$ clearly $j_{v_1} = \mu j_{v_0}$, where $\mu = \sqrt{q(v_0)/q(v_1)} \in K$. Hence, the k-subspace $j_{v_0}(V^{\perp}) \subset K$ is independent of v_0 up to $K^{\times}$-scaling, so the k-subalgebra

$$F = \{\lambda \in K \,|\, \lambda \cdot j_{v_0}(V^{\perp}) \subset j_{v_0}(V^{\perp})\}$$

of K is independent of v_0. This is called the *root field* of $(V^{\perp}, q|_{V^{\perp}})$. Observe that $V^{\perp}$ thereby acquires a structure of F-vector space over its k-linear structure, and this is independent of v_0; it is characterized by the property $q(\lambda v) = \lambda^2 q(v)$ for $\lambda \in F$ and $v \in V^{\perp}$.

The formation of F commutes with separable extension on k, and $[F : k]$ is a power of 2, so if $\dim_k V^{\perp}$ is odd then $F = k$. In the non-degenerate case $F = k$, but in the degenerate case F may be larger than k (e.g., for $V^{\perp} = K$ with $q|_{V^{\perp}}$ the squaring map $K \to k$ we have $F = K$).

It is obvious that the (long) root field of G is k. In general we claim that F coincides with the short root field of G. It suffices to check this equality of purely inseparable extensions of k after scalar extension to k_s. In the proof of Proposition 7.1.3 it has been shown that if $k = k_s$ then for any short root $b \in \Phi(G, T)$ we have $G_b \simeq \mathrm{P}H_{j_{v_0}(V^{\perp}), K/k}$ for some $v_0 \in V^{\perp} - \{0\}$.

Definition 7.1.9. For regular (V, q) satisfying $0 < \dim V^{\perp} < \dim V$, the *spin group* $\mathrm{Spin}(q)$ is the universal smooth k-tame central extension of $\mathrm{SO}(q)$.

The k-group $\mathrm{SO}(q)$ in Definition 7.1.9 is of minimal type since $Z_{\mathrm{SO}(q)} = 1$, so by Proposition 5.3.3 the k-group $\mathrm{Spin}(q)$ is of minimal type. This recovers the usual spin group in the non-degenerate case (i.e., when $\dim V^{\perp} = 1$).

7.2 Conformal isometries

We next turn our attention to k_s/k-forms of $\mathrm{SO}(q)$ for finite-dimensional regular quadratic spaces (V, q) over k such that $0 < \dim V^{\perp} < \dim V$; our main interest will be in the degenerate case (i.e., $\dim V^{\perp} > 1$). A *conformal isometry* between quadratic spaces over a field is a linear isomorphism that respects the quadratic forms up to a nonzero scaling factor (which is uniquely determined when the quadratic forms are nonzero).

Let $\mathrm{CIsom}(q', q)$ be the set of conformal isometries onto (V, q) from a finite-dimensional regular quadratic space (V', q') satisfying $0 < \dim V'^{\perp} < \dim V'$. Let $\mathrm{CO}(q) \subset \mathrm{GL}(V)$ be the smooth k-subgroup that is the Galois descent of the Zariski closure of $\mathrm{CIsom}(q_{k_s}, q_{k_s})$ inside $\mathrm{GL}(V_{k_s})$. The (possibly non-smooth) *scheme* of conformal isometries is the closed subgroup scheme of $\mathrm{GL}(V)$ assigning to any k-algebra A the group of pairs (L, μ) consisting of an A-linear automorphism L of V_A and a unit $\mu \in A^{\times}$ such that $q_A \circ L = \mu \cdot q_A$; its maximal smooth k-subgroup is $\mathrm{CO}(q)$ by [CGP, Lemma C.4.1], so if E/k is an arbitrary separable extension field then $\mathrm{CO}(q)(E)$ is the group of conformal isometries of $(V, q)_E$ onto itself.

Remark 7.2.1. If q is degenerate then the k-group $\mathrm{CO}(q)$ that is smooth by definition generally does *not* contain the group scheme $\mathrm{O}(q)$; see Example 7.2.3.

Let F be the root field of $(V^{\perp}, q|_{V^{\perp}})$ as in Example 7.1.8, so it is also the short root field of $\mathrm{SO}(q)$ and $V^{\perp}$ has a canonical structure of F-vector space over its given k-linear structure (with $q(\lambda v) = \lambda^2 q(v)$ for $v \in V^{\perp}$ and $\lambda \in F$). When $G := \mathrm{SO}(q)$ contains a split maximal k-torus T, we can build a copy of $\mathrm{R}_{F/k}(\mathrm{GL}_1)$ inside $\mathrm{CO}(q)$ centralizing T, as follows. Consider the unique T-equivariant decomposition $V = V_0 \oplus V^{\perp}$ as in Proposition 7.1.5, and decompose V_0 into a direct sum of T-weight lines $\ell_{\pm 1}, \dots, \ell_{\pm n}$ for pairs of opposite weights (the short roots in $\Phi(G, T)$). For any $\lambda \in F^{\times}$ we have $\lambda^2 \in k^{\times}$ and define $[\lambda] \in \mathrm{CO}(q)$ via $[\lambda]|_{V^{\perp}} = \lambda$, $[\lambda]|_{\ell_i} = \lambda^2$ for $i = 1, \dots, n$, and $[\lambda]|_{\ell_i} = 1$ for $i = -1, \dots, -n$. This construction defines a k-subgroup inclusion

$$\mathrm{R}_{F/k}(\mathrm{GL}_1) \hookrightarrow \mathrm{CO}(q) \tag{7.2.1.1}$$

resting on a choice within each pair of opposite nontrivial T-weights on V. The set of such choices is permuted transitively under the action of $N_{\mathrm{SO}(q)}(T)(k)$ on T through its Weyl group quotient $W(\Phi(G, T))$, so the composite map

$$\theta_q : \mathrm{R}_{F/k}(\mathrm{GL}_1) \to \mathrm{CO}(q)/\mathrm{SO}(q) \tag{7.2.1.2}$$

is intrinsic to the k-split T. It follows from the $\mathrm{SO}(q)(k_s)$-conjugacy of maximal k_s-tori in $\mathrm{SO}(q)_{k_s}$ that (7.2.1.2) is independent of T and so may be constructed in general over k via k_s/k-descent (not depending on an initial choice of maximal k-torus).

Using k-split T, the restriction of (7.2.1.1) to the maximal k-torus $\mathrm{GL}_1 \subset$

$\mathrm{R}_{F/k}(\mathrm{GL}_1)$ carries λ to the product of λ-multiplication and

$$(\lambda, 1/\lambda, \ldots, \lambda, 1/\lambda, \mathrm{id}) \in T \subset \mathrm{GL}(V_0) \times \mathrm{GL}(V^{\perp}), \qquad (7.2.1.3)$$

so if $F = k$ (as when q is non-degenerate) then (7.2.1.1) provides nothing in $\mathrm{CO}(q)$ beyond what is already obtained from T and the central GL_1. In general $\mathrm{R}_{F/k}(\mathrm{GL}_1) \cap T = 1$ schematically inside $\mathrm{GL}(V)$, so since $Z_{\mathrm{SO}(q)}(T) = T$ it follows that the k-homomorphism

$$\mathrm{R}_{F/k}(\mathrm{GL}_1) \ltimes \mathrm{SO}(q) \to \mathrm{CO}(q)$$

provided by the split k-torus T has trivial kernel (equivalently, $\ker \theta_q = 1$). The k-subgroup $\mathrm{R}_{F/k}(\mathrm{GL}_1) \times T$ of $\mathrm{CO}(q)$ contains the central GL_1 due to (7.2.1.3).

The k-subgroup $\mathrm{SO}(q) \subset \mathrm{GL}(V)$ is insensitive to $k^{\times}$-scaling of q (and likewise after any separable extension on k), so there is an evident action of $\mathrm{CO}(q)/\mathrm{GL}_1$ on $\mathrm{SO}(q)$ and the natural map

$$\mathrm{CIsom}(q', q) \to \mathrm{Isom}_k(\mathrm{SO}(q'), \mathrm{SO}(q)) \qquad (7.2.1.4)$$

is equivariant with respect to $\mathrm{CO}(q)(k) \to \mathrm{CO}(q)(k)/k^{\times}$.

Proposition 7.2.2. *Consider (V, q) and (V', q') as above (with no pseudo-split hypothesis on $\mathrm{SO}(q)$).*

(i) *The map (7.2.1.4) is the quotient by the $k^{\times}$-action on $\mathrm{CIsom}(q', q)$. In particular, $\mathrm{SO}(q') \simeq \mathrm{SO}(q)$ if and only if (V', q') is conformal to (V, q), and $\mathrm{CO}(q)(k)/k^{\times} \simeq \mathrm{Aut}_k(\mathrm{SO}(q))$.*

(ii) *The map θ_q in (7.2.1.2) is an isomorphism, so $\mathrm{CO}(q)$ is connected. In particular, $\mathrm{SO}(q)(k) \simeq \mathrm{Aut}_k(\mathrm{SO}(q))$ if and only if $F = k$, in which case the k_s/k-forms of $\mathrm{SO}(q)$ are the k-groups $\mathrm{SO}(q')$ for (V', q') that becomes isometric to (V, q) over k_s. Moreover, in general $\mathrm{SO}(q')$ is a k_s/k-form of $\mathrm{SO}(q)$ if and only if $\dim V'/V'^{\perp} = 2n$ and $q'|_{V'^{\perp}}$ becomes conformal to $q|_{V^{\perp}}$ over k_s.*

Proof. The first assertion in (i) clearly implies the rest of (i), and both sides of (7.2.1.4) are compatible with Galois descent. By Hilbert's Theorem 90, to prove (i) it suffices to consider the case $k = k_s$. Granting (ii) when $k = k_s$, let us deduce (ii) in general. The general k-isomorphism assertion in (ii) is immediate via Galois descent, so if $F = k$ then the central GL_1 in $\mathrm{CO}(q)$ is a complement to $\mathrm{SO}(q)$ (due to how this central GL_1 lies inside $\mathrm{R}_{F/k}(\mathrm{GL}_1) \times T$ via θ_q), so in such cases the k_s/k-forms of $\mathrm{SO}(q)$ are classified up to k-isomorphism by the cohomology set

$$\begin{aligned}\mathrm{H}^1(k_s/k, \mathrm{Aut}_{k_s}(\mathrm{SO}(q_{k_s}))) &= \mathrm{H}^1(k_s/k, \mathrm{CO}(q)(k_s)/k_s^{\times})\\ &= \mathrm{H}^1(k_s/k, \mathrm{CO}(q)(k_s))\end{aligned}$$

since $\mathrm{H}^1(k_s/k, k_s^{\times}) = 1$. Hence, if $F = k$ then this cohomology set classifies conformal isometry classes (V', q') that become conformal to (V, q) over k_s (forcing $\dim V'/V'^{\perp} = \dim V/V^{\perp}$), and the form of $\mathrm{SO}(q)$ associated to such a (V', q') is $\mathrm{SO}(q')$. Likewise, if $F = k$ then $\mathrm{Aut}_{k_s}(\mathrm{SO}(q_{k_s})) = \mathrm{SO}(q)(k_s) = \mathrm{O}(q)(k_s)$, so in such cases the k_s/k-forms of $\mathrm{SO}(q)$ are classified by the set $\mathrm{H}^1(k_s/k, \mathrm{O}(q)(k_s))$ of isometry classes of (V', q') over k that are isometric to (V, q) over k_s. Explicitly, if $F = k$ then the k_s/k-form of $\mathrm{SO}(q)$ classified by (V', q') is the k-group $\mathrm{SO}(q')$.

Clearly if (V', q') and (V, q) become conformal over k_s then the same holds for $q'|_{V'^{\perp}}$ and $q|_{V^{\perp}}$. Conversely, suppose these defect spaces are conformal over k_s and $\dim V'/V'^{\perp} = 2n$. Each of (V, q) and (V', q') is isometric to the orthogonal sum of its defect space and a non-degenerate quadratic space of dimension $2n$, yet there is only one isometry class of $2n$-dimensional non-degenerate quadratic spaces over k_s, so (V, q) and (V', q') become conformal over k_s. This completes the reduction of the proof of (ii) to the case $k = k_s$.

For the remainder of this proof we may and do assume $k = k_s$. We shall first prove that if $\mathrm{SO}(q') \simeq G := \mathrm{SO}(q)$ then (V', q') is conformal to (V, q). The isomorphism of root systems and equality of dimensions of short root groups imply that $\dim V'/V'^{\perp} = \dim V/V^{\perp} = 2n$ and $\dim V^{\perp} = \dim V'^{\perp}$ (see Proposition 7.1.3). We have just seen that (V', q') is conformal to (V, q) if and only if $q'|_{V'^{\perp}}$ is conformal to $q|_{V^{\perp}}$. We will construct a conformal isometry of the latter type via root groups.

Choose a maximal k-torus $T \subset G$ and a short root $b \in \Phi(G, T)$, with $U_b \subset G$ the associated root group. Let K/k be the minimal field of definition for the geometric unipotent radical of G. By Proposition 7.1.2 we may choose a Levi k-subgroup $L \subset G$ containing T, so $L_K \to G' := G_K^{\mathrm{red}}$ is an isomorphism; explicitly, by Proposition 7.1.5, $L = \mathrm{SO}(q')$ with $q' := q|_{V' \oplus \ell}$ for the unique T-stable linear complement V' to $V^{\perp}$ in V and a unique line $\ell \subset V^{\perp}$. The choice of L provides a k-structure on G' compatible with the natural one on its maximal K-torus $T' := T_K$. For $\overline{q}_K$ as in (7.1.4), the canonical K-isogeny $G' = \mathrm{SO}(\overline{q}_K) \to \mathrm{Sp}(B_{\overline{q}_K}) =: \overline{G}'$ has unipotent infinitesimal kernel (it is the classical isogeny $\mathrm{SO}_{2n+1} \to \mathrm{Sp}_{2n}$ in characteristic 2). This identifies T' with a split maximal K-torus $\overline{T}' \subset \overline{G}'$, and the quotient map $(G', T') \to (\overline{G}', \overline{T}')$ over

K descends to the classical unipotent k-isogeny $L = \mathrm{SO}(q') \to \mathrm{Sp}(\overline{B}_{q'}) = \overline{L}$ carrying T isomorphically onto its image $\overline{T}$. Under this isogeny, the long roots in $\Phi(\overline{G}', \overline{T}') = \Phi(\overline{L}, \overline{T})$ are twice the short roots in $\Phi(G,T) = \Phi(G',T') = \Phi(L,T)$ via the identification $\mathrm{X}(T) = \mathrm{X}(T')$.

Let $\overline{b} = 2b \in \Phi(\overline{L}, \overline{T})$, so the associated root group $U_{\overline{b}}$ of $(\overline{L}, \overline{T})$ is k-isomorphic to $\mathbf{G}_{\mathrm{a}}$ (isomorphism unique up to $k^{\times}$-scaling) and provides a k-structure on the $\overline{b}$-root group of $(\overline{G}', \overline{T}')$. The quotient map $G_K \twoheadrightarrow \overline{G}'$ carries $U_b(K)$ into $U_{\overline{b}}(K)$ and carries $U_b(k)$ into $U_{\overline{b}}(k) \simeq \mathbf{G}_{\mathrm{a}}(k) = k$. The description of the orthogonal automorphism $[v]$ in the proof of Proposition 7.1.3 provides a T-equivariant isomorphism between U_b and the vector group associated to $V^{\perp}$ (equipped with T-action through b). Combining this with Remark 7.1.4, we see that the composite map $U_b(k) \to U_{\overline{b}}(k) = k$ is a $k^{\times}$-multiple of $q|_{V^{\perp}}$ because on $V_K^{\perp} := K \otimes_k V^{\perp}$ we have

$$1 \otimes v \equiv \sqrt{q(v)/q(v_0)} \otimes v_0 \bmod \mathscr{V}$$

for any $v \in V^{\perp}$ and a fixed nonzero $v_0 \in V^{\perp}$, where $\mathscr{V}$ is the K-hyperplane $\ker(q_K|_{V_K^{\perp}}) \subset V_K^{\perp}$. Therefore, $(V'^{\perp}, q'|_{V'^{\perp}})$ is conformal to $(V^{\perp}, q|_{V^{\perp}})$, and hence (V', q') is conformal to (V, q).

The preceding argument proves that the target of (7.2.1.4) is non-empty if and only if the source of (7.2.1.4) is non-empty, so (i) is reduced to showing that the natural map

$$h_q : \mathrm{CO}(q)(k)/k^{\times} \to \mathrm{Aut}_k(G)$$

is bijective (with $k = k_s$). For non-degenerate quadratic spaces with odd dimension $\geqslant 3$ this is a special case of the Isomorphism Theorem for connected semisimple groups (and is due to Dieudonné away from characteristic 2). In the non-degenerate odd-dimensional case, a scheme-theoretic proof that h_q is bijective in all characteristics is given in [C2, Lemma C.3.12], and a characteristic-free proof via classical algebraic geometry rather than via schemes is given in [KMRT, VI, 26.12, 26.15, 26.17]. This special case will be applied over K in our general proof below that h_q is bijective.

As a preliminary step in the general case (with $k = k_s$), we check that $\ker h_q = 1$; i.e., any conformal isometry f of (V, q) centralizing G inside $\mathrm{GL}(V)$ is a scalar. For a maximal torus $T \subset G$, let V_0 be the unique T-equivariant linear complement to $V^{\perp}$ in V, so we may identify $q_0 := q|_{V_0}$ with

$$x_1 x_2 + \cdots + x_{2n-1} x_{2n}$$

making $T = \mathrm{GL}_1^n$ act on V_0 with weight space ke_{2i-1} for $t \mapsto t_i$ and weight space ke_{2i} for $t \mapsto t_i^{-1}$. Let $\mu \in k^\times$ be the scalar for which $q \circ f = \mu q$ (so $B_q \circ (f \times f) = \mu B_q$). Since f centralizes T, each e_j is an eigenvector for f with eigenvalue λ_j satisfying $\lambda_{2i-1}\lambda_{2i} = \mu$ for all i. Likewise, f carries $V^\perp = V^T$ into itself preserving $q|_{V^\perp}$ up to multiplication by μ.

For every line $\ell \subset V^\perp$, the Levi k-subgroup $L_\ell := \mathrm{SO}(q|_{V_0 \oplus \ell}) \subset G$ containing T is centralized by f, so f preserves $L_\ell.V_0 = V_0 \oplus \ell$. Hence, f preserves $V^\perp \cap (V_0 \oplus \ell) = \ell$, so the linear automorphism $f|_{V^\perp}$ preserves every line and thus is scaling by some $c \in k^\times$. But $q|_{V^\perp} \circ f = \mu q|_{V^\perp}$, so $\mu = c^2$. Replacing f with $(1/c)f$ thereby reduces us to the case $\mu = 1$, so f arises from $T(k) \subset G(k)$. But T is arbitrary in G, so the triviality of Z_G (or of the center of any Levi k-subgroup of G containing T) then forces $f = 1$; i.e., $\ker h_q = 1$.

Now we address the surjectivity of h_q. Since $Z_G = 1$, $G(k)$ is a normal subgroup of $\mathrm{Aut}_k(G)$ via conjugation; this subgroup is obviously contained inside the image of $\mathrm{CO}(q)(k)/k^\times$ under h_q. For a given $f \in \mathrm{Aut}_k(G)$, to determine if f lies in the image of h_q it is harmless to replace f with a $G(k)$-translate. Thus, we may assume that f preserves a chosen maximal k-torus T of G and a minimal pseudo-parabolic k-subgroup B of G containing T, so f centralizes T (as the diagram of $\Phi(G,T)$ has no nontrivial automorphism).

Consider the induced automorphism $\overline{f}_K$ of $\mathrm{SO}(\overline{q}_K) = G_K^{\mathrm{ss}}$. By the settled case of non-degenerate quadratic spaces of odd dimension $2n+1$, $\overline{f}_K$ arises from some $t \in T(K)$. Let Δ be the basis of $\Phi(G,T) = \Phi(G_K^{\mathrm{ss}}, T_K)$ associated to B, so the natural map $T \to \mathrm{GL}_1^\Delta$ is an isomorphism (as we may check over K, or over k using that $Z_G = 1$). This identifies t with an ordered n-tuple $(t_1, \dots, t_n) \in (K^\times)^n$ where $t_i = a_i(t)$ for the vertices $a_i \in \Delta$ denoted according to the Dynkin diagram of (G,B,T):

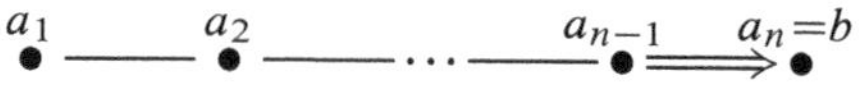

Here, b denotes the unique short root in Δ (where we use the convention for $n = 1$ that the unique root in Δ is short).

Writing U_a and U'_a to denote the respective a-root groups of (G,T) and $(G' := G_K^{\mathrm{ss}}, T_K)$ for any $a \in \Delta$, we have $U_a(k) \subset U'_a(K)$ since G is pseudo-reductive. For long a, the quotient map $G_K \to G' = \mathrm{SO}(\overline{q}_K)$ carries the 1-dimensional $(U_a)_K$ isomorphically onto U'_a, so $U_a(k)$ is a k-line in the K-line $U'_a(K)$. Hence, if a is long then the K-linear t-action on $U'_a(K)$ preserves this k-line, so the scalar $a(t) \in K^\times$ lies in $k^\times$ for such a.

For the unique short root $b \in \Delta$, multiplication on the K-line $U'_b(K)$ by $b(t) \in K^\times$ must preserve the image of the additive injection $j : V^\perp = U_b(k) \to U'_b(K)$. If we use a nonzero $v_0 \in V^\perp$ as a K-basis of $U'_b(K)$ then the map j is identified with $v \mapsto \sqrt{q(v)/q(v_0)}$ (due to the definition of the isomorphism $V^\perp \simeq U_b(k)$ and the congruence $v \equiv \sqrt{q(v)/q(v_0)} \cdot v_0 \bmod \mathscr{V}$ with $\mathscr{V}$ as in (7.1.4)). Hence, $b(t)$ lies in the short root field F, so we can translate f against a k-point of the k-subgroup $\mathrm{R}_{F/k}(\mathrm{GL}_1) \subset \mathrm{CO}(q)$ defined as in (7.2.1.1) to reduce to the case $b(t) = 1$ without affecting the property that f centralizes T. But then $\overline{f}_K \in \prod_{a \in \Delta_>} k^\times \subset T(k)$, so f arises from $T(k)$ by [CGP, Prop. 1.2.2]. This proves (i), and via (7.2.1.3) the same reasoning shows that $\mathrm{CO}(q)$ is generated by G and the k-subgroup $\mathrm{R}_{F/k}(\mathrm{GL}_1)$ provided by (7.2.1.1). In other words, θ_q is an isomorphism. □

Example 7.2.3. Suppose $F = k$, so the isomorphism property of θ_q implies that the inclusion $\mathrm{GL}_1 \times \mathrm{SO}(q) \subset \mathrm{CO}(q)$ is an equality. We claim that if q is degenerate then the closed subgroup scheme $\mathrm{O}(q) \subset \mathrm{GL}(V)$ is *not* contained in the smooth k-subgroup $\mathrm{CO}(q) \subset \mathrm{GL}(V)$.

Suppose such a containment holds. Since $\mathrm{SO}(q) \subset \mathrm{CO}(q)$ and $\mathrm{O}(q)$ meets the central GL_1 in μ_2, we obtain that $\mathrm{O}(q) = \mu_2 \times \mathrm{SO}(q)$. Hence, the identification of $V^\perp_{k_s}$ with each of the $2n$ short root spaces in $\mathrm{Lie}(\mathrm{SO}(q))_{k_s}$ for the action of a maximal k_s-torus in $\mathrm{SO}(q)_{k_s}$ implies that

$$\begin{aligned} \dim \mathrm{Lie}(\mathrm{O}(q)) = 1 + \dim \mathrm{SO}(q) &= 1 + \dim \mathrm{SO}_{2n+1} + 2n(\dim V^\perp - 1) \\ &= 1 + n(2n+1) + 2n(\delta - 1) \\ &= n(2n-1) + (2n\delta + 1) \end{aligned}$$

where $\delta := \dim V^\perp$. However, by direct calculation we see that the Lie subalgebra $\mathrm{Lie}(\mathrm{O}(q)) \subset \mathfrak{gl}(V) = \mathrm{End}(V)$ consists of those endomorphisms ϕ that satisfy $B_q(v, \phi(v)) = 0$ for all $v \in V$, which is to say that the bilinear form $B_q(v, \phi(w))$ on V factors through an alternating form on $V/V^\perp$. In particular, there is a short exact sequence

$$0 \to \mathrm{Hom}(V, V^\perp) \to \mathrm{Lie}(\mathrm{O}(q)) \to \wedge^2(V/V^\perp)^* \to 0,$$

so $\dim \mathrm{Lie}(\mathrm{O}(q)) = n(2n-1) + (2n+\delta)\delta$.

Hence, if $F = k$ then $\mathrm{O}(q)$ is not contained in $\mathrm{CO}(q)$ whenever $\delta \neq 1$. The case $\delta = 1$ is the non-degenerate case, so if (V, q) is degenerate and $\mathrm{SO}(q)$ has short root field equal to k then $\mathrm{O}(q) \not\subset \mathrm{CO}(q)$. (Note that the short root field of an absolutely pseudo-simple basic exotic k-group is a *nontrivial* purely inseparable extension of k, unlike for $\mathrm{SO}(q)$ when $F = k$ and q is degenerate.)

In contrast with the adjoint *semisimple* case, the above $\mathrm{SO}(q)$-construction with $n > 2$ does not account for all centerless absolutely pseudo-simple k-groups of type B_n whose Cartan k-subgroups are tori. This is relevant even over non-archimedean local fields k of characteristic 2:

Example 7.2.4. Let k be a non-archimedean local field of characteristic 2. For $n \geqslant 2$, the proof of [CGP, Prop. 7.3.6] provides basic exotic absolutely pseudo-simple k-groups G of type B_n such that G has k-rank $\lfloor n/2 \rfloor$. Thus, the groups G/Z_G are absolutely pseudo-simple of type B_n with trivial center (Proposition 5.3.1(ii)) and their Cartan k-subgroups are tori (by inspection). But any $\mathrm{SO}(q)$ as above has k-rank at least $n-1$ due to the existence of a Levi k-subgroup and the isotropicity of non-degenerate quadratic spaces of dimension > 4 over k. (See [EKM, Lemma 36.8] for an elementary proof of such isotropicity in dimension $> 2^{e+1}$ over fields F of characteristic 2 such that $[F : F^2] = 2^e$.)

The following proposition provides sufficient conditions for an absolutely pseudo-simple k-group to be isomorphic to $\mathrm{SO}(q)$ for a regular quadratic form q; it is a partial converse to Proposition 7.1.2.

Proposition 7.2.5. *For $n \geqslant 1$, consider absolutely pseudo-simple k-groups G with root system* B_n *over k_s such that the Cartan k-subgroups are tori and $Z_G = 1$.*

If G is pseudo-split then it is isomorphic to $\mathrm{SO}(q)$ *for a regular quadratic space (V, q) of Witt index n with nonzero $V^{\perp}$ of codimension $2n$ in V. If instead G merely admits a pseudo-split k_s/k-form but its short root field is equal to k then it is isomorphic to* $\mathrm{SO}(q')$ *for a regular quadratic space (V', q') with nonzero $V'^{\perp}$ of codimension $2n$ in V' such that $(V'^{\perp}, q'|_{V'^{\perp}})$ has root field k in the sense of Example* 7.1.8.

Proof. As the root system over k_s is reduced and the center is trivial, it follows from the centrality of the kernel (of i_G) in Proposition 2.3.4 that all such groups G are of minimal type.

To prove the first assertion, assume G is pseudo-split and let K/k be the minimal field of definition for its geometric unipotent radical and let T be a fixed split maximal k-torus in G. Note that i_G maps G isomorphically onto its image in $\mathrm{R}_{K/k}(G_K^{\mathrm{ss}})$ since G is of minimal type and its root system is reduced. As G has trivial center and is generated by its root groups, we see that $\Phi := \Phi(G, T)\ (= \Phi(G_K^{\mathrm{ss}}, T_K)$ under the natural identification of the character groups

$X(T)$ and $X(T_K)$) spans $X(T)$. This implies that G_K^{ss} is of adjoint type. We fix a basis Δ of the root system Φ.

Let G_c be the absolutely pseudo-simple k-subgroup of G generated by the $\pm c$-root groups for any $c \in \Phi$. In view of the construction of these subgroups as derived groups of centralizers of codimension-1 tori, the hypothesis that the Cartan subgroups of G are tori is inherited by each G_c. Likewise, each G_c is of minimal type. Also, recall from 3.2.2 that the natural maps

$$h_c : (G_c)_K^{ss} \to (G_K^{ss})_c$$

between split connected semisimple K-groups of rank 1 are isomorphisms.

Since G_K^{ss} has root system B_n (understood to have only short roots when $n = 1$) and its center is trivial, for any long root $a \in \Delta$ (if $n > 1$) and the unique short root $b \in \Delta$ we conclude via the isomorphisms h_a and h_b that $(G_a)_K^{ss}$ is simply connected (so it is isomorphic to SL_2) and $(G_b)_K^{ss}$ is adjoint (so it is isomorphic to PGL_2). As the Cartan subgroups of the pseudo-split absolutely pseudo-simple k-group G_a of minimal type are tori, and the root system of G_a is A_1, Proposition 3.1.8 implies (using the description of Cartan subgroups provided by Proposition 3.1.4) that $G_a \simeq SL_2$ for any long $a \in \Delta$. Hence, by Proposition 3.2.5 (and the trivial equality $G_b = G$ when $n = 1$), the minimal field of definition for the geometric unipotent radical of G_b is K. Now using Proposition 3.1.8 again we see that $G_b \simeq PH_{V_0,K/k}$ for a nonzero kK^2-subspace V_0 of K that contains 1 and generates K as a k-algebra. In particular, $Z_{G_b} = 1$.

The Cartan subgroups of G_b are (1-dimensional) tori, so the description of Cartan subgroups in Proposition 3.1.4 implies that $K^2 \subset k$. Now we can define a k-valued anisotropic quadratic form on V_0 by $q_0 : v \mapsto v^2$, so Proposition 7.1.3 implies $G_b \simeq SO(H \perp q_0)$ for a hyperbolic plane H. This settles the case $n = 1$. In general, let $\mathfrak{q}$ be the orthogonal sum of n hyperbolic planes and the anisotropic (V_0, q_0), so if $n \geqslant 2$ then the Isomorphism Theorem (i.e., Theorem 6.1.1) and the description of G_b imply $G \simeq SO(\mathfrak{q})$. This proves the first assertion.

To prove the second assertion, suppose instead that G admits a pseudo-split k_s/k-form $\mathscr{G}$ and has short root field equal to k. Then $\mathscr{G}$ inherits the hypotheses from G, so $\mathscr{G} \simeq SO(q)$ for some q as above. Since the short root field of $\mathscr{G}$ is k, by Proposition 7.2.2 applied to $SO(q)$ we have $G \simeq SO(q')$ for a regular quadratic space (V', q') as desired. (The end of Example 7.1.8 relates the short root field of $SO(q')$ to that of $(V'^{\perp}, q'|_{V'^{\perp}})$.) □

Remark 7.2.6. In the second assertion of Proposition 7.2.5, the hypothesis that G admits a pseudo-split k_s/k-form cannot be dropped: this property is a consequence of the conclusion by Proposition 7.1.2, and examples satisfying all of the other hypotheses but lacking a pseudo-split k_s/k-form are given by the k-groups G/Z_G for G of type B_n in Example C.4.1 with many k admitting a quadratic Galois extension and satisfying $[k:k^2] \geqslant 8$.

7.3 Severi–Brauer varieties

For imperfect fields k of characteristic 2, an exhaustive description of all *pseudo-split* absolutely pseudo-simple k-groups of type-B with trivial center and tori as Cartan k-subgroups is provided by the first assertion in Proposition 7.2.5; this applies in particular over k_s. To give an exhaustive description without a pseudo-split hypothesis (and without assuming that there exists a pseudo-split k_s/k-form) requires a description of k-descents of $\mathrm{SO}(q')$ for degenerate regular (V', q') with $V'^{\perp} \neq V'$ over finite Galois extensions k'/k.

Since isomorphisms among such groups $\mathrm{SO}(q')$ generally arise from *conformal* isometries of quadratic spaces rather than from isometries of quadratic spaces (Proposition 7.2.2), such k'/k-descents generally do not have the form $\mathrm{SO}(q)$ for regular (V, q) over k. (The same phenomenon is seen in the classical semisimple case for non-degenerate quadratic spaces with even dimension $\geqslant 4$.) Remark 7.2.6 provides such descents with no pseudo-split k_s/k-form (and with short root field equal to k) for many k admitting a quadratic Galois extension and satisfying $[k:k^2] \geqslant 8$. We now prepare to describe all descents in terms of automorphisms of certain quadrics in Severi–Brauer varieties over k.

7.3.1. Recall that a *Severi–Brauer variety* of dimension $N > 0$ over a field F is a smooth proper F-scheme X such that $X_{\overline{F}} \simeq \mathbf{P}^N_{\overline{F}}$, or equivalently $X_{F_s} \simeq \mathbf{P}^N_{F_s}$. Since the latter isomorphism is well-defined up to the action of $\mathrm{PGL}_{N+1}(F_s)$, for a geometrically reduced hypersurface $D \subset X$ we can speak of its *degree* $\delta \geqslant 1$: this means the degree of D_{F_s} as a hypersurface in $X_{F_s} \simeq \mathbf{P}^N_{F_s}$ (or equivalently, the index of the subgroup of $\mathrm{Pic}(X_{F_s})$ generated by the invertible ideal sheaf of D in $\mathscr{O}_X$). For instance, when $\delta = 2$ we call D a *quadric* in X. Every Severi–Brauer variety X over F is projective, and by descent theory the automorphism functor $\underline{\mathrm{Aut}}_{X/F}$ is represented by an F-form $\mathrm{Aut}_{X/F}$ of PGL_{N+1}. For any F-algebra A, an A-automorphism f of X_A that carries D_A into itself restricts to an automorphism of D_A [EGA, IV_4, 17.9.6], so the subfunctor

$$\underline{\mathrm{Aut}}_{(X,D)/F} : A \rightsquigarrow \{f \in \mathrm{Aut}_A(X_A) \mid f_A(D_A) \subset D_A\}$$

of $\underline{\mathrm{Aut}}_{X/F}$ is a subgroup functor.

Since D is geometrically reduced, or equivalently reduced and generically smooth, the subset $D(F_s) \subset D_{F_s}$ is schematically dense (in the sense of [EGA, IV$_3$, 11.10.2]). Thus, $\underline{\mathrm{Aut}}_{(X,D)/F}$ is represented by a closed subgroup scheme

$$\mathrm{Aut}_{(X,D)/F} \subset \mathrm{Aut}_{X/F},$$

namely by the Galois descent of the closed subscheme of $\mathrm{Aut}_{X_{F_s}/F_s}$ defined by the intersection of the closed conditions "$f(d) \in D_{F_s}$" for all $d \in D(F_s)$.

Remark 7.3.2. The case of most interest to us is imperfect F of characteristic 2 with $X_{F_s} = \mathbf{P}(V^*) := \mathrm{Proj}(\mathrm{Sym}(V^*))$ and $D_{F_s} = (q = 0)$ for a regular quadratic form $q : V \to F_s$ such that $V^\perp$ is a nonzero proper subspace of V. Since $V^\perp$ has even codimension in V, necessarily $\dim V \geqslant 3$ and when $\dim V \leqslant 4$ we can identify $D_{\overline{F}} \hookrightarrow X_{\overline{F}}$ with $(x^2 + yz = 0) \subset \mathbf{P}^{\dim V - 1}$.

These are instances of the general context above with D a geometrically integral quadric (so $N \geqslant 2$), subject to the further condition that when $N = 3$ it is non-smooth at a unique geometric point. In such cases, we claim that the restriction map $\mathrm{Aut}_{(X,D)/F} \to \mathrm{Aut}_{D/F}$ to the automorphism scheme of D is an isomorphism, so $\mathrm{Aut}_{(X,D)/F}$ is intrinsic to D (and $\mathrm{Aut}_{D/F}$ is affine). This is not needed below, but for the interested reader we now sketch the proof.

We may assume F is algebraically closed, and we claim that $\mathrm{Pic}(D) \simeq \mathbf{Z}$. The case $\dim D = 1$ is obvious (as D is an integral plane conic, hence smooth), the case $\dim D = 2$ is a classical fact about the quadric cone in $\mathbf{P}^3$, and the case $\dim D \geqslant 3$ is a consequence of the global Lefschetz theorem [SGA2, XII, Cor. 3.6] (which gives that $\mathbf{Z} = \mathrm{Pic}(X) \simeq \mathrm{Pic}(D)$ in such cases). The Picard scheme $\mathrm{Pic}_{D/F}$ is constant since $\mathrm{H}^1(D, \mathscr{O}_D) = 0$, so the action of $\mathrm{Aut}_{D/F}$ on $\mathrm{Pic}_{D/F} = \mathbf{Z}$ is trivial (as it must preserve the unique ample generator). Thus, this action extends to a projective representation of $\mathrm{Aut}_{D/F}$ on $\Gamma(D, \mathscr{L})$ for any line bundle $\mathscr{L}$ on the integral hypersurface D. The projective embedding $j : D \hookrightarrow X = \mathbf{P}^N$ is given by the space of global sections of $\mathscr{L} = j^*(\mathscr{O}_X(1))$ since D has degree > 1, so the desired isomorphism property follows.

Proposition 7.3.3. *Let X be a Severi–Brauer variety over k with* $\dim X > 0$, *and let $D \subset X$ be a geometrically reduced quadric that is regular (equivalently, smooth) at all k_s-points. If D is smooth then assume* $\dim X$ *is even. Let $G_{X,D}$ be the maximal smooth k-subgroup scheme $\mathrm{Aut}^{\mathrm{sm}}_{(X,D)/k}$ of $\mathrm{Aut}_{(X,D)/k}$.*

(i) *The k-group $G_{X,D}$ is connected. Its derived group $\mathscr{D}(G_{X,D})$ is absolutely pseudo-simple of type* B *with trivial center, and its Cartan k-subgroups are tori.*

(ii) *If (X', D') is a second such pair then*

$$\mathrm{Isom}_k((X', D'), (X, D)) \to \mathrm{Isom}_k(\mathscr{D}(G_{X',D'}), \mathscr{D}(G_{X,D}))$$

is bijective. In particular, the k-isomorphism class of any of (X, D), $G_{X,D}$, or $\mathscr{D}(G_{X,D})$ determines the others.

(iii) *The k-groups $\mathscr{D}(G_{X,D})$ for (X, D) with $X(k) \neq \emptyset$ are precisely $\mathrm{SO}(q)$ for regular (V, q) with $0 < \dim V^{\perp} < \dim V$, and in such cases $G_{X,D} = \mathscr{D}(G_{X,D})$ if and only if $(V^{\perp}, q|_{V^{\perp}})$ has root field k in the sense of Example 7.1.8.*

By Remark 7.3.2, we have $G_{X,D} = \mathrm{Aut}^{\mathrm{sm}}_{D/k}$; this will not be used.

Proof. First we consider (i) and (ii), so we may and do assume $k = k_s$. Thus, we identify X with the projective space $\mathbf{P}(V^*) = \mathrm{Proj}(\mathrm{Sym}(V^*))$ covariantly attached to a k-vector space V of finite dimension at least 2, and D is thereby the zero scheme of a nonzero quadratic form $q \in \mathrm{Sym}^2(V^*)$. The regularity of D at its k-points is exactly the regularity of q (i.e., the injectivity of the 2-linear $q|_{V^{\perp}}$). The generic smoothness expresses that $q_{\overline{k}}$ is reduced, which is to say that q is not a sum of squares; i.e., $V^{\perp} \neq V$. Since smoothness of the quadric corresponds to non-degeneracy of the quadratic form, and $\dim(V/V^{\perp})$ is even, non-smoothness of the quadric expresses exactly the condition that $\dim V^{\perp} > 1$. If instead the quadric is smooth then (V, q) is non-degenerate and by hypothesis $\dim X$ is even and positive in such cases, so $\dim V$ is odd and $\geqslant 3$ (forcing $V^{\perp}$ to be a line in V).

The smooth part of $\mathrm{Aut}_{(X,D)/k}$ is the Zariski closure in $\mathrm{PGL}(V)$ of the group of projective linear automorphisms preserving $(q = 0)$. These arise from exactly the linear automorphisms f of V that preserve q up to a $k^{\times}$-scaling factor, which is to say that f is a conformal automorphism of (V, q). Hence, $G_{X,D}$ is the k-subgroup $\mathrm{CO}(q)/\mathrm{GL}_1 \hookrightarrow \mathrm{PGL}(V)$ that is connected (by Proposition 7.2.2(ii)). Since $\mathrm{SO}(q)$ has trivial center, the natural maps

$$\mathrm{SO}(q) \to \mathrm{PGL}(V), \;\; \mathrm{SO}(q) \to \mathrm{CO}(q)/\mathrm{GL}_1$$

have trivial kernel. But $\mathrm{CO}(q)/\mathrm{SO}(q)$ is commutative since (7.2.1.2) is an isomorphism (Proposition 7.2.2(ii)), so the perfect $\mathrm{SO}(q)$ coincides with $\mathscr{D}(G_{X,D})$. This proves (i), and (ii) is exactly Proposition 7.2.2(ii).

Consider (iii). If $X(k) \neq \emptyset$ then we may identify X with a projective space over k and hence D with a quadric. Thus, the preceding considerations work over k to show that $\mathscr{D}(G_{X,D}) = \mathrm{SO}(q)$ for (V, q) as in (iii). Conversely, every such $\mathrm{SO}(q)$ has the desired form by defining $X = \mathbf{P}(V^*)$ and $D = (q = 0)$ since $\dim X = \dim V - 1$ is even and positive when D is smooth (because $V^\perp \neq 0$ by hypothesis and smoothness of D expresses non-degeneracy for q). This proves the first part of (iii), and for the equivalence at the end we may assume $k = k_s$. Thus, $G_{X,D} = \mathrm{CO}(q)/\mathrm{GL}_1$. This has derived group $\mathrm{SO}(q)$ and maximal commutative affine quotient $\mathrm{R}_{F/k}(\mathrm{GL}_1)/\mathrm{GL}_1$, where F is the root field of $(V^\perp, q|_{V^\perp})$, so perfectness of $G_{X,D}$ is indeed the condition $F = k$. □

Remark 7.3.4. For any (X, D) as above, X is classified up to k-isomorphism by an element $[X] \in \mathrm{Br}(k)$ called the *Brauer invariant* of $\mathscr{D}(G_{X,D})$. By Proposition 7.3.3(iii), $[X] = 1$ if and only if $\mathscr{D}(G_{X,D}) \simeq \mathrm{SO}(q)$ for a regular (V, q) with $0 < \dim V^\perp < \dim V$. We claim that $[X]$ is 2-torsion, and more generally that a Severi–Brauer variety X containing a geometrically reduced hypersurface D of degree d (over k_s) corresponds to a d-torsion class in $\mathrm{Br}(k)$. This is proved via an argument of Artin [Ar, 5.2] that we now review.

Let V be a k-vector space of dimension $N + 1 \geqslant 3$ such that X is a k-form of the projective space $\mathbf{P}(V^*) := \mathrm{Proj}(\mathrm{Sym}(V^*))$; this projective space classifies lines in V (equivalently, hyperplanes in V^*) and has a left action by $\mathrm{PGL}(V)$ under which the action of $L \in \mathrm{PGL}(V)$ carries the hypersurface $(h = 0)$ for nonzero $h \in \mathrm{Sym}^d(V^*)$ to the zero scheme of $(\mathrm{Sym}^d(L^*))^{-1}(h)$. For every $r \geqslant 1$, the natural representation

$$\mathrm{GL}(V) \to \mathrm{GL}((\mathrm{Sym}^r(V^*))^*)$$

induces a corresponding k-homomorphism

$$f_r : \mathrm{PGL}(V) \to \mathrm{PGL}((\mathrm{Sym}^r(V^*))^*).$$

The f_r-pushforward of the 1-cocycle descent datum for X valued in $\mathrm{PGL}(V)$ is the 1-cocycle descent datum valued in $\mathrm{PGL}((\mathrm{Sym}^r(V^*))^*)$ for a Severi–Brauer variety $X(r)$ equipped with an inclusion $X \hookrightarrow X(r)$ encoding a descent of the r-fold Segre map. But $X(d)$ contains a *degree*-1 hypersurface H corresponding to a descent of $D_{k_s} \subset X_{k_s}$, so the ideal sheaf $\mathscr{I}_H$ of H in $\mathscr{O}_{X(d)}$ is a k-descent of $\mathscr{O}(-1)$ on P'_{k_s}, where $P' = \mathbf{P}((\mathrm{Sym}^d(V^*)))$ (so $\mathscr{I}_H$ is invertible).

By computing over k_s, we see that $\mathscr{L} := \mathscr{I}_H^{-1}$ is generated by its global sections and that the natural map

$$X(d) \to \mathrm{Proj}(\mathrm{Sym}(\Gamma(X(d), \mathscr{L}))) = \mathbf{P}(\Gamma(X(d), \mathscr{L}))$$

(assigning to each $x \in X(d)$ the hyperplane of global sections of $\mathscr{L}$ that vanish at x) is an isomorphism; i.e., $X(d)$ is a trivial Severi–Brauer variety. Hence, the map of pointed sets

$$\mathrm{H}^1(f_d):\mathrm{H}^1(k,\mathrm{PGL}(V))\to\mathrm{H}^1(k,\mathrm{PGL}((\mathrm{Sym}^d(V^*))^*))$$

kills $[X]$. For any $r\geqslant 1$ we have a commutative diagram of central extensions of affine algebraic k-groups

$$\begin{array}{ccccccccc} 1 & \longrightarrow & \mathrm{GL}_1 & \longrightarrow & \mathrm{GL}(V) & \longrightarrow & \mathrm{PGL}(V) & \longrightarrow & 1 \\ & & \downarrow{\scriptstyle t^r} & & \downarrow & & \downarrow{\scriptstyle f_r} & & \\ 1 & \longrightarrow & \mathrm{GL}_1 & \longrightarrow & \mathrm{GL}((\mathrm{Sym}^r(V^*))^*) & \longrightarrow & \mathrm{PGL}((\mathrm{Sym}^r(V^*))^*) & \longrightarrow & 1 \end{array}$$

from which it follows that $\mathrm{H}^1(f_r)$ is compatible with multiplication by r on $\mathrm{H}^2(k,\mathrm{GL}_1)=\mathrm{Br}(k)$ via the connecting map

$$\delta:\mathrm{H}^1(k,\mathrm{PGL}(V))\to\mathrm{H}^2(k,\mathrm{GL}_1).$$

Thus, $[X]\in\mathrm{Br}(k)[d]$.

Proposition 7.3.5. *Let $G=\mathscr{D}(G_{(X,D)})$ for (X,D) as in Proposition 7.3.3. Then G contains a Levi k-subgroup if and only if $G\simeq\mathrm{SO}(q)$ for a regular (V,q) over k with $0<\dim V^{\perp}<\dim V$.*

Proof. The implication "$\Leftarrow$" is immediate from Proposition 7.1.5. For the converse, suppose G has a Levi k-subgroup L. Let $T\subset L$ be a maximal k-torus. By Proposition 7.3.3(iii) over k_s, we have $G_{k_s}\simeq\mathrm{SO}(\mathfrak{q})$ for a regular quadratic space $(\mathscr{V},\mathfrak{q})$ over k_s with $0<\dim\mathscr{V}^{\perp}<\dim\mathscr{V}$. Concretely, viewing the quadratic form $\mathfrak{q}:\mathscr{V}\to k_s$ as a nonzero element of $\mathrm{Sym}^2(\mathscr{V}^*)$, the pair (X,D) is a k_s/k-descent of $(\mathbf{P}(\mathscr{V}^*),(\mathfrak{q}=0))$ via a descent datum valued in conformal isometries between $(\mathscr{V},\mathfrak{q})$ and its k_s/k-twists.

Letting $\mathscr{V}_0$ be the unique T_{k_s}-equivariant complement to $\mathscr{V}^{\perp}$, Proposition 7.1.5 gives that $L_{k_s}=\mathrm{SO}(\mathfrak{q}|_{\mathscr{V}_0\oplus\ell})$ for a unique k_s-line ℓ in $\mathscr{V}^{\perp}$. In view of the uniqueness of ℓ, the conformal isometries of $(\mathscr{V},\mathfrak{q})$ encoding the k_s/k-descent datum on $(\mathbf{P}(\mathscr{V}^*),(\mathfrak{q}=0))$ must preserve the k_s-line ℓ in $\mathscr{V}$ because L and T are k-subgroups of the k_s/k-descent G of $\mathrm{SO}(\mathfrak{q})$. Hence, the closed subscheme $\mathbf{P}(\ell^*)\subset\mathbf{P}(\mathscr{V}^*)$ is preserved by the k_s/k-descent datum that defines X. But $\mathbf{P}(\ell^*)=\mathrm{Spec}(k_s)$ is a k_s-point of $\mathbf{P}(\mathscr{V}^*)=X_{k_s}$, so its k_s/k-descent to a closed subscheme of X must be a k-point of X. Thus, by Proposition 7.3.3(iii) it follows that $G\simeq\mathrm{SO}(q)$ for some regular (V,q) with $0<\dim V^{\perp}<\dim V$. $\square$

Now we finally obtain the result we have been after in this section:

Theorem 7.3.6. *The absolutely pseudo-simple k-groups G such that the Cartan k-subgroups are tori, $Z_G = 1$, and the root system over k_s is B_n for some $n \geqslant 1$ are precisely the k-groups $\mathscr{D}(G_{X,D})$ for (X, D) as in Proposition 7.3.3.*

Moreover, the pair (X, D) is uniquely functorial with respect to isomorphisms in the associated k-group, and for $G := \mathscr{D}(G_{X,D})$ the following three conditions are equivalent: $X(k) \neq \emptyset$, $G \simeq \mathrm{SO}(q)$ for a regular (V, q) over k such that $0 < \dim V^{\perp} < \dim V$, and G admits a Levi k-subgroup.

Proof. Proposition 7.3.3(i) gives that the k-groups $\mathscr{D}(G_{X,D})$ satisfy the initial list of properties. For a general k-group G satisfying those properties, we shall prove that it has the form $\mathscr{D}(G_{X,D})$ for some such pair (X, D); once that is shown, Propositions 7.3.3 and 7.3.5 give everything else.

We may choose a finite Galois extension k'/k such that the k'-group $G_{k'}$ is pseudo-split, so by the first assertion in Proposition 7.2.5 we have $G_{k'} \simeq \mathrm{SO}(q')$ for a regular quadratic space (V', q') over k' such that $0 < \dim V'^{\perp} < \dim V'$ and $\dim(V'/V'^{\perp}) = 2n$. By Proposition 7.3.3(iii), $G_{k'} \simeq \mathscr{D}(G_{X',D'})$ for some pair (X', D') over k' (with $X'(k') \neq \emptyset$). This k'-isomorphism transfers the k'/k-descent datum on $G_{k'}$ over to such descent datum on the k'-group $\mathscr{D}(G_{X',D'})$, and by Proposition 7.3.3(ii) this arises from a uniquely determined k'/k-descent datum on (X', D'). By effectivity of Galois descent for (quasi-)projective schemes over fields, we obtain (X, D) over k equipped with a k'-isomorphism $(X_{k'}, D_{k'}) \simeq (X', D')$ such that the resulting composite k'-isomorphism

$$f' : \mathscr{D}(G_{X,D})_{k'} = \mathscr{D}(G_{X_{k'},D_{k'}}) \simeq \mathscr{D}(G_{X',D'}) \simeq G_{k'}$$

is compatible with the natural k'/k-descent datum on each side. By Galois descent, f' descends to a k-isomorphism $\mathscr{D}(G_{X,D}) \simeq G$. □

Proposition 7.3.7. *Let k be a field. Non-reductive pseudo-reductive k-groups G whose Cartan k-subgroups are tori exist if and only if k is imperfect of characteristic 2, in which case such groups are precisely $H \times \mathrm{R}_{k'/k}(G')$ for a connected reductive k-group H, nonzero finite étale k-algebra k', and smooth affine k'-group G' whose fiber G'_i over each factor field k'_i of k' is a group $\mathscr{D}(G_{X'_i,D'_i})$ as in Theorem 7.3.6 for which the quadric D'_i is not k'_i-smooth.*

That such G only exist over imperfect fields of characteristic 2 is [CGP, Thm. 11.1.1], but our proof of the above more precise result is simpler.

Proof. We first observe that when k is imperfect of characteristic 2, such a triple $(H, k'/k, G')$ describing G is unique up to unique isomorphism if it exists. Indeed, to prove this we may (by Galois descent) assume $k = k_s$, and then it suffices to show that if $H, \mathcal{H}$ are connected reductive k-groups and $\{H_i\}_{i \in I}, \{\mathcal{H}_j\}_{j \in J}$ are non-empty finite collections of non-reductive pseudo-simple k-groups with trivial center then any k-isomorphism

$$f : H \times \prod_i H_i \simeq \mathcal{H} \times \prod_j \mathcal{H}_j$$

arises uniquely from a k-isomorphism $\varphi : H \simeq \mathcal{H}$, a bijection $\tau : I \simeq J$, and k-isomorphisms $\varphi_i : H_i \simeq \mathcal{H}_{\tau(i)}$ for $i \in I$. The groups H_i are precisely the non-reductive pseudo-simple normal k-subgroups of the left side, and H is the centralizer of $\prod_i H_i$ in the left side (since Z_{H_i} is trivial for all i). This intrinsic characterization of $(H, \{H_i\})$ in terms of the k-group $H \times \prod H_i$ has an evident analogue on the right side, so the (unique) description of f is immediate.

By Galois descent, now we may assume $k = k_s$ and that k is imperfect with $p = \mathrm{char}(k)$. Let $T \subset G$ be a maximal k-torus, so $T = Z_G(T)$ and hence $G = T \cdot \mathscr{D}(G)$. By [CGP, Lemma 1.2.5(ii)], T is an almost direct product of the maximal central k-torus Z and the maximal k-torus $T \cap \mathscr{D}(G)$ of $\mathscr{D}(G)$, so $G = Z \cdot \mathscr{D}(G)$. The collection $\{G_i\}_{i \in I}$ of pseudo-simple normal k-subgroups of G is finite and non-empty, the G_i's pairwise commute, and $\prod G_i \to \mathscr{D}(G)$ is a surjection with central kernel (see [CGP, Prop. 3.1.8]). The Cartan k-subgroups of any smooth connected normal k-subgroup of G are tori, and G is non-reductive if and only if some G_i is non-reductive, so we may (and will) assume that G is pseudo-simple (and non-reductive).

By Theorem 7.3.6, it suffices to show that $\mathrm{char}(k) = 2$, Z_G is trivial, and the root system is of type B. Consider the k-subgroup $\mathscr{C}_G = (\ker i_G)^T$ as in §2.3. By centrality, $\mathscr{C}_G$ is contained in $Z_G(T) = T$, yet it is unipotent since $\ker i_G$ is unipotent, so $\mathscr{C}_G = 1$. Thus, G is of minimal type. Hence, its universal smooth k-tame central extension $\widetilde{G}$ is also of minimal type, by Proposition 5.3.3.

Assume G has a non-reduced root system, so $\widetilde{G}$ has trivial center by [CGP, Prop. 9.4.9] and hence $G = \widetilde{G}$. The possibilities for such G are given in [CGP, Thm. 9.8.6], and an explicit description of the Cartan k-subgroups of G is given in [CGP, (9.7.6)] when $n \neq 2$ and in [CGP, (9.8.2)] when $n = 2$ (see [CGP, Thm. 9.8.1(2), Prop. 9.8.4(1)] for this determination of Cartan k-subgroups). By inspection, these k-subgroups are not tori (because $\dim V_0 \geqslant 2$ for V_0 as defined there). Hence, the root system $\Phi := \Phi(G, T)$ of G is reduced.

Let K/k be the minimal field of definition for $\mathscr{R}_u(G_{\overline{k}}) \subset G_{\overline{k}}$, so $K \neq k$

since G is not reductive. By Proposition 3.2.6, the extension K/k attached to G coincides with the analogue for $\widetilde{G}$. Since $\widetilde{G}$ is of minimal type with a reduced root system, $\ker i_{\widetilde{G}}$ is trivial by Proposition 2.3.4. Let $\widetilde{G}' = \widetilde{G}_K/\mathscr{R}_{u,K}(\widetilde{G}_K)$, so naturally $\widetilde{G} \subset \mathrm{R}_{K/k}(\widetilde{G}')$ and $Z_{\widetilde{G}} \subset \mathrm{R}_{K/k}(Z_{\widetilde{G}'})$. Assume that Φ is not of type B, so $c_K^\vee(\mathrm{GL}_1) \cap Z_{\widetilde{G}'} = 1$ for every $c \in \Phi$. Thus, $\widetilde{G}_c$ maps isomorphically onto its image G_c in the central quotient G of $\widetilde{G}$ for all $c \in \Phi$, so the Cartan k-subgroup of each $\widetilde{G}_c$ is GL_1. But each $\ker i_{\widetilde{G}_c}$ is trivial, so the possibilities for each $\widetilde{G}_c$ are given by Lemma 3.2.3 (in the "SL_2-case") in terms of K_c/k (and auxiliary data in characteristic 2). It follows that $K_c = k$ for all c. This contradicts Proposition 3.2.5 since $K \neq k$, so Φ is of type B.

Consider the rank-1 case; i.e., $\Phi = \mathrm{B}_1$. By Proposition 3.1.8, if $\mathrm{char}(k) \neq 2$ then $G = \mathrm{R}_{K/k}(G')$ with G' equal to SL_2 or PGL_2, but then its Cartan k-subgroups are not tori since $K \neq k$. Suppose $\mathrm{char}(k) = 2$, so we just need to show Z_G is trivial. If Z_G is nontrivial then Proposition 3.1.8 implies $G \simeq H_{V,K/k}$ for a nonzero kK^2-subspace $V \subset K$ satisfying $k\langle V\rangle = K$ (so $\dim_k V > 1$). The Cartan k-subgroup $V_{K/k}^*$ has dimension larger than 1, so it is not a k-torus. The B_1-case is settled.

Suppose $\Phi = \mathrm{B}_n$ with $n \geqslant 2$, and consider long $a \in \Phi$, so $a_K^\vee(\mathrm{GL}_1) \cap Z_{\widetilde{G}'}$ is trivial. The proof that Φ is of type B shows that $\widetilde{G}_a \simeq \mathrm{SL}_2$. If $p \neq 2$ then $\widetilde{G} \simeq \mathrm{R}_{K/k}(\widetilde{G}')$ by Theorem 3.4.1(i), so $\widetilde{G}_a \simeq \mathrm{R}_{K/k}(\widetilde{G}'_{a_K}) = \mathrm{R}_{K/k}(\mathrm{SL}_2)$, a contradiction. Thus, $p = 2$ (so we just need to show Z_G is trivial) and in the notation of Theorem 3.4.1(iii) we have shown $V_> = k$ (even if $n = 2$). Hence, $\widetilde{G}$ is classified by a nonzero k-subspace $V \subset K$ satisfying $k\langle V\rangle = K$.

By Proposition 4.3.3, Cartan k-subgroups of $\widetilde{G}$ are $\mathrm{GL}_1^{n-1} \times V_{K/k}^*$; the factor $V_{K/k}^*$ is associated to the unique short root b in a basis of Φ. Since $Z_{\widetilde{G}} \subset \mathrm{R}_{K/k}(Z_{\widetilde{G}'}) = \mathrm{R}_{K/k}(b_K^\vee(\mu_2))$, clearly $Z_{\widetilde{G}} = V_{K/k}^*[2]$. The central quotient G_b of $\widetilde{G}_b = H_{V,K/k}$ is of minimal type (as G is), so the map $\widetilde{G}_b \twoheadrightarrow G_b$ is either an isomorphism or projection onto $\mathrm{P}H_{V,K/k}$, and the latter case is the quotient by $Z_{\widetilde{G}_b} = Z_{\widetilde{G}}$. Thus, $G = \widetilde{G}$ or $G = \widetilde{G}/Z_{\widetilde{G}}$. Since $\dim_k V \geqslant 2$ (as $k\langle V\rangle = K \neq k$), so $V_{K/k}^*$ is not a torus, the case $G = \widetilde{G}$ cannot occur. Hence, $G = \widetilde{G}/Z_{\widetilde{G}}$, so Z_G is trivial by Proposition 5.3.1(ii). □

8

Constructions when Φ has a double bond

In §8.1–§8.3 *we assume* $\mathrm{char}(k) = 2$.

8.1 Additional constructions for type B

Inspired by Proposition 7.3.3, we are led to the following construction that goes beyond $\mathrm{SO}(q)$'s and will be seen to provide the right generalization of the basic exotic construction for type-B_n when $n \neq 2$. (The case $n = 2$ admits additional constructions; see §8.3.)

Definition 8.1.1. A *type-*B *adjoint generalized basic exotic k-group* is a k-group G of the form $\mathscr{D}(\mathrm{Aut}^{\mathrm{sm}}_{(X,D)/k})$ for (X, D) as in Proposition 7.3.3 with non-smooth D, subject to the addition requirement in the rank-1 case that G has root field equal to k.

A *type-*B *generalized basic exotic k-group* is the universal smooth k-tame central extension of a type-B adjoint generalized basic exotic k-group.

The root-field hypothesis in the rank-1 adjoint type-B case is automatic in the higher-rank case (as we may verify over k_s). Moreover, by Theorem 7.3.6, for any separable extension k'/k the k-group G is in one of the above two classes of k-groups if and only if $G_{k'}$ is in the analogous class of k'-groups.

The type-B adjoint generalized basic exotic k_s-groups are precisely $\mathrm{SO}(q)$ for regular (V, q) over k_s satisfying $1 < \dim V^{\perp} < \dim V$, subject to the requirement for rank 1 that $(V^{\perp}, q|_{V^{\perp}})$ has root field k_s in the sense of Example 7.1.8. This extra condition for rank 1 serves two purposes: it prevents the intervention of Weil restriction (since $\mathrm{P}H_{V^{\perp},K/k_s} = \mathscr{D}(\mathrm{R}_{F/k_s}(\mathrm{P}H_{V^{\perp},K/F}))$, where F is the root field), and it ensures that the Cartan subgroups are tori (automatic in the higher-rank cases) due to the relation "$kK^2 \subset F$" in (3.3.8).

By inspection of SO(q)'s over k_s, all k-groups in Definition 8.1.1 are absolutely pseudo-simple of minimal type, non-standard (by consideration of dimensions of root spaces over k_s), have root system over k_s of type B, and have trivial center in the adjoint case. In particular, if G is a type-B generalized basic exotic k-group then G/Z_G is a type-B adjoint generalized basic exotic k-group and moreover G has root field equal to k (inherited from G/Z_G by Remark 3.3.3).

8.1.2. Let G be a type-B generalized basic exotic k-group with minimal field of definition K/k for its geometric unipotent radical, so $K^2 \subset k$ since the root field is k. To justify the terminology in Definition 8.1.1, we now explain how G lies inside a canonically associated basic exotic k-group $\mathscr{G}$ with the same associated invariant K/k and same root system over k_s.

Since $(G/Z_G)_{k_s} \simeq \mathrm{SO}(q)$ for (V, q) over k_s as above (so $k_s\langle V^\perp\rangle = K_s$ and $V^\perp$ is a nonzero proper k_s-subspace of K_s), we expect that $\mathscr{G}_{k_s}$ corresponds to enlarging $V^\perp$ to K_s. However, we want to describe $\mathscr{G}$ directly over k and to make its construction functorial (with respect to isomorphisms) in G.

The connected semisimple K-group $G' := G_K^{\mathrm{ss}}$ is simply connected and absolutely simple of type B_n. As G is of minimal type, the kernel of $i_G : G \to \mathrm{R}_{K/k}(G')$ has trivial intersection with any Cartan k-subgroup of G, yet (by Proposition 2.3.4) $\ker i_G$ is central since G_{k_s} has a reduced root system, so $\ker i_G = 1$. Via i_G, G embeds into a type-B basic exotic k-group when $n \geqslant 2$:

Proposition 8.1.3. *For G and $(K/k, G')$ as above, let $\pi : G' \to \overline{G}'$ be the very special K-isogeny and define $f = \mathrm{R}_{K/k}(\pi)$. The image $\overline{G} = f(G)$ is a Levi k-subgroup of $\mathrm{R}_{K/k}(\overline{G}')$, and if G has k_s-rank $n \geqslant 2$ then $\mathscr{G} := f^{-1}(\overline{G})$ is a basic exotic pseudo-simple k-group of type B_n containing G.*

Thus, any type-B generalized basic exotic k-group with k_s-rank $n \geqslant 2$ is contained in a *functorially associated* basic exotic k-group of type B_n.

Proof. Once it is shown that $\overline{G}$ is a Levi k-subgroup of $\mathrm{R}_{K/k}(\overline{G}')$, it is immediate from the definitions that $\mathscr{G}$ is a basic exotic pseudo-semisimple k-group if the k_s-rank of G is $\geqslant 2$. To verify that $\overline{G}$ is a Levi k-subgroup of $\mathrm{R}_{K/k}(\overline{G}')$ we may assume $k = k_s$, so G contains a split maximal k-torus T and there exists a Levi k-subgroup L of G containing T [CGP, Thm. 3.4.6].

Since $L_K \to G'$ is an isomorphism, π is the scalar extension to K of the very special isogeny $L \to \overline{L}$. In particular, via the inclusion $\overline{L} \hookrightarrow \mathrm{R}_{K/k}(\overline{L}_K) = \mathrm{R}_{K/k}(\overline{G}')$ we have $f(L) = \overline{L}$, and $\overline{L}$ is a Levi k- subgroup of $\mathrm{R}_{K/k}(\overline{L}_K)$ [CGP, Cor. A.5.16]. By Proposition 2.1.2(i), $\overline{G}$ is pseudo-semisimple because it lies between $\overline{L}$ and $\mathrm{R}_{K/k}(\overline{L}_K)$.

It now suffices to check that the inclusion $\overline{L} \subset \overline{G}$ of smooth connected k-groups is an equality. By considering the product structure on an open cell for (G, T), it suffices to check that the inclusion $\overline{T} := f(T) \subset f(Z_G(T))$ is an equality and that $f(U_c)$ is 1-dimensional for all $c \in \Phi := \Phi(G, T)$. The direct product structure on $Z_G(T)$ (Proposition 4.3.3) further reduces the task to showing that $f(G_c)$ is 3-dimensional for all c. (We shall follow the convention that in the rank-1 case the roots in Φ are short and the roots in $\overline{\Phi}$ are long.)

The finite-index inclusion $\mathrm{X}(\overline{T}) \subset \mathrm{X}(T)$ identifies each long root $a \in \Phi = \Phi(G', T')$ with a short root of $\overline{\Phi} := \Phi(\overline{G}', \overline{T}') = \Phi(\overline{G}, \overline{T})$ and identifies each short root $b \in \Phi$ with $(1/2)\overline{b}$ for a long root $\overline{b} \in \overline{\Phi}$. In this way, π carries G'_a isomorphically onto $\overline{G}'_a$ for $a \in \Phi_>$ and carries G'_b onto $\overline{G}'_{\overline{b}}$ via a Frobenius isogeny for $b \in \Phi_<$. From the description of G we see that $G_a = \mathrm{SL}_2$ for $a \in \Phi_>$, so we just have to check that the inclusion $\overline{L}_{\overline{b}} \subset f(G_b)$ is an equality for $b \in \Phi_<$. But $K^2 \subset k$, so the restriction $f : \mathrm{R}_{K/k}(G_b) \to \mathrm{R}_{K/k}(\overline{G}'_{\overline{b}})$ has image equal to $\overline{L}_{\overline{b}}$ for dimension reasons. $\square$

We would like to describe maximal k-tori and pseudo-parabolic k-subgroups of type-B generalized basic exotic k-groups in terms of the diagram

$$\begin{array}{ccc} & & \mathrm{R}_{K/k}(G') \\ & & \big\downarrow f \\ \overline{G} & \longrightarrow & \mathrm{R}_{K/k}(\overline{G}') \end{array}$$

As motivation, for a basic exotic k-group G arising from a triple $(k'/k, G', \overline{G})$, [CGP, Prop. 11.1.3] provides a natural bijective correspondence between the set of maximal k-tori T in G and the set of pairs $(\overline{T}, T')$ consisting of a maximal k-torus $\overline{T} \subset \overline{G}$ and a maximal k'-torus $T' \subset G'$ such that the very special k'-isogeny $G' \to \overline{G}' = \overline{G}_{k'}$ carries T' onto $\overline{T}_{k'}$. Likewise, in [CGP, Prop. 11.4.6] a similar bijection is established for pseudo-parabolic k-subgroups P, as well as an equivalence between the conditions that $T \subset P$ and that $T' \subset P'$.

For a type-B generalized basic exotic k-group G with k_s-rank $n \geqslant 2$, we generally cannot establish an analogous dictionary for the sets of maximal k-tori of G (in terms of the associated triple $(K/k, G', \overline{G})$). The difficulty is that since perfect smooth connected affine k-groups are generated by their maximal k-tori [CGP, Prop. A.2.11], the basic exotic k-group $\mathscr{G} = f^{-1}(\overline{G})$ that is strictly larger than G always contains maximal k-tori that are not inside G. Consequently, we establish the following dictionary for pseudo-parabolic k-subgroups:

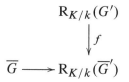

Proposition 8.1.4. *Let G be a type-B generalized basic exotic k-group with k_s-rank $n \geqslant 2$ and associated invariants $(K/k, G', \overline{G})$. If P is a pseudo-parabolic k-subgroup of G then $P' := \mathrm{im}(P_K \to G')$ is a parabolic K-subgroup of G' and $G \cap \mathrm{R}_{K/k}(P') = P$. Moreover, if $T \subset G$ is a maximal k-torus then $P \mapsto P'$ is a bijection between the set of pseudo-parabolic k-subgroups of G containing T and the set of parabolic K-subgroups of G' containing the associated T'.*

In particular, a parabolic K-subgroup P' of G' arises in this way from a pseudo-parabolic k-subgroup P of G if and only if P' contains a maximal K-torus of G' arising from a maximal k-torus of G.

Proof. Given P, we can choose a 1-parameter k-subgroup $\lambda : \mathrm{GL}_1 \to P$ such that $P = P_G(\lambda)$. Thus, the image P' of P_K under the surjective map $\pi : G_K \twoheadrightarrow G_K^{\mathrm{ss}} =: G'$ is $P_{G'}(\lambda')$ for the composition $\lambda' = \pi \circ \lambda_K$ [CGP, Cor. 2.1.9]. To establish that the inclusion $P \subset G \cap \mathrm{R}_{K/k}(P')$ is an equality, we first note that $\mathrm{R}_{K/k}(P') = P_{\mathrm{R}_{K/k}(G')}(\lambda)$ by [CGP, Prop. 2.1.13], where λ is valued in the k-subgroup G of $\mathrm{R}_{K/k}(G')$. Hence, $G \cap \mathrm{R}_{K/k}(P') = P_G(\lambda) = P$ as desired.

It remains to show for a maximal k-torus T of G that every parabolic K-subgroup P' of G' containing the associated T' arises from a pseudo-parabolic k-subgroup P of G containing T. Such a P is unique if it exists (since it must equal $G \cap \mathrm{R}_{K/k}(P')$), so for existence we may assume $k = k_s$. By [CGP, Prop. 3.5.4], there exists a unique pseudo-parabolic k-subgroup P of G containing T such that $P_{\overline{k}}$ has image in $G_{\overline{k}}^{\mathrm{ss}} = G'_{\overline{k}}$ coinciding with the parabolic subgroup $P'_{\overline{k}}$ containing $T'_{\overline{k}}$. Thus, the parabolic K-subgroups $\mathrm{im}(P_K \to G')$ and P' containing T' agree after scalar extension to $\overline{k}$, so they coincide. $\square$

An exceptional rank-2 construction in §8.3 will require an analogue, for type-B generalized basic exotic k-groups, of the classical very special isogeny

$$\mathrm{Spin}(q) \to \mathrm{SO}(q) \to \mathrm{Sp}(\overline{B}_q)$$

associated to non-degenerate (V, q) with $\dim V^{\perp} = 1$. Although such an analogue is only needed in rank-2 cases, we shall now define and study this analogue in any rank since that clarifies the construction and entails no extra effort.

Let us first address a version of the very special isogeny for k-groups $\mathrm{SO}(q)$ with regular quadratic spaces (V, q) satisfying $0 < \dim V^{\perp} < \dim V$. For such (V, q), let $\overline{B}_q$ be the symplectic form on $V/V^{\perp}$ obtained from B_q, and let

$$\pi_q : \mathrm{SO}(q) \to \mathrm{Sp}(\overline{B}_q)$$

be the natural map; this is surjective since its restriction to a Levi k_s-subgroup is the unipotent isogeny from an absolutely simple k_s-group of adjoint type B to an absolutely simple and simply connected k_s-group of type C. (In Proposition 8.1.8(i) we will see that $\dim \ker \pi_q = 2n(\dim V^{\perp} - 1)$.)

For $c \in k^{\times}$ we have $\mathrm{SO}(cq) = \mathrm{SO}(q)$ inside $\mathrm{GL}(V)$ and $\mathrm{Sp}(\overline{B}_{cq}) = \mathrm{Sp}(\overline{B}_q)$ inside $\mathrm{GL}(V/V^{\perp})$ (since $\overline{B}_{cq} = c\overline{B}_q$); in this way we have $\pi_q = \pi_{cq}$. Thus, π_q is functorial with respect to *conformal isometries* in (V, q), so by Proposition 7.2.2(i) the k-homomorphism π_q is *intrinsic* to the k-group $G = \mathrm{SO}(q)$ in the sense that it is independent of the choice of (V, q) and isomorphism $G \simeq \mathrm{SO}(q)$. The following definition is therefore well-posed:

Definition 8.1.5. Let G be a type-B generalized basic exotic k-group. The *very special quotient* $G \twoheadrightarrow \overline{G}$ is the composition of $G \to G/Z_G$ and the quotient map $\pi_u : G/Z_G \to \overline{G}$ that descends the k_s-homomorphism $\pi_q : \mathrm{SO}(q) \twoheadrightarrow \mathrm{Sp}(\overline{B}_q)$ for the conformal isometry class of (V, q) over k_s such that $(G/Z_G)_{k_s} \simeq \mathrm{SO}(q)$. In particular, $\overline{G}$ is absolutely simple and simply connected of type C.

Remark 8.1.6. Very special quotient maps $\pi : G \to \overline{G}$ always have connected kernel. Indeed, we may assume $k = k_s$, so the problem is to show that the quotient map $G/(\ker \pi)^0 \to \overline{G} \simeq \mathrm{Sp}_{2n}$ is an isomorphism. This latter map has finite étale kernel $E = (\ker \pi)/(\ker \pi)^0$, so $G/(\ker \pi)^0$ is a smooth connected central extension of Sp_{2n} by the finite étale E. But Sp_{2n} has no such *nontrivial* central extension since it is simply connected, so E is trivial.

Lemma 8.1.7. *Let G be a pseudo-split absolutely pseudo-simple group over a field k of characteristic 2, and for a split maximal k-torus T assume the root system $\Phi = \Phi(G, T)$ is of type B_n with $n \geqslant 2$. If $b, b' \in \Phi$ are linearly independent short roots then the root groups U_b and $U_{b'}$ commute inside G.*

This lemma expresses a well-known degeneration of Chevalley commutation relations for B_2 in characteristic 2 [Hum, 33.4(b)], but we give a proof avoiding the use of this degeneration.

Proof. By [CGP, Lemma 1.2.1], to prove the triviality of $(U_b, U_{b'})$ it suffices to prove the triviality of the image of $(U_b, U_{b'})_{\overline{k}}$ in $G_{\overline{k}}^{\mathrm{red}}$. Since Φ is reduced, it follows from [CGP, Rem. 2.3.6, Thm. 2.3.10] that the quotient map $G_{\overline{k}} \to G_{\overline{k}}^{\mathrm{red}}$ carries root groups onto root groups identifying the root systems. Hence, it suffices to treat the case of semisimple G over $k = \overline{k}$. For the classical purely inseparable isogeny $\varpi : G = \mathrm{SO}_{2n+1} \to \mathrm{Sp}_{2n}$ in characteristic 2, the long roots of the target are identified with twice the short roots of the source and ϖ carries each short root group onto the corresponding long root group [CGP, Prop. 7.1.5(2)].

Thus, it suffices to prove the commutation of root groups attached to a pair of linearly independent long roots a, a' for a connected semisimple k-group H of type C_n. But in the root system C_n the sum of two linearly independent long roots is never a root (seen by inspection, or in other ways), so we are done. □

The very special quotient map $\pi : G \to \overline{G}$ associated to a type-B generalized basic exotic group G satisfies properties analogous to very special isogenies between simply connected semisimple groups as in [CGP, Prop. 7.1.5] (such as swapping long and short root groups). This is primarily an assertion about the intermediate map $\pi_u : G/Z_G \to \overline{G}$ that is analogous to the (unipotent) isogeny $\pi_Q : \mathrm{SO}(Q) \to \mathrm{Sp}(\overline{B}_Q)$ for non-degenerate odd-dimensional (W, Q):

Proposition 8.1.8. *Let G be a type-B generalized basic exotic k-group and let T be a maximal k-torus in G/Z_G, with $n = \dim T$.*

(i) *The k-group $\ker \pi_u$ is commutative with no nontrivial k_s-points, and if $G/Z_G = \mathrm{SO}(q)$ for a regular (V, q) then $\ker \pi_q$ is a k_s/k-form of $(\ker \underline{q}^{\perp})^{2n}$ for $\underline{q}^{\perp} : \underline{V}^{\perp} \to \mathbf{G}_a$ defined by the 2-linear restriction $q^{\perp} : V^{\perp} \to k$ of q (so $\dim(\ker \pi_q) = 2n(\dim V^{\perp} - 1)$).*

(ii) *The map π_u carries T isomorphically onto a maximal k-torus $\overline{T} \subset \overline{G}$.*

(iii) *Assume T is split (so $G/Z_G \simeq \mathrm{SO}(q)$ for (V, q) as in (i)), and define $\Phi = \Phi(G/Z_G, T)$ and $\overline{\Phi} = \Phi(\overline{G}, \overline{T})$. Let $\Phi_<, \Phi_> \subset \Phi$ be the respective subsets of short and long roots, with both roots understood to be short when $n = 1$, and similarly define $\overline{\Phi}_<, \overline{\Phi}_> \subset \overline{\Phi}$ with both roots understood to be long when $n = 1$.*
The induced isomorphism $\mathrm{X}(\overline{T}) \simeq \mathrm{X}(T)$ identifies $\overline{\Phi}_<$ with $\Phi_>$ and identifies $\overline{\Phi}_>$ with $2 \cdot \Phi_<$. If a root $a \in \Phi_>$ coincides with a root $\overline{a} \in \overline{\Phi}_<$ then $\pi_u : U_a \to \overline{U}_{\overline{a}}$ is an isomorphism, and if a root $b \in \Phi_<$ coincides with $\overline{b}/2$ for a root $\overline{b} \in \overline{\Phi}_>$ then $\pi_u : U_b \to \overline{U}_{\overline{b}}$ is identified with $\underline{q}^{\perp}$.

Proof. We may assume $G/Z_G = \mathrm{SO}(q)$ for some (V, q) as in (i), so $\pi_u = \pi_q$. Let $\ell \subset V^{\perp}$ a line, and $L \subset \mathrm{SO}(q)$ the associated Levi k-subgroup containing T as in Proposition 7.1.5. By design, for the unique T-equivariant complement V_0 of $V^{\perp}$ we have $L = \mathrm{SO}(q|_W)$ for $W := V_0 \oplus \ell$ of dimension $2n + 1$. In particular, $q|_W$ is non-degenerate with defect line $W^{\perp}$ equal to $\ell = W \cap V^{\perp}$.

The resulting equality $W/W^{\perp} = V/V^{\perp}$ is compatible with the symplectic forms arising from $q|_W$ and q respectively, so $\pi_q|_L$ is the classical (unipotent) isogeny and its composition with the simply connected central cover $\widetilde{L} = \mathrm{Spin}(q|_W) \to L$ is the very special isogeny for $\widetilde{L}$ [CGP, Rem. 7.1.6]. In particular, π_q carries T isomorphically onto its image $\overline{T}$, and for split T the map π_q

has the asserted behavior between root systems (via the natural identification $\Phi(L, T) = \Phi$) as well as from long root groups onto short root groups (since the long root groups of $(\mathrm{SO}(q), T)$ coincide with those of (L, T)).

Since T_{k_s} is k_s-split, for the analysis of $\ker \pi_q$ and the description of $\pi_q|_{U_a} : U_a \to \overline{U}_{2a}$ for $a \in \Phi_<$ we may and do assume T is k-split. The kernel $\ker \pi_q$ is connected by Remark 8.1.6. Since T is its own centralizer in $\mathrm{SO}(q)$, it follows from consideration of compatible open cells with respect to π_q that $\ker \pi_q$ is a direct product of $2n$ copies of $\ker \underline{q}^\perp$ provided that (1) the k-groups $\ker \pi_q|_{U_b}$ for $b \in \Phi_<$ pairwise commute, and (2) $\pi_q|_{U_b} : U_b \to \overline{U}_{2b}$ is identified with $\underline{q}^\perp$ for all such b. For linearly independent $b, b' \in \Phi_<$ the associated root groups U_b and $U_{b'}$ commute by Lemma 8.1.7, so our problem is reduced to the study of π_q on $G_b := \langle U_b, U_{-b} \rangle$.

The k-subgroup G_b coincides with $\mathscr{D}(Z_{\mathrm{SO}(q)}(S))$ for the codimension-1 subtorus $S := (\ker b)^0_{\mathrm{red}} \subset T$, and the computations in the proof of Proposition 7.1.3 identify G_b with $\mathrm{SO}(q|_{W_b})$ where W_b denotes the span $V_b \oplus V_{-b} \oplus V^\perp$ of $V^\perp$ and the $\pm b$-weight lines $V_{\pm b}$ for T acting on V. The restriction $q_b = q|_{W_b}$ is regular with defect space of codimension 2, and $\pi_q|_{G_b}$ is thereby identified with the composition of π_{q_b} and the natural identification of $\mathrm{Sp}(\overline{B}_{q_b})$ with the copy of SL_2 inside $\mathrm{Sp}(\overline{B}_q)$ generated by the $\overline{T}$-root groups for $\pm 2b$. Hence, we may replace (V, q) with (W_b, q_b) for each such b, thereby reducing our problem to the case $n = 1$.

The restriction of q to the T-equivariant complement of $V^\perp$ in V may be identified with $(k^2, x_1 x_2)$, and we may k-linearly embed $V^\perp$ into K via $v \mapsto \sqrt{q(v)/q(v_0)}$ for a nonzero $v_0 \in V^\perp$; this embedding identifies $q^\perp$ with the restriction to $V^\perp$ of the 2-linear squaring map $K \to k$. In this way, as we saw in the proof of Proposition 7.1.3, the map π_q is identified with the natural map $\mathrm{P}H_{V^\perp, K/k} \to \mathrm{SL}_2$ induced by applying $\mathrm{R}_{K/k}$ to the natural unipotent K-isogeny $\Pi : \mathrm{PGL}_2 \to \mathrm{SL}_2$. The restriction of Π between each pair of corresponding root groups (relative to the diagonal tori) is identified with the squaring endomorphism of $\mathbf{G}_{\mathrm{a}}$ over K, and its kernels α_2 in the standard root groups over K commute inside PGL_2 by inspection. Hence, the restriction π_q of $\mathrm{R}_{K/k}(\Pi)$ to the k-subgroup $\mathrm{P}H_{V^\perp, K/k} \subset \mathrm{R}_{K/k}(\mathrm{PGL}_2)$ has the asserted form between compatible root groups and its kernel admits the desired description. □

8.2 Constructions for type C

We shall next build a large class of absolutely pseudo-simple k-groups of type C via fiber products using type-B generalized basic exotic groups. To motivate this, recall that the construction of basic exotic k-groups G of type C_n for $n \geqslant 2$ is given by fiber products

$$\begin{array}{ccc} G & \longrightarrow & R_{K/k}(G') \\ \downarrow & & \downarrow f \\ \overline{L} & \xrightarrow[j]{} & R_{K/k}(\overline{G}') \end{array}$$

where K/k is a nontrivial finite extension satisfying $K^2 \subset k$ and $f = R_{K/k}(\pi)$ for the very special isogeny $\pi : G' \to \overline{G}'$ from absolutely simple type C_n onto absolutely simple type B_n, with $\overline{L}$ a Levi k-subgroup of $R_{K/k}(\overline{G}')$ (equivalently, $\overline{L}$ is a k-descent of $\overline{G}'$) such that $\overline{L} \subset \text{im}(f)$. If G' contains a split maximal K-torus T' and we define $\overline{T}' = \pi(T')$ then π carries long T'-root groups isomorphically onto short $\overline{T}'$-root groups and it carries short T'-root groups onto long $\overline{T}'$-root groups via a Frobenius isogeny. In particular, f carries long root groups *isomorphically* onto short root groups.

If $\overline{L}$ contains the split maximal k-torus $\overline{T} \subset R_{K/k}(\overline{T}')$ then f carries short root groups $R_{K/k}(\mathbf{G}_a)$ into the 1-dimensional long $\overline{T}$-root groups of $\overline{L}$ since $K^2 \subset k$. Hence, the idea is that we try to fatten $\overline{L}$ into a type-B generalized basic exotic k-subgroup $\overline{G} \subset R_{K/k}(\overline{G}')$ by enlarging the short root groups of $\overline{L}$ to become the k-subgroup $\underline{V}_> \subset R_{K/k}(\mathbf{G}_a)$ associated to a k-subspace $V_> \subset K$ strictly containing k. Indeed, by inspection of open cells we expect that $f^{-1}(\overline{G})$ should be pseudo-semisimple of minimal type and that if $k = k_s$ then the root system of $f^{-1}(\overline{G})$ is type C_n with short root groups equal to $R_{K/k}(\mathbf{G}_a)$ (as $K^2 \subset k$) and long root groups equal to $\underline{V}_>$. If $k = k_s$ then this should account for the non-standard possibilities of type C_n in Theorem 3.4.1(iii) when $n \geqslant 3$. (In the special case $V_> = k$, which is to say $\overline{G}$ of type B is semisimple rather than basic exotic, we recover basic exotic k-groups of type C.)

In view of Proposition 3.3.6, to avoid the intervention of nontrivial Weil restrictions we want $f^{-1}(\overline{G})$ to have root field k, and that is the same as the long root field, so we should only consider those $\overline{G}$ of type B whose *short* root field is k. (For pseudo-split $\overline{G}$, this says $\{\lambda \in K \mid \lambda \cdot V_> \subset V_>\} = k$.) The presence of two vector spaces V and $V_>$ in Theorem 3.4.1(iii) for the case $n = 2$ shows that such a fiber product idea cannot be exhaustive for type $B_2 = C_2$ over k_s.

Although $C_2 = B_2$, when the above idea is specialized to the case $n = 2$ it does not generally coincide with the type-B generalized basic exotic construction for $n = 2$ if $[k : k^2] > 2$. Thus, to avoid confusion, the reader should regard the notion of "type-C generalized (basic) exotic" defined below (without reference to rank) as always distinct from “type-B generalized (basic) exotic”.

8.2.1. We now turn to the actual construction. Let K/k be a nontrivial finite extension satisfying $K^2 \subset k$, and let G' be a connected semisimple K-group that is absolutely simple and simply connected of type C_n with $n \geqslant 2$ (i.e., G' is a K-form of Sp_{2n}). Let $\pi : G' \to \overline{G}'$ be the very special isogeny, so $\overline{G}'$ is a K-form of Spin_{2n+1}. Let $\overline{G}$ be either a simply connected and absolutely simple connected semisimple k-group of type B_n or a type-B generalized basic exotic k-group with k_s-rank n. Such a $\overline{G}$ is of minimal type, and the minimal field of definition K'/k for its geometric unipotent radical is a (possibly trivial) finite extension of k contained inside $k^{1/2}$. Consider such $\overline{G}$ for which $K' \subset K$ over k and there is *given* a K-isomorphism $\overline{G}_K^{\mathrm{ss}} \simeq \overline{G}'$; we can encode this K-isomorphism via a k-homomorphism

$$j : \overline{G} \to \mathrm{R}_{K/k}(\overline{G}')$$

by composing $\xi_{\overline{G}} = i_{\overline{G}}$ with $\mathrm{R}_{K'/k}(\overline{G}_{K'}^{\mathrm{ss}}) \hookrightarrow \mathrm{R}_{K/k}(\overline{G}_K^{\mathrm{ss}}) \simeq \mathrm{R}_{K/k}(\overline{G}')$. (The case of semisimple $\overline{G}$ is included above so that the “type-C generalized basic exotic” construction below includes as a special case the type-C “basic exotic” construction in §2.2.)

Since $\overline{G}_{k_s}$ has a reduced root system, $\ker i_{\overline{G}}$ is central (Proposition 2.3.4). But $\overline{G}$ is of minimal type, so $\ker i_{\overline{G}}$ is trivial; i.e., j is a closed immersion. We will only be interested in those $(\overline{G}, j)$ for which $j(\overline{G})$ lies inside the image of

$$f = \mathrm{R}_{K/k}(\pi) : \mathrm{R}_{K/k}(G') \to \mathrm{R}_{K/k}(\overline{G}').$$

Proposition 8.2.2. *Let $(K/k, G')$ and $(\overline{G}, j)$ be as above (so $K' \subset K$) with $j(\overline{G}) \subset \mathrm{im}(f)$, and define G to be the fiber product*

$$\begin{array}{ccc} G & \longrightarrow & \mathrm{R}_{K/k}(G') \\ \downarrow & & \downarrow{\scriptstyle f} \\ \overline{G} & \xrightarrow[j]{} & \mathrm{R}_{K/k}(\overline{G}') \end{array}$$

(so $\overline{G} = f(G)$). The k-group G is absolutely pseudo-simple of minimal type with $G_{\overline{k}}^{\mathrm{ss}}$ simply connected of type C_n, K/k is the minimal field of definition for $\mathscr{R}_u(G_{\overline{k}}) \subset G_{\overline{k}}$, and the inclusion $G \hookrightarrow \mathrm{R}_{K/k}(G')$ is identified with i_G (so the corresponding K-homomorphism $G_K \to G'$ is surjective and identifies G' with G_K^{ss}). In particular, the data $(K/k, K'/k, G', \overline{G}, j)$ satisfying $K' \subset K$ is uniquely functorial with respect to k-isomorphisms in G.

Proof. We assume (as we may) after replacing k with k_s that $k = k_s$. The quotient scheme $G/(\ker f) = \overline{G}$ is connected, so to prove that G is connected it suffices to show that $\ker f$ is connected. As a K-scheme (ignoring the K-group structure), the infinitesimal group scheme $\ker\pi$ is a direct product of copies of the K-scheme $\mathrm{Spec}(K[x]/(x^2)) = \alpha_2 = \mu_2$. Thus, the k-scheme $\ker f = \mathrm{R}_{K/k}(\ker\pi)$ is a direct product of copies of the k-scheme $\mathrm{R}_{K/k}(\alpha_2)$ that is a quadric in the affine space $\mathrm{R}_{K/k}(\mathbf{G}_\mathrm{a})$ (since $K^2 \subset k$). This quadric hypersurface is geometrically irreducible since the underlying reduced scheme of its $\overline{k}$-fiber is a hyperplane, so $\ker f$ is connected and thus so is G.

Since k is separably closed, by [CGP, Thm. 3.4.6] there exists a Levi k-subgroup $\overline{L}$ in the pseudo-simple k-group $\overline{G}$, so $\overline{L}_K \to \overline{G}_K^{\mathrm{ss}}$ is an isomorphism. The inclusion $j : \overline{G} \hookrightarrow \mathrm{R}_{K/k}(\overline{G}')$ corresponds to a K-isomorphism $\overline{G}_K^{\mathrm{ss}} \simeq \overline{G}'$, so the composite inclusion $\overline{L} \hookrightarrow \mathrm{R}_{K/k}(\overline{G}')$ corresponds to a K-isomorphism $\overline{L}_K \simeq \overline{G}'$. Thus, $\overline{L}$ is a Levi k-subgroup of $\mathrm{R}_{K/k}(\overline{G}')$. Clearly $\overline{L} \subset \mathrm{im}(f)$, so the triple $(K/k, G', \overline{L})$ is exactly the setup for a basic exotic k-group of type C_n (see [CGP, Prop. 7.2.7, Prop. 7.3.1]). The schematic preimage $f^{-1}(\overline{L})$ is therefore a basic exotic k-group of type C_n.

Let L be a Levi k-subgroup of $f^{-1}(\overline{L})$ (see [CGP, Thm. 3.4.6]), so L is also a Levi k-subgroup of $\mathrm{R}_{K/k}(G')$ (since the natural map $f^{-1}(\overline{L})_K \to G'$ is a K-descent of the maximal geometric reductive quotient [CGP, Prop. 7.2.7(2)]). The connected k-subgroup scheme G of $\mathrm{R}_{K/k}(G')$ contains the Levi k-subgroup L of $\mathrm{R}_{K/k}(G')$, so if G is *smooth* then by Proposition 2.1.2(i) the k-group G is pseudo-reductive with L a Levi k-subgroup of G.

Assume G is smooth, so $L_{\overline{k}} \to G_{\overline{k}}^{\mathrm{red}} = G'_{\overline{k}}$ is an isomorphism. Thus, $G_{\overline{k}}^{\mathrm{red}}$ is semisimple, simply connected, and absolutely simple of type C_n (so G has an irreducible root system and therefore is pseudo-simple if it is pseudo-semisimple). By Proposition 2.1.2(i), $G_K \cap \mathscr{R}_{u,K}(\mathrm{R}_{K/k}(G')_K)$ is a K-descent of $\mathscr{R}_u(G_{\overline{k}})$, so it coincides with $\mathscr{R}_{u,K}(G_K)$ and the natural map $G_K \to G'$ (corresponding to the given inclusion $G \hookrightarrow \mathrm{R}_{K/k}(G')$) is the quotient modulo $\mathscr{R}_{u,K}(G_K)$. Hence, K/k is a field of definition $\mathscr{R}_u(G_{\overline{k}})$. In fact, K/k is the *minimal* such field because it is the minimal field of definition for the geometric unipotent radical of the basic exotic $f^{-1}(\overline{L})$ and

$$f^{-1}(\overline{L})_{\overline{k}} \cap \mathscr{R}_u(G_{\overline{k}}) = f^{-1}(\overline{L})_{\overline{k}} \cap \mathscr{R}_u(\mathrm{R}_{K/k}(G')_{\overline{k}}) = \mathscr{R}_u(f^{-1}(\overline{L})_{\overline{k}}).$$

It follows that the inclusion $G \hookrightarrow \mathrm{R}_{K/k}(G')$ is i_G, and the consequent triviality of $\ker i_G$ implies that G is of minimal type.

It remains to prove that the connected k-group G is smooth and perfect.

For this purpose we shall study compatible open cells of $\mathrm{R}_{K/k}(G')$, $\mathrm{R}_{K/k}(\overline{L})$, and $\overline{G}$ built as follows. Pick a maximal k-torus T of $f^{-1}(\overline{L})$, so T is also a maximal k-torus of $\mathrm{R}_{K/k}(G')$. Letting $\overline{T}$ be the maximal k-torus image of T in $\overline{L} = f(f^{-1}(\overline{L}))$, we may naturally identify T_K and $\overline{T}_K$ with respective maximal K-tori of G' and $\overline{G}'$. Define the root systems

$$\Phi = \Phi(\mathrm{R}_{K/k}(G'), T) = \Phi(G', T_K),$$

$$\overline{\Phi} = \Phi(\overline{G}, \overline{T}) = \Phi(\mathrm{R}_{K/k}(\overline{G}'), \overline{T}) = \Phi(\overline{G}', \overline{T}_K),$$

so the inclusion $\mathrm{X}(\overline{T}) \subset \mathrm{X}(T)$ identifies $\overline{\Phi}_<$ with $\Phi_>$ and identifies $\overline{\Phi}_>$ with $2 \cdot \Phi_<$, and the map $\pi : G' \to \overline{G}'$ carries long root groups isomorphically onto short root groups and carries short root groups onto long root groups via a Frobenius isogeny. For $c \in \Phi$, let $\overline{c} \in \overline{\Phi}$ be the corresponding root (i.e., $\overline{c} = c$ if c is long and $\overline{c} = 2c$ if c is short).

Choose compatible positive systems of roots Φ^+ and $\overline{\Phi}^+$ in Φ and $\overline{\Phi}$ respectively, so we get compatible open cells

$$\Omega \subset G', \ \overline{\Omega} = \pi(\Omega) \subset \overline{G}'.$$

By [CGP, Prop. 2.1.8(3)], $\overline{L} \cap \mathrm{R}_{K/k}(\overline{\Omega})$ and $\overline{G} \cap \mathrm{R}_{K/k}(\overline{\Omega})$ are the corresponding open cells of $\overline{L}$ and $\overline{G}$ respectively. Since $\ker \pi$ is infinitesimal, the inclusion of open subschemes $\Omega \subset \pi^{-1}(\overline{\Omega})$ is an equality. Thus,

$$G \cap \mathrm{R}_{K/k}(\Omega) = f^{-1}(\overline{G} \cap \mathrm{R}_{K/k}(\overline{\Omega})).$$

By expressing $\pi : \Omega \to \overline{\Omega}$ as a direct product of maps between maximal tori and root groups (arranged in whatever compatible order we wish among the positive roots, and among their negatives), we get an analogous expression for $f : \mathrm{R}_{K/k}(\Omega) \to \mathrm{R}_{K/k}(\overline{\Omega})$ as direct product of maps. Since the open cell $\overline{G} \cap \mathrm{R}_{K/k}(\overline{\Omega})$ associated to $(\overline{G}, \overline{T}, \overline{\Phi}^+)$ is likewise such a direct product, the f-preimage of $\overline{G} \cap \mathrm{R}_{K/k}(\overline{\Omega})$ may be computed within each factor separately. Hence, G is smooth if and only if (i) the restriction $f : \mathrm{R}_{K/k}(T_K) \to \mathrm{R}_{K/k}(\overline{T}_K)$ has smooth preimage of $\overline{C} := Z_{\overline{G}}(\overline{T})$, and (ii) the restriction $f : \mathrm{R}_{K/k}(U_c) \to \mathrm{R}_{K/k}(U_{\overline{c}})$ has smooth preimage for the $\overline{c}$-root group of $(\overline{G}, \overline{T})$. Note that (i) is a direct product of maps $\mathrm{R}_{K/k}(c_K^\vee(\mathrm{GL}_1)) \to \mathrm{R}_{K/k}(\overline{c}_K^\vee(\mathrm{GL}_1))$ for roots c in the basis Δ of Φ^+ (with the associated $\overline{c}$'s constituting the basis $\overline{\Delta}$ of $\overline{\Phi}^+$), and $\overline{C}$ is given by a compatible direct product due to Proposition 4.3.3 (applied to $\overline{G}$).

If c is long then $U_c \to U_{\overline{c}}$ is an isomorphism and if also $c \in \Delta_>$ then the

c-factor of (i) is an isomorphism, so (i) and (ii) are clear for long roots of Φ (and associated coroots). For short $c \in \Phi$ we have $\overline{c} = 2c \in \overline{\Phi}_>$ and the 1-dimensional $\overline{c}$-root group of $(\overline{G}, \overline{T})$ coincides with that of $(\overline{L}, \overline{T})$, so its f-preimage meets $\mathrm{R}_{K/k}(U_c)$ in the c-root group of $f^{-1}(\overline{L})$. If also $c \in \Delta_<$ then $\overline{c} = 2c \in \overline{\Delta}_>$ and the c-factor of (i) is the squaring endomorphism of $\mathrm{R}_{K/k}(\mathrm{GL}_1)$ whose image is $\mathrm{GL}_1 \subset \mathrm{R}_{K/k}(\mathrm{GL}_1)$ (since $K^2 \subset k$). The $\overline{c}$-factor of $\overline{C}$ is $\overline{c}^\vee(\mathrm{GL}_1)$, and its preimage in the c-factor $\mathrm{R}_{K/k}(c_K^\vee(\mathrm{GL}_1))$ of $\mathrm{R}_{K/k}(T_K)$ is the entire c-factor. Thus, G is smooth and therefore pseudo-reductive.

This argument proves: $G_a = \overline{G}_{\overline{a}}$ inside $\mathrm{R}_{K/k}(G'_a) = \mathrm{R}_{K/k}(\overline{G}'_{\overline{a}})$ for $a \in \Phi_>$, $G_b = \mathrm{R}_{K/k}(G'_b)$ for $b \in \Phi_<$, and $Z_G(T)$ is generated by a Cartan k-subgroup of G_{a_0} for the unique $a_0 \in \Delta_>$ and a Cartan k-subgroup of the pseudo-simple subgroup $f^{-1}(\overline{L}) \subset G$, so $Z_G(T) \subset \mathscr{D}(G)$. But $G = \mathscr{D}(G) \cdot Z_G(T)$, so $G = \mathscr{D}(G)$; i.e., G is perfect. □

Definition 8.2.3. A *type-*C *generalized basic exotic group* is a k-group G as in Proposition 8.2.2 (so G is absolutely pseudo-simple of minimal type with $G^{\mathrm{ss}}_{\overline{k}}$ simply connected of type $\mathrm{C}_n, n \geqslant 2$) such that the root field of G (or equivalently, the short root field of $\overline{G}$) is k.

By Proposition 8.2.2, if k'/k is a separable extension then a k-group G is type-C generalized basic exotic if and only if $G_{k'}$ is such a k'-group.

8.2.4. It is instructive to compare the root groups for pseudo-split "type-B" and "type-C" generalized basic exotic groups. Let G be a group of either type with rank $n \geqslant 2$, and let K/k the minimal field of definition for the geometric unipotent radical, so $K \neq k$ and $K^2 \subset k$.

The type-B construction has long root groups of dimension 1 and short root groups $\underline{V}$ for a nonzero k-subspace $V \subset K$ satisfying $k\langle V\rangle = K$ (so $1 < \dim_k V \leqslant [K:k]$); the case $V = K$ coincides with "basic exotic of type B_n".

The proof of Proposition 8.2.2 shows that (as expected from the motivation given near the beginning of §8.2) the type-C construction has as its short root groups $\mathrm{R}_{K/k}(\mathbf{G}_\mathrm{a})$ and as its long root groups either $\mathbf{G}_\mathrm{a}$ (when $\overline{G}$ is semisimple) or the k-group $\underline{V}_>$ (when $\overline{G}$ is a type-B generalized basic exotic k-group) with associated nontrivial field extension $k\langle V_>\rangle/k$, where $V_> \subset k\langle V_>\rangle \subset K$ and $V_>$ has root field k (so $1 < \dim_k(V_>) < [K:k]$). The type-C case with semisimple $\overline{G}$ is the basic exotic construction of type C, so the new cases of interest are when $\overline{G}$ is a type-B generalized basic exotic k-group with k_s-rank $n \geqslant 2$.

When $k = k_s$, the generalized basic exotic groups that are *not* basic exotic rest on a choice of a nonzero k-subspace of K whose k-dimension is strictly between 1 and $[K:k]$. The type-B generalized basic exotic construction of rank $n \geqslant 2$ is obtained from the basic exotic construction of type B_n relative to K/k by shrinking short root groups from $\mathrm{R}_{K/k}(\mathbf{G}_\mathrm{a})$ to $\underline{V}$ for V satisfying $k\langle V\rangle = K$, whereas the type-C_n generalized basic exotic construction for $n \geqslant 2$ is obtained from a basic exotic construction of type C_n relative to K/k by enlarging long root groups from $\mathbf{G}_\mathrm{a}$ to $\underline{V}_>$ where the nonzero proper k-subspace $V_> \subset K$ has root field k. By Theorem 3.4.1(iii), the isomorphism class of each such k-group depends only on the root system over k_s and on the $K^\times$-homothety class of the indicated k-subspace $V_>$ of K. Such k-groups that are not basic exotic can only exist when $[K:k] > 2$, and since $K \subset k^{1/2}$ this can only occur if $[k:k^2] > 2$.

For $n = 2$ over $k = k_s$ with $[K:k] > 2$ (so $[k:k^2] > 2$), we therefore have two entirely different classes of new absolutely pseudo-simple k-groups G for which $G^{\mathrm{ss}}_{\overline{k}}$ is simply connected of type $\mathrm{B}_2 = \mathrm{C}_2$: the "type-B" case has long root groups of dimension 1 and short root groups of dimension strictly between 1 and $[K:k]$ whereas the "type-C" case has short root groups of dimension $[K:k] > 1$ and long root groups of dimension strictly between 1 and $[K:k]$.

Here is the classification of *pseudo-split* possibilities for the generalized basic exotic constructions beyond rank 1; it is established by the preceding considerations along with Theorem 3.4.1(iii) and Proposition 4.3.3.

Proposition 8.2.5. *Fix $n \geqslant 2$ and let K/k be a nontrivial finite extension satisfying $K^2 \subset k$. Consider pseudo-split type-B or type-C generalized basic exotic k-groups G of rank n that have K/k as the minimal field of definition for $\mathscr{R}_u(G_{\overline{k}}) \subset G_{\overline{k}}$. Let T be a split maximal k-torus in G, and $\Phi = \Phi(G,T)$.*

The set of k-isomorphism classes of such G in the "type-B" case is in bijective correspondence with the set of $K^\times$-homothety classes of nonzero k-subspaces $V \subset K$ satisfying $k\langle V\rangle = K$ by assigning to G the homothety class of V for which $G_b \simeq H_{V,K/k}$ for short roots $b \in \Phi$. Moreover, if Δ is a basis of Φ then naturally

$$Z_G(T) = \Big(\prod_{a\in\Delta_>} a^\vee(\mathrm{GL}_1)\Big) \times (\mathrm{R}_{K/k}(b_K^\vee))(V_{K/k}^*)$$

for the unique $b \in \Delta_<$.

The set of k-isomorphism classes of such G in the "type-C" case is in bijective correspondence with the set of $K^\times$-homothety classes of nonzero k-subspaces $V_> \subset K$ with root field k such that $G_a \simeq H_{V_>,K/k}$ for long $a \in \Phi$ (so $G_a \simeq H_{V_>,K_>/k}$ for $K_> := k\langle V_>\rangle$). Moreover, if Δ is a basis of Φ then *naturally*

$$Z_G(T) = (\mathrm{R}_{K/k}(a_K^\vee))((V_>)^*_{K/k}) \times \prod_{b \in \Delta_<} \mathrm{R}_{K/k}(\mathrm{GL}_1)$$

for the unique $a \in \Delta_>$.

In Proposition 8.2.5, the k-group G is basic exotic precisely when $V = K$ in the "type-B" case and precisely when $V_>$ is a k-line in the "type-C" case. Note that we allow $k\langle V_> \rangle$ to be a proper subfield of K over k for the "type-C" case (it is precisely the subfield $K_>$ in such cases).

In the spirit of Proposition 8.1.4, here is a description of the pseudo-parabolic k-subgroups of a type-C generalized basic exotic k-group G in terms of an associated type-B generalized basic exotic k-group $\overline{G}$:

Proposition 8.2.6. *Let G be a type-C generalized basic exotic k-group with k_s-rank $n \geqslant 2$ and associated triple $(K/k, G', \overline{G})$ as in Proposition 8.2.2. Define $f = \mathrm{R}_{K/k}(\pi)$ for the very special K-isogeny $\pi : G' \to \overline{G}'$. For any pseudo-parabolic k-subgroup P of G, define*

$$\overline{P} = f(P) \subset f(G) = \overline{G}, \;\; P' = \mathrm{im}(P_K \to G')$$

(so $\overline{P}' := \pi(P')$ is the image of $\overline{P}_K \hookrightarrow \overline{G}_K \twoheadrightarrow \overline{G}_K^{\mathrm{ss}} \simeq \overline{G}'$).

(i) *The k-subgroup $\overline{P} \subset \overline{G}$ is pseudo-parabolic, the K-subgroup $P' \subset G'$ is pseudo-parabolic, and $P = G \cap \mathrm{R}_{K/k}(P')$. Moreover, the natural map*

$$G/P \to (\overline{G}/\overline{P}) \times_{\mathrm{R}_{K/k}(\overline{G}'/\overline{P}')} \mathrm{R}_{K/k}(G'/P')$$

is an isomorphism.

(ii) *For any pseudo-parabolic k-subgroup $\overline{Q} \subset \overline{G}$ and pseudo-parabolic K-subgroup $Q' \subset G'$ such that $\pi(Q') = \mathrm{im}(\overline{Q}_K \to \overline{G}')$, there is a unique pseudo-parabolic k-subgroup $Q \subset G$ giving rise to Q' and $\overline{Q}$.*

(iii) *Let $T \subset G$ be a maximal k-torus, and let $T' := \mathrm{im}(T_K \hookrightarrow G')$. Then $P \supset T$ if and only if $P' \supset T'$, and every pseudo-parabolic K-subgroup of G' containing T' comes from a unique such P.*

Since G' and $\overline{G}'$ are connected reductive, we remind the reader that over a field F the pseudo-parabolic F-subgroups of a connected reductive F-group are precisely its parabolic F-subgroups [CGP, Prop. 2.2.9].

Proof. The proof of (iii) and the proof of (i) apart from the assertions concerning G/P proceed exactly as in the proof of Proposition 8.1.4. In particular, the uniqueness in (ii) is proved.

It remains to establish the description of G/P in (i) and to prove the existence of Q in (ii). For both assertions we may assume $k = k_s$, and the arguments will rest on calculations with the Bruhat decomposition for the pairs

$$(G'(K), P'(K)), \ (\overline{G}'(K), \overline{P}'(K)), \ (\overline{G}(k), \overline{P}(k))$$

similar to the basic exotic case in the proof of [CGP, Prop. 11.4.6]. However, although G' and $\overline{G}'$ are reductive, now $\overline{G}$ is *not* reductive (in contrast with the basic exotic case). This causes two complications:

(1) the proof of the Bruhat decomposition for $\overline{G}(k)$ relative to a choice of pseudo-parabolic k-subgroup of $\overline{G}$ lies deeper than in the reductive case,
(2) the map $\overline{G}_K \rightarrow \overline{G}'$ corresponding to the tautological inclusion $j : \overline{G} \hookrightarrow \mathrm{R}_{K/k}(\overline{G}')$ is the quotient map $\overline{G}_K \twoheadrightarrow \overline{G}_K^{\mathrm{ss}}$ rather than an isomorphism.

Consider the existence part of (ii): given a pseudo-parabolic K-subgroup $Q' \subset G'$ and a pseudo-parabolic k-subgroup $\overline{Q} \subset \overline{G}$ such that

$$\pi(Q') = \mathrm{im}(\overline{Q}_K \rightarrow \overline{G}_K^{\mathrm{ss}} \simeq \overline{G}'),$$

we seek a (necessarily unique) pseudo-parabolic k-subgroup $Q \subset G$ giving rise to Q' and $\overline{Q}$. As in the basic exotic case in [CGP, Prop. 11.4.6], we first pick a maximal k-torus $T \subset G$ and let T' denote the maximal K-torus $T_K \hookrightarrow G_K^{\mathrm{ss}} = G'$, so $G'(K)$-conjugacy of maximal K-tori in G' provides $g' \in G'(K)$ such that the pseudo-parabolic K-subgroup $P' := g'Q'g'^{-1}$ of G' contains T'. Hence, by (iii) there is a unique pseudo-parabolic k-subgroup $P \subset G$ containing T such that P' is the image of $P_K \rightarrow G_K^{\mathrm{ss}} \simeq G'$.

Let $\overline{P}$ denote the pseudo-parabolic image of P under the quotient map $G \twoheadrightarrow \overline{G}$. It is harmless to change g' by left $P'(K)$-multiplication and also by right $G(k)$-multiplication (since for $g \in G(k)$ there is no harm in replacing Q' and $\overline{Q}$ with their respective conjugates against the images of g in $G'(K)$ and $\overline{G}(k)$ respectively). Thus, it suffices to show that necessarily $g' \in P'(K)G(k)$.

For the pseudo-parabolic k-subgroups $\overline{P}$ and $\overline{Q}$ in $\overline{G}$, the scalar extensions $\overline{P}_K$ and $\overline{Q}_K$ in $\overline{G}_K$ have respective images in $\overline{G}_K^{\mathrm{ss}} = \overline{G}'$ equal to $\overline{P}'$ and $\overline{Q}'$. These images are $\overline{G}'(K)$-conjugate by design, so $\overline{P}$ and $\overline{Q}$ are $\overline{G}(k)$-conjugate due to:

Lemma 8.2.7. *Let H be a pseudo-reductive group over a field F, and consider a pair of pseudo-parabolic F-subgroups $P, Q \subset H$ and the respective parabolic images $P', Q' \subset H' =: H_{\overline{F}}^{\mathrm{red}}$ of $P_{\overline{F}}$ and $Q_{\overline{F}}$. If P' is $H'(\overline{F})$-conjugate to Q' then P is $H(F)$-conjugate to Q.*

This is a strengthening of [CGP, Cor. C.2.6] in which a separability condition has been removed (but otherwise the proof is essentially the same).

Proof. By the $H(F)$-conjugacy of minimal pseudo-parabolic F-subgroups of H [CGP, Thm. C.2.5], we may replace Q with an $H(F)$-conjugate so that P and Q contain a common (minimal) pseudo-parabolic F-subgroup. But then P' and Q' are parabolic subgroups of H' containing a common parabolic subgroup, and hence containing a common Borel subgroup. By hypothesis P' and Q' are $H'(\overline{F})$-conjugate, and in a connected reductive $\overline{F}$-group there is no conjugacy among distinct parabolic subgroups containing a common Borel subgroup. Thus, $P' = Q'$, so $P = Q$ by [CGP, Cor. 3.5.11]. ☐

We shall now complete the proof of Proposition 8.2.6. Since $\overline{P}$ and $\overline{Q}$ are now known to be $\overline{G}(k)$-conjugate, by exactly the same reasoning as in the treatment of the basic exotic case (see the proof of [CGP, Prop. 11.4.6(1)]) it suffices for the existence aspect of (ii) to show that under the map

$$G'(K)/P'(K) \to \overline{G}'(K)/\overline{P}'(K)$$

the preimage of the $\overline{P}'(K)$-coset of any point of $\overline{G}(k)$ admits a representative in $G(k)$. In contrast with the basic exotic case, $\overline{G}_K \to \overline{G}'$ is now merely a surjection (it is the maximal reductive quotient) rather than an isomorphism. Nonetheless, the map $\overline{G}(k) \to \overline{G}'(K)$ is injective since it is induced by the inclusion $\overline{G} \hookrightarrow \mathrm{R}_{K/k}(\overline{G}')$.

Choose a positive system of roots Φ^+ in $\Phi := \Phi(G,T) = \Phi(G',T')$, and let $\overline{\Phi}^+$ be the corresponding positive system of roots in $\overline{\Phi} := \Phi(\overline{G},\overline{T}) = \Phi(\overline{G}',\overline{T}')$. Via the equality $\mathrm{X}(T)_{\mathbf{Q}} = \mathrm{X}(\overline{T})_{\mathbf{Q}}$ and the length-swapping correspondence $c \mapsto \overline{c}$ between Φ and $\overline{\Phi}$, the reflections r_c and $r_{\overline{c}}$ coincide. Thus, $W := W(\Phi)$ coincides with $W(\overline{\Phi})$ as reflection groups on the common rational character space. Letting Δ be the basis of Φ corresponding to Φ^+, the parabolic set of roots $\Phi(P,T)$ is labelled by a unique subset $J \subset \Delta$ (with $J = \emptyset$ when P is a pseudo-Borel K-subgroup).

Inspection of the behavior of the very special K-isogeny $\pi : G' \to \overline{G}'$ on root groups relative to T' and $\overline{T}'$ shows that $\overline{P}$ and P' are also labelled by J via the natural bijection between Δ and the basis of $\overline{\Phi}$ corresponding to $\overline{\Phi}^+$. Thus, via the injection $\overline{G}(k) \hookrightarrow \overline{G}'(K)$ we can conclude by the exact same argument as in the basic exotic case (in the proof of [CGP, Prop. 11.4.6(1)]) provided that the following two properties are proved:

(a) the natural map

$$\coprod_{w \in W^J} \overline{U}_w(k) \cdot \overline{n}_w \cdot \overline{P}(k) \to \overline{G}(k)$$

is bijective,

(b) each constituent on the left side of the map in (a) is a direct product set.

Here we use standard notation as in the reductive case [Bo, 21.29]: W^J is the set of unique shortest-length representatives in $W = W(G', T')$ of the cosets in W/W_J where W_J is generated by the reflections associated to elements of J, $\overline{n}_w \in N_{\overline{G}}(\overline{T})(k)$ is a representative of w, and $\overline{U}_w$ is the k-subgroup of $\overline{G}$ directly spanned in any order by the root groups of $(\overline{G}, \overline{T})$ associated to the roots $c \in \overline{\Phi}^+$ such that $w^{-1}(c) \notin \overline{\Phi}^+$. (See [CGP, Cor. 3.3.13(1)] for the existence of the k-subgroup $\overline{U}_w$.)

Using the "open cell" structure associated to pseudo-parabolic subgroups of pseudo-reductive groups in [CGP, Prop. 2.1.8(2),(3)], properties (a) and (b) are straightforward consequences of the Bruhat decomposition for pseudo-reductive groups [CGP, Thm. C.2.8] (applied to $(\overline{G}, \overline{P})$) exactly as in the reductive case. This completes the proof of (ii), and the proof of the formula for G/P in (i) goes exactly as in the basic exotic case (using calculations with the bijection as in (a) above and its better-known analogue for the pairs $(G'(K), P'(K))$ and $(\overline{G}'(K), \overline{P}'(K))$ with reductive G' and $\overline{G}'$). □

8.3 Exceptional construction for rank 2

Let k be imperfect with $\mathrm{char}(k) = 2$. Among the absolutely pseudo-simple k-groups G of minimal type with a reduced root system over k_s such that $G^{\mathrm{ss}}_{\overline{k}}$ is simply connected of type $\mathrm{B}_2 = \mathrm{C}_2$, Theorem 3.4.1(iii) provides pseudo-split non-standard possibilities beyond the type-B and type-C generalized basic exotic constructions in §8.1–§8.2. These additional G are those for which $(G_{k_s})_c$ is non-standard for *all* roots c (rather than only for short c as in §8.1, or only for long c as in §8.2). We aim to go beyond the pseudo-split case and describe such G in terms of a variant of the fiber product construction in Proposition 8.2.2.

The pseudo-split pseudo-simple groups with root system B_2 are classified by field-theoretic and linear-algebraic data: the minimal field of definition K/k

for the geometric unipotent radical (a nontrivial purely inseparable finite extension), the subextension $K_>/k$ from Theorem 3.3.8(ii) that contains kK^2 by (3.3.8), and the nonzero kK^2-subspace $V_> \subset K_>$ and nonzero $K_>$-subspace $V \subset K$ satisfying $k\langle V_>\rangle = K_>$ and $k\langle V\rangle = K$ (see Theorem 3.3.8(iii)). Since $G_a \simeq H_{V_>,K_>/k}$ for long roots a and $G_b \simeq H_{V,K/k}$ for short roots b, non-standardness of G_c for all roots c says exactly $V \neq K$ and $V_> \neq K_>$ (so $[K : K_>] > 2$ and $[K_> : kK^2] > 2$).

Remark 8.3.1. We claim that such pseudo-split G exist with non-standard G_c for all c if and only if $[k : k^2] \geqslant 16$. Suppose $[k : k^2] \geqslant 16$, so we can choose $K/K_>/k$ inside $k^{1/2}$ with $[K : K_>] \geqslant 4$ and $[K_> : k] \geqslant 4$. Thus, we can choose $V_>$ to be a k-hyperplane in $K_>$ and V to be a $K_>$-hyperplane in K. Conversely, non-standardness forces $[K_> : kK^2] > 2$ and $[K : K_>] > 2$, and these degrees are powers of 2, so both degrees are multiples of 4 and $[K : K^2] \geqslant [K : kK^2] \geqslant 16$.

It remains to show that $[k : k^2] = [K : K^2]$, or more generally that if E is any field of characteristic $p > 0$ then $[E' : E'^p] = [E : E^p]$ for any finite extension E' of E. If E'/E is separable then $E \otimes_{E^p} E'^p$ is a field and $E \otimes_{E^p} E'^p \to E'$ is a map of fields that is both separable and purely inseparable (hence an equality). Thus, the general case reduces to $E' = E(\alpha)$ where $\alpha^p = a \in E - E^p$. This special case amounts to an elementary computation given in the proof of [Mat, Thm. 26.10] since $[E : E^p] = \dim_E \Omega^1_{E/E^p}$ (by the relationship between p-bases and differential bases [Mat, Thm. 26.5]) and $\Omega^1_{E/E^p} = \Omega^1_{E/\mathbf{F}_p}$.

Now consider G as above *without* a pseudo-split hypothesis, and assume $(G_{k_s})_c$ is non-standard for all c and that G has root field k. The next two lemmas show how to construct G in terms of a fiber product analogous to the generalized basic exotic construction in Proposition 8.2.2, after which we will show that such fiber products account for all G in a canonical manner.

Let K/k be the minimal field of definition for $\mathscr{R}_u(G_{\overline{k}}) \subset G_{\overline{k}}$, and let $K_>/k$ be the subextension that is the k_s/k-descent of the analogous subextension of K_s/k_s associated to G_{k_s} in Theorem 3.4.1(iii). Let $F_>, F_<$ denote the long and short root fields of G (as in Definition 3.3.7), so $kK^2 \subset F_> \subset K_> \subset F_<$ by Galois descent of (3.3.8) applied to G_{k_s}. The root field of G is k by hypothesis, but $F_>$ is equal to the root field of G by (3.3.2), so $F_> = k$ and hence $K^2 \subset k$.

Lemma 8.3.2. *Let* $\mathscr{G} = G^{\mathrm{prmt}}_{F_<}$.

(i) *The $F_<$-group $\mathscr{G}$ is type-*B *generalized basic exotic of rank* 2 *but not basic exotic, and the natural map $G \to \mathrm{R}_{F_</k}(\mathscr{G})$ has trivial kernel.*

(ii) *Let $\pi : \mathscr{G} \to \overline{\mathscr{G}}$ be the very special quotient as in Definition* 8.1.5 *(so $\overline{\mathscr{G}}$ is semisimple and simply connected of type* $\mathrm{C}_2 = \mathrm{B}_2$*), and define* $f =$

$\mathrm{R}_{F_</k}(\pi)$ *and* $\overline{G} := f(G)$. *The* k*-group* $\overline{G}$ *is type-*B *generalized basic exotic of rank* 2 *but not basic exotic, and the natural map* $\overline{G}_{F_<} \to \overline{\mathscr{G}}$ *is an* $F_<$*-descent of the maximal geometric reductive quotient of* $\overline{G}$.

(iii) *If* G *is pseudo-split and is classified by the data* $(K/k, V, V_>)$ *in Theorem* 3.4.1(iii) *then in Proposition* 8.2.5 *the* $F_<$*-group* $\mathscr{G}$ *is classified by* $(K/F_<, V)$ *and the* k*-group* $\overline{G}$ *is classified by* $(K_>/k, V_>)$.

Before we prove this lemma, for the convenience of the reader we display the groups and maps in the form of a commutative diagram

$$\begin{array}{ccccc} G & \longrightarrow & & & \mathrm{R}_{F_</k}(\mathscr{G}) \\ \downarrow & & & & \downarrow f \\ \overline{G} & \xrightarrow[i_{\overline{G}}]{} & \mathrm{R}_{K_>/k}(\overline{G}^{\mathrm{ss}}_{K_>}) & \longrightarrow & \mathrm{R}_{F_</k}(\overline{\mathscr{G}}) \end{array} \tag{8.3.2}$$

in which the top map is an inclusion by (i) and the lower-right horizontal map is induced by the k-inclusion $K_> \subset F_<$ and the identification $\overline{G}^{\mathrm{ss}}_{F_<} \simeq \overline{\mathscr{G}}$ arising from (ii).

Proof. For the proofs of (i) and (ii) it is harmless to extend scalars to k_s, so for the proof of the entire lemma we may assume G has a split maximal k-torus T and associated linear algebra data $(V, V_>)$ as in Theorem 3.4.1(iii). Let $\Phi = \Phi(G, T)$. The maximal geometric reductive quotient of G coincides with that of $\mathsf{G} := G^{\mathrm{pred}}_{F_<}$ and hence with that $\mathscr{G}$, so $\mathscr{G}^{\mathrm{ss}}_{\overline{k}}$ is simply connected of type $\mathrm{B}_2 = \mathrm{C}_2$. Identify $\mathsf{T} := T_{F_<}$ with a maximal $F_<$-torus of G, so $\Phi(\mathsf{G}, \mathsf{T}) = \Phi$.

For every $c \in \Phi$, the quotient map $G_{F_<} \twoheadrightarrow \mathscr{G}$ restricts to a quotient map $\Pi_c : (G_c)_{F_<} \twoheadrightarrow \mathscr{G}_c$ whose target is pseudo-reductive of minimal type (see Lemma 2.3.10). We claim that Π_c identifies $\mathscr{G}_c$ with $(G_c)^{\mathrm{prmt}}_{F_<}$ as quotients of $(G_c)_{F_<}$. Since the geometric unipotent radical of G_c is defined over K, by Remark 2.3.12 it suffices to identify $\mathscr{G}_c$ (as a quotient of $(G_c)_{F_<}$) with the image of the natural map $j_c : (G_c)_{F_<} \to \mathrm{R}_{K/F_<}((G_c)^{\mathrm{ss}}_K)$. But $(G_c)^{\mathrm{ss}}_K$ is naturally a K-subgroup of G^{ss}_K by 3.2.2, so the image of j_c is identified with that of the composite map

$$(G_c)_{F_<} \hookrightarrow G_{F_<} \to \mathrm{R}_{K/F_<}(G^{\mathrm{ss}}_K).$$

The second map in this composition has image $\mathscr{G}$ by Remark 2.3.12, and the resulting map $(G_c)_{F_<} \to \mathscr{G}$ clearly has image $\mathscr{G}_c$, as required.

For long roots $a \in \Phi$, since $K_> \subset F_<$ over k we conclude that

$$\mathscr{G}_a \simeq (H_{V_>, K_>/k})^{\mathrm{prmt}}_{F_<} \simeq \mathrm{SL}_2$$

because $(H_{V_>,K_>/k})^{\rm pred}_{K_>} \simeq \mathrm{SL}_2$. Likewise, for short roots $b \in \Phi$, since $F_< \subset K$ over k and V is an $F_<$-subspace of K we have

$$\mathscr{G}_b \simeq (H_{V,K/k})^{\rm prmt}_{F_<} \simeq H_{V,K/F_<}$$

via (3.1.6) and (2.3.13). Thus, by Theorem 3.4.1(iii), the $F_<$-group $\mathscr{G}$ is classified by the data $(K/F_<, V, F_<)$, so it is a type-B generalized basic exotic $F_<$-group of rank 2 and is not basic exotic since $V \neq K$ (as G_c is non-standard for short c). This settles the first assertions in (i) and (iii).

To establish the second assertion in (i) we note that the quotient map $G_K \twoheadrightarrow G_K^{\rm ss}$ factors through the quotient $\mathscr{G}_K$ of $G_K = (G_{F_<}) \otimes_{F_<} K$ since $G_K^{\rm ss}$ has no nontrvial central unipotent K-subgroup scheme. Thus, we can factor i_G through the map of interest $G \to \mathrm{R}_{F_</k}(\mathscr{G})$, so this latter map has trivial kernel since $\ker i_G = 1$ (by Proposition 2.3.4, since G is of minimal type).

Turning to (ii), for a Levi k-subgroup $L \subset G$ containing T (which exists by [CGP, Thm. 3.4.6]) clearly $L_{F_<}$ is a Levi $F_<$-subgroup of $\mathscr{G}$ containing $T_{F_<}$. The restriction $\pi : L_{F_<} \to \overline{\mathscr{G}}$ is the very special isogeny for $L_{F_<}$ (by Definition 8.1.5 and Proposition 7.1.5), so it identifies $\overline{\mathscr{G}}$ with $\overline{L}_{F_<}$ for the very special quotient $\overline{L}$ of L. Hence, $f(L) = \overline{L}$; this is a Levi k-subgroup of $\mathrm{R}_{F_</k}(\overline{L}_{F_<}) = \mathrm{R}_{F_</k}(\overline{\mathscr{G}})$. It follows from Proposition 2.1.2(i) that $\overline{G}$ is pseudo-semisimple with $\overline{L}$ as a Levi k-subgroup and with the same root system as $\overline{L}$ relative to the split maximal k-torus $\overline{T} := f(T)$. Since $\overline{L}$ is simultaneously a Levi k-subgroup of $\mathrm{R}_{F_</k}(\overline{\mathscr{G}})$ and $\overline{G}$, we have established the final assertion in (ii).

Consider the split maximal $F_<$-torus $\overline{T}_{F_<}$ in $\overline{\mathscr{G}}$ and associated root system $\overline{\Phi} = \Phi(\overline{L}, \overline{T}) = \Phi(\overline{G}, \overline{T})$. By [CGP, Prop. 7.1.5] applied to the very special isogeny $L \to \overline{L}$, the isogeny $T \to \overline{T}$ identifies each long $c \in \Phi$ with a short $\overline{c} \in \overline{\Phi}$ and identifies each short $c \in \Phi$ with $\overline{c}/2$ for a long $\overline{c} \in \overline{\Phi}$. Moreover, the definition of π implies that the restriction $\pi_c : \mathscr{G}_c \twoheadrightarrow \overline{\mathscr{G}}_{\overline{c}}$ is an $F_<$-isomorphism for long c whereas if c is short then π_c corresponds to the natural map $H_{V,K/F_<} \to \mathrm{SL}_2$ induced by $\mathrm{R}_{K/F_<}$ applied to the Frobenius isogeny of $(L_c)_K \simeq \mathrm{SL}_2$. Thus, $f_c : G_c \twoheadrightarrow \overline{G}_{\overline{c}}$ (induced by $\mathrm{R}_{F_</k}(\pi_c)$) is an isomorphism if c is long, but if c is short then f_c has 1-dimensional image on the $\pm c$-root groups since even on

$$\mathrm{R}_{F_</k}(\mathrm{R}_{K/F_<}(\mathrm{SL}_2)) = \mathrm{R}_{K/k}(\mathrm{SL}_2)$$

the effect of the Frobenius isogeny of the K-group SL_2 carries each root group of $\mathrm{R}_{K/k}(\mathrm{SL}_2)$ onto a 1-dimensional image (as $K^2 \subset k$).

By our description of $\overline{G}_{\overline{c}}$ for roots $\overline{c} \in \overline{\Phi}$, Theorem 3.4.1(iii) now classifies

$\overline{G}$ via the data $(K_>/k, V_>, k)$, so $\overline{G}$ is a type-B generalized basic exotic k-group of rank 2 and it is not basic exotic since $V_> \neq K_>$ (as G_c is non-standard for long c). Parts (ii) and (iii) are now proved. □

Lemma 8.3.3. *The inclusion $G \subset f^{-1}(\overline{G})$ induced by* (8.3.2) *is an equality.*

Proof. We can replace k with k_s so that $k = k_s$. We first check that $f^{-1}(\overline{G})$ is smooth by computing where it meets an open cell of $\mathrm{R}_{F_</k}(\mathscr{G})$. Let T be a maximal k-torus in G, and define $\overline{T} = f(T)$. Note that $\Phi := \Phi(G,T) = \Phi(\mathrm{R}_{F_</k}(\mathscr{G}), T) = \Phi(\mathscr{G}, T_{F_<})$, so upon fixing a positive system of roots $\Phi^+ \subset \Phi$ and using the corresponding positive system of roots for the common root system $\overline{\Phi}$ of $(\overline{G}, \overline{T})$, $(\mathrm{R}_{F_</k}(\overline{\mathscr{G}}), \overline{T})$, and $(\overline{\mathscr{G}}, \overline{T}_{F_<})$ we get open cells $\Omega \subset \mathscr{G}$ and $\overline{\Omega} \subset \overline{\mathscr{G}}$ such that $G \cap \mathrm{R}_{F_</k}(\Omega)$ and $\overline{G} \cap \mathrm{R}_{F_</k}(\overline{\Omega})$ are the corresponding open cells of G and $\overline{G}$ (by [CGP, Prop. 2.1.8(3)]). Let $\Phi_>, \Phi_<$ be the respective subsets of long and short roots in Φ, and define the subsets $\overline{\Phi}_>$ and $\overline{\Phi}_<$ of $\overline{\Phi}$ similarly.

The restriction $f : \mathrm{R}_{F_</k}(\Omega) \to \mathrm{R}_{F_</k}(\overline{\Omega})$ is a direct product of copies of $\mathrm{R}_{F_</k}$ applied to maps between root groups and between coroot subgroups of Cartan subgroups of $\mathscr{G}_c$ and $\overline{\mathscr{G}}_{\overline{c}}$ (with $\overline{c}$ short precisely when c is long). By Lemma 8.3.2(iii) for $\mathscr{G}$, the maps between corresponding direct factors for long c arise from the identity map of $\mathrm{R}_{F_</k}(\mathrm{SL}_2)$ and such maps for short c arise from $\mathrm{R}_{F_</k}(\varphi) : \mathrm{R}_{F_</k}(H_{V_<,K/F_<}) \to \mathrm{R}_{F_</k}(\mathrm{SL}_2)$ where $\varphi : H_{V_<,K/F_<} \to \mathrm{SL}_2$ is the "Frobenius" over $F_<$.

The description of $\mathrm{R}_{F_</k}(\Omega)$ as a direct product yields a description of the open cell $G \cap \mathrm{R}_{F_</k}(\Omega)$ of G as well as of the open cell $\overline{G} \cap \mathrm{R}_{F_</k}(\overline{\Omega})$ of $\overline{G}$ and the map induced between them by f. By computing separately with root groups for long roots and short roots in Φ as well as for direct factors of Cartan subgroups in accordance with Proposition 4.3.3, we see (using that $K^2 \subset k$ when working with short roots in Φ) that $f^{-1}(\overline{G}) \cap \mathrm{R}_{F_</k}(\Omega) = G \cap \mathrm{R}_{F_</k}(\Omega)$. This establishes the smoothness of $f^{-1}(\overline{G})$ and that $G = f^{-1}(\overline{G})^0$, so G is *normal* in $f^{-1}(\overline{G})$.

It remains to show $f^{-1}(\overline{G})$ is connected, and it suffices to check that every k-point $\gamma \in f^{-1}(\overline{G})$ lies in G. By $G(k)$-conjugacy of maximal k-tori of G and the equality $N_G(T)(k)/Z_G(T)(k) = W(\Phi)$, we may replace γ with a $G(k)$-translate so that conjugation by γ stabilizes T and preserves Φ^+. Since the diagram of Φ has no nontrivial automorphisms, the effect of γ-conjugation on G fixes Δ pointwise. Hence, γ centralizes T. But the map

$$\mathrm{R}_{F_</k}(T_{F_<}) \to \mathrm{R}_{F_</k}(\overline{T}_{F_<})$$

induced by $f = \mathrm{R}_{F_</k}(\pi)$ is identified with the map

$$V^*_{K/k} \times \mathrm{R}_{F_</k}(\mathrm{GL}_1) \to \mathrm{R}_{F_</k}(\mathrm{GL}_1) \times \mathrm{R}_{F_</k}(\mathrm{GL}_1)$$

given by squaring between the first factors and the identity between the second factors. Inside the target, $Z_{\overline{G}}(\overline{T})$ is compatibly identified with $\mathrm{GL}_1 \times (V_>)^*_{K_>/k}$ by Lemma 8.3.2(iii) for $\overline{G}$ (recall $k\langle V_> \rangle = K_> \subset F_<$). But $K^2 \subset k$, so $\gamma \in V^*_{K/k} \times (V_>)^*_{K_>/k} = Z_G(T)$. □

8.3.4. The two preceding lemmas motivate the following construction. Let $\overline{G}$ be a type-B generalized basic exotic k-group of rank 2 that is *not* basic exotic, with $K_>/k$ the minimal field of definition for $\mathscr{R}_u(\overline{G}_{\overline{k}}) \subset \overline{G}_{\overline{k}}$ (so $K_>^2 \subset k$ and $[K_> : k] \geqslant 4$). Let $F'/K_>$ be a finite-degree subfield of $k^{1/2}$ and $\mathscr{G}$ a type-B generalized basic exotic F'-group of rank 2 with short root field F' such that $\mathscr{G}$ is *not* basic exotic and the minimal field of definition K/F' for $\mathscr{R}_u(\mathscr{G}_{\overline{F'}}) \subset \mathscr{G}_{\overline{F'}}$ satisfies $K^2 \subset k$. Note that $[K : F'] \geqslant 4$. Let $\pi : \mathscr{G} \to \overline{\mathscr{G}}$ be the (semisimple) very special quotient over F' as in Definition 8.1.5, and define $f = \mathrm{R}_{F'/k}(\pi)$.

Proposition 8.3.5. *Let there be given an F'-isomorphism $\theta : \overline{G}^{\mathrm{ss}}_{F'} \simeq \overline{\mathscr{G}}$, and assume that the inclusion $j : \overline{G} \stackrel{i_{\overline{G}}}{\hookrightarrow} \mathrm{R}_{K_>/k}(\overline{G}^{\mathrm{ss}}_{K_>}) \hookrightarrow \mathrm{R}_{F'/k}(\overline{\mathscr{G}})$ defined via θ lands inside* $\mathrm{im}(f)$. *The fiber product*

$$\begin{array}{ccc} G & \xrightarrow{\iota} & \mathrm{R}_{F'/k}(\mathscr{G}) \\ \downarrow & & \downarrow f \\ \overline{G} & \xrightarrow[j]{} & \mathrm{R}_{F'/k}(\overline{\mathscr{G}}) \end{array}$$

(with $\ker j = 1$*) is pseudo-reductive and satisfies the following properties:*

(i) *G is absolutely pseudo-simple with root system* $\mathrm{B}_2 = \mathrm{C}_2$,
(ii) *G has short root field F' and $G^{\mathrm{ss}}_{\overline{k}}$ is simply connected,*
(iii) *ι corresponds to an isomorphism $G^{\mathrm{prmt}}_{F'} \simeq \mathscr{G}$ (so G is of minimal type since* $\ker \iota = 1$*),*
(iv) *$(G_{k_s})_c$ is non-standard for all roots c relative to a maximal k-torus of G.*

In particular, since $\overline{G} = (f \circ \iota)(G)$, the 4-tuple $(\overline{G}, F'/K_>, \mathscr{G}, \theta)$ is uniquely functorial with respect to k-isomorphisms in G.

Proof. We may and do assume without loss of generality that $k = k_s$. The main work is to prove that G is smooth. To prove such smoothness we will build a basic exotic k-group H of type B_2 inside G, so let us first construct such an H.

Since $k = k_s$, by [CGP, Thm. 3.4.6] there exists a Levi F'-subgroup $\mathscr{L} \subset \mathscr{G}$ (so $\mathscr{L} \simeq \mathrm{Spin}_5$ and the natural map $\mathscr{L}_K \to \mathscr{G}_K^{\mathrm{red}}$ is an isomorphism). A (split) maximal F'-torus $\mathscr{T}$ of $\mathscr{L}$ is also one for $\mathscr{G}$, and $\Phi(\mathscr{L}, \mathscr{T}) = \Phi(\mathscr{G}, \mathscr{T})$ since both coincide with $\Phi := \Phi(\mathscr{G}_K^{\mathrm{red}}, \mathscr{T}_K)$. Hence, by consideration of open cells we see that the schematic center $Z_{\mathscr{L}}$ is contained in the center of $\mathscr{G}$, so the containment $\mathscr{L} \cap Z_{\mathscr{G}} \subset Z_{\mathscr{L}}$ is an equality. In particular, $\mathscr{L}/Z_{\mathscr{L}}$ is identified with a Levi F'-subgroup of the type-B adjoint generalized basic exotic k-group $\mathscr{G}/Z_{\mathscr{G}}$, so (via Proposition 7.1.5) the restriction of the very special quotient map $\pi : \mathscr{G} \to \overline{\mathscr{G}}$ to $\mathscr{L}$ is the very special isogeny $\pi_{\mathscr{L}}$ for $\mathscr{L}$. Let $f_{\mathscr{L}} = \mathrm{R}_{F'/k}(\pi_{\mathscr{L}})$.

Let V be a nonzero proper F'-subspace of K that classifies $\mathscr{G}$ via the "type-B" case of Proposition 8.2.5 (so $F'\langle V \rangle = K$). For $b \in \Phi_<$ the natural map $\pi_b : \mathscr{G}_b \to \overline{\mathscr{G}}_{\overline{b}}$ (with $\overline{b} = 2b \in \overline{\Phi}_>$) is identified with the natural "Frobenius" quotient map $H_{V,K/F'} \to \mathrm{SL}_2$ (recall $K^2 \subset k$), whereas for $a \in \Phi_>$ the natural map $\pi_a : \mathscr{G}_a \to \overline{\mathscr{G}}_{\overline{a}}$ (with $\overline{a} = a \in \overline{\Phi}_<$) is an isomorphism (with source and target isomorphic to SL_2). Hence, by treating long and short roots in Φ separately and using that $K^2 \subset k$ when considering short roots, we see that the images $f(\mathrm{R}_{F'/k}(\mathscr{G}_c))$ and $f(\mathrm{R}_{F'/k}(\mathscr{L}_c))$ coincide for all $c \in \Phi$. The Cartan F'-subgroup $Z_{\mathscr{G}}(\mathscr{T})$ has a natural product decomposition as in the "type-B" case of Proposition 8.2.5, so $\mathrm{R}_{F'/k}(\mathscr{G})$ is generated by the groups $\mathrm{R}_{F'/k}(\mathscr{G}_c)$ for $c \in \Phi$. This proves that $f(\mathrm{R}_{F'/k}(\mathscr{G})) = f_{\mathscr{L}}(\mathrm{R}_{F'/k}(\mathscr{L}))$, hence $\overline{G} \subset f_{\mathscr{L}}(\mathrm{R}_{F'/k}(\mathscr{L}))$.

If $\overline{L}$ denotes a Levi k-subgroup of $\overline{G}$ (so it is a Levi k-subgroup of $\mathrm{R}_{F'/k}(\overline{\mathscr{G}})$, as $\overline{G}_{F'}^{\mathrm{ss}} \to \overline{\mathscr{G}}$ is an isomorphism) then the triple $(\mathscr{L}, F'/k, \overline{L})$ is data as required in the construction of a basic exotic k-group of type B_2. Thus, the scheme-theoretic preimage

$$H := f_{\mathscr{L}}^{-1}(\overline{L}) \subset G$$

is a basic exotic k-group of type B_2. In particular, this preimage is smooth (even absolutely pseudo-simple) over k and $f(H) = \overline{L}$.

Let $T \subset H$ be a (split) maximal k-torus, so for rank reasons T is maximal in $\mathrm{R}_{F'/k}(\mathscr{G})$ and hence $T_{F'}$ is a maximal F'-torus in $\mathscr{G}$. Likewise the image $\overline{T} = f(T)$ is a maximal k-torus in $f(H) = \overline{L}$ and so also in $\overline{G}$. Clearly $\overline{T}_{F'}$ is identified with a maximal F'-torus of $\overline{\mathscr{G}}$. Let Δ be a basis of $\Phi(H,T) = \Phi(\mathscr{L}, T_{F'}) = \Phi(\mathscr{G}, T_{F'}) = \Phi(\mathrm{R}_{F'/k}(\mathscr{G}), T)$. The basis Δ defines a positive system of roots Φ^+ for the common root system Φ of $(\mathrm{R}_{F'/k}(\mathscr{G}), T)$ and $(\mathscr{G}, T_{F'})$. Let $\overline{\Phi}^+$ be the corresponding positive system of roots for $(\overline{G}, \overline{T})$, $(\mathrm{R}_{F'/k}(\overline{\mathscr{G}}), \overline{T})$, and $(\overline{G}, \overline{T}_{F'})$. For the corresponding open cells $\Omega \subset \mathscr{G}$ and

$\overline{\Omega} = \pi(\Omega) \subset \overline{\mathscr{G}} = \overline{G}_{F'}$, clearly $\mathrm{R}_{F'/k}(\Omega)$ and $\mathrm{R}_{F'/k}(\overline{\Omega})$ are respective open cells of $\mathrm{R}_{F'/k}(\mathscr{G})$ and $\mathrm{R}_{F'/k}(\overline{\mathscr{G}})$, and

$$\overline{G} \cap \mathrm{R}_{F'/k}(\overline{\Omega})$$

is the associated open cell of $\overline{G}$. The smoothness of $G := f^{-1}(\overline{G})$ is equivalent to the smoothness of

$$X = f^{-1}(\overline{G} \cap \mathrm{R}_{F'/k}(\overline{\Omega})) \cap \mathrm{R}_{F'/k}(\Omega).$$

The compatible product decompositions of all of these open cells (including product decompositions of Cartan factors via Proposition 8.2.5) decomposes X into a direct product of analogous constructions indexed by the elements of Φ and $\Delta^\vee$. Thus, the smoothness of X can be studied separately on each factor of the direct product. For roots in $\Phi_>$ and the coroot $a^\vee$ for the long root $a \in \Delta$, the corresponding factors are identified with those of $\overline{G}_{\overline{a}} \simeq H_{V_>,K_>/k}$ since $\mathscr{G}_a \to \overline{\mathscr{G}}_{\overline{a}}$ is an isomorphism. For roots in $\Phi_<$ and the coroot $b^\vee$ for the short root $b \in \Delta$, the corresponding factors are identified with those of $\mathrm{R}_{F'/k}(\mathscr{G}_b) \simeq H_{V,K/k}$ since the inclusion $\overline{L}_b \subset \pi_b(\mathrm{R}_{F'/k}(\mathscr{G}_b))$ is an equality for dimension reasons (as $K^2 \subset k$). This completes the proof that G is smooth.

Next we prove that the smooth connected G^0 is pseudo-reductive. Note that $\mathscr{R}_{u,k}(G^0) \subset \ker f$ since $G^0 \to \overline{G}$ is surjective with pseudo-reductive target. But $\ker f = \mathrm{R}_{F'/k}(\ker \pi)$, so it suffices to show that $\ker \pi$ has no nontrivial F'-points. That $(\ker \pi)/Z_{\mathscr{G}}$ does not have nontrivial F'-points follows from Proposition 8.1.8(i), and $Z_{\mathscr{G}}(F')$ is trivial since $Z_{\mathscr{G}} \subset \mathrm{R}_{K/F'}(Z_{\mathscr{G}_K^{\mathrm{ss}}}) = \mathrm{R}_{K/F'}(\mu_2)$. Rank considerations and pseudo-semisimplicity of $\overline{G} = f(G) = f(G^0)$ imply that the maximal tori of G lie inside $\mathscr{D}(G^0)$.

Since $(\overline{G}, F'/K_>, \mathscr{G})$ and $\theta : \overline{G}_{F'}^{\mathrm{ss}} \simeq \overline{\mathscr{G}}$ are the data arising from applying Lemma 8.3.2 to $\mathscr{D}(G^0)$, via the compatible open cells considered above we see that $\mathscr{D}(G^0)$ satisfies all of the desired properties (i)–(iv) for G (using the inclusion $i_{\mathscr{G}} : \mathscr{G} \hookrightarrow \mathrm{R}_{K/F'}(G')$ and $E = K$ in Remark 2.3.12 for (iii)). Hence, Lemma 8.3.3 gives that $\mathscr{D}(G^0) = f^{-1}(\overline{G})$, so the inclusion $\mathscr{D}(G^0) \subset G$ inside $f^{-1}(\overline{G})$ is an equality. □

Definition 8.3.6. A k-group G is (rank-2) *basic exceptional* if it arises as in Proposition 8.3.5 (so G is absolutely pseudo-simple of rank 2) with root field k.

By Remark 8.3.1 over k_s, such k-groups exist if and only if $[k : k^2] \geqslant 16$. The following analogue of Proposition 8.2.5 is now immediate from Theorem 3.4.1(iii), Proposition 4.3.3, and Lemmas 8.3.2 and 8.3.3.

Corollary 8.3.7. *Let k be imperfect with* $\mathrm{char}(k) = 2$ *such that* $[k : k^2] \geqslant 16$, *and let $K/K_>/k$ be finite subextensions of $k^{1/2}/k$ with $[K : K_>] \geqslant 4$, $[K_> : k] \geqslant 4$.*

A pseudo-split basic exceptional k-group G with associated invariants K and $K_>$ is classified up to isomorphism by a pair $(V_<, V_>)$ consisting of a $K^\times$-homothety class of nonzero proper $K_>$-subspaces $V_< \subset K$ such that $k\langle V_<\rangle = K$ and a $K_>^\times$-homothety class of nonzero proper k-subspaces $V_> \subset K_>$ satisfying $k\langle V_>\rangle = K_>$ and $\{\lambda \in K_> \mid \lambda \cdot V_> \subset V_>\} = k$, and all such pairs arise. For a split maximal k-torus T in such a G, we have $Z_G(T) \simeq (V_>)^*_{K_>/k} \times (V_<)^*_{K/k}$.

We finish our discussion of the rank-2 basic exceptional construction with an analogue of Proposition 8.2.6:

Proposition 8.3.8. *Let G be a basic exceptional k-group with associated 4-tuple $(\overline{G}, F_</K_>, \mathscr{G}, \theta)$. Let $\pi : \mathscr{G} \to \overline{\mathscr{G}}$ be the very special quotient over $F_<$, and let $f = \mathrm{R}_{F_</k}(\pi)$. For any pseudo-parabolic k-subgroup $P \subset G$, define*

$$\overline{P} = f(P) \subset \overline{G},\ \ P' = \mathrm{im}(P_{F_<} \to \mathscr{G})$$

(so $\overline{P}' := \pi(P')$ is the image of $\overline{P}_{F_<} \hookrightarrow \overline{G}_{F_<} \twoheadrightarrow \overline{G}^{\mathrm{ss}}_{F_<} \overset{\theta}{\simeq} \overline{\mathscr{G}}$).

Assertions (i), (ii), (iii) *of Proposition* 8.2.6 *hold in the present situation upon replacing $(G', \overline{G}', K)$ with $(\mathscr{G}, \overline{\mathscr{G}}, F_<)$.*

Proof. We may carry over the proof of Proposition 8.2.6 essentially without change except that we work with the Cartesian diagram as in Proposition 8.3.5 and now $\mathscr{G}$ is not reductive (so the Bruhat decomposition for $(G'(K), P'(K))$ in the proof of Proposition 8.2.6 is replaced with the pseudo-reductive Bruhat decomposition for $(\mathscr{G}(F_<), P'(F_<))$ provided by [CGP, Thm. C.2.8], and we replace the K-isomorphism $G_K^{\mathrm{ss}} \simeq G'$ from the proof of Proposition 8.2.6 with the $F_<$-isomorphism $G_{F_<}^{\mathrm{prmt}} \simeq \mathscr{G}$). Proposition 8.1.8 provides the required properties of the very special quotient $\mathscr{G} \to \overline{\mathscr{G}}$ on root groups, akin to very special isogenies in the semisimple case. □

8.4 Generalized exotic groups

Over imperfect fields of characteristic 2, the generalized basic exotic groups of types B and C, as well as the rank-2 basic exceptional groups, underlie a construction beyond the standard case that is exhaustive under a locally minimal type hypothesis. To explain this, we need to introduce auxiliary Weil restrictions:

Definition 8.4.1. Let k be an imperfect field of characteristic $p \in \{2, 3\}$. A *generalized exotic* k-group is any $G \simeq \mathscr{D}(\mathrm{R}_{k'/k}(\mathscr{G}'))$ for a nonzero finite reduced

k-algebra k' and k'-group $\mathscr{G}'$ whose fiber over each factor field of k' is type-G_2 basic exotic when $p = 3$ or is either type-F_4 basic exotic, type-B or type-C generalized basic exotic, or rank-2 basic exceptional when $p = 2$.

Any such k-group is pseudo-semisimple and is clearly of minimal type with a reduced root system over k_s. By Proposition 3.4.4 it is also clear that such G are never standard. If $\{k'_i\}$ is the set of factor fields of k' and $\mathscr{G}'_i$ denotes the k'_i-fiber of $\mathscr{G}'$ then applying [CGP, Prop. A.4.8] over $\overline{k}$ gives that

$$G^{\mathrm{ss}}_{\overline{k}} \simeq \prod_i \prod_{\sigma_i} (\mathscr{G}'_i \otimes_{k'_i, \sigma_i} \overline{k})^{\mathrm{ss}} \tag{8.4.1}$$

where σ_i varies through the set of k-embeddings of k'_i into $\overline{k}$. In particular, $G^{\mathrm{ss}}_{\overline{k}}$ is simply connected.

Remark 8.4.2. By (8.4.1), G as in Definition 8.4.1 is absolutely pseudo-simple if and only if k' is a purely inseparable field extension of k (see [CGP, Lemma 3.1.2]). In such cases with $p = 2$ and absolutely pseudo-simple G with root system of type B or C, the next result implies that G is generalized basic exotic if and only if $k' = k$; for rank 1 (arising for type-B cases) this requires the root field condition that we included in Definition 8.1.1.

It is important that the data $(\mathscr{G}', k'/k)$ used in the construction of a generalized exotic k-group G is functorial with respect to isomorphisms in G:

Proposition 8.4.3. *Let G and H be generalized exotic k-groups arising from respective pairs $(\mathscr{G}', k'/k)$ and $(\mathscr{H}', \ell'/k)$. Any k-isomorphism $\sigma : G \simeq H$ has the form $\mathscr{D}(\mathrm{R}_{\alpha/k}(\varphi))$ for a unique pair (φ, α) consisting of a k-algebra isomorphism $\alpha : k' \simeq \ell'$ and a group isomorphism $\varphi : \mathscr{G}' \simeq \mathscr{H}'$ over α.*

Proof. In view of the uniqueness assertion, by Galois descent we may and do assume $k = k_s$. Writing $k' = \prod_{i \in I} k'_i$ for finite extensions k'_i/k (which must be purely inseparable) and $\mathscr{G}'_i$ for the k'_i-fiber of $\mathscr{G}'$, we have $G = \prod \mathscr{D}(\mathrm{R}_{k'_i/k}(\mathscr{G}'_i))$. Each of the factors in this direct product is absolutely pseudo-simple by Remark 8.4.2, so these factors are the minimal nontrivial smooth connected normal k-subgroups of G due to [CGP, Prop. 3.1.6]. Hence, if $\{\ell'_j, \mathscr{H}'_j\}_{j \in J}$ is the analogous such data for H, any isomorphism $G \simeq H$ arises from a bijection $\tau : I \simeq J$ and a collection of k-isomorphisms

$$\mathscr{D}(\mathrm{R}_{k'_i/k}(\mathscr{G}'_i)) \simeq \mathscr{D}(\mathrm{R}_{\ell'_{\tau(i)}/k}(\mathscr{H}'_{\tau(i)})).$$

In this way we are reduced to the special case where k' and ℓ' are fields.

By definition of generalized basic exotic groups, the root field of $\mathscr{G}'$ is k'. Thus, by Proposition 3.1.8 and (3.1.6), the (rank-1) minimal-type k-subgroup of G generated by a pair of opposite long root groups has root field equal to k'. In particular, the long root field of G is k'. Likewise, the long root field of H is ℓ'. Hence, the existence of a k-isomorphism $G \simeq H$ implies that there is a k-isomorphism $\alpha : k' \simeq \ell'$ between the long root fields (and α is unique since $k = k_s$). Now identifying ℓ' with k' via α, the inclusions $G \hookrightarrow \mathrm{R}_{k'/k}(\mathscr{G}')$ and $H \hookrightarrow \mathrm{R}_{k'/k}(\mathscr{H}')$ correspond to k'-homomorphisms $G_{k'} \to \mathscr{G}'$ and $H_{k'} \to \mathscr{H}'$ that are maximal among pseudo-reductive quotients of minimal type over k' by Proposition 2.3.13 (since k'/k is purely inseparable). Thus, $\sigma_{k'}$ induces a k'-isomorphism $\varphi : \mathscr{G}' \simeq \mathscr{H}'$, and it is clear that this is the unique φ which does the job. □

By Galois descent, we obtain:

Corollary 8.4.4. *A k-group G is generalized exotic if and only if G_{k_s} is generalized exotic.*

For an imperfect field k of characteristic $p \in \{2,3\}$, among all pseudo-semisimple k-groups of minimal type with a reduced root system over k_s there is a simple intrinsic characterization of the generalized exotic k-groups:

Theorem 8.4.5. *Let G be a nontrivial pseudo-semisimple k-group of minimal type with a reduced root system over k_s. Then G is generalized exotic if and only if G satisfies the following two properties:*

(i) *$G^{\mathrm{ss}}_{\overline{k}}$ is simply connected,*
(ii) *the minimal normal pseudo-simple k_s-subgroups of G_{k_s} are non-standard.*

Note that non-standardness of a pseudo-reductive k-group is insensitive to scalar extension to k_s [CGP, Cor. 5.2.3].

Proof. By Corollary 8.4.4, we can (and we will) assume that $k = k_s$. The implication "$\Rightarrow$" follows from the preceding discussion. For the converse, we may assume $k = k_s$ and first reduce to considering pseudo-simple G.

Let $\{G_j\}$ be the collection of minimal nontrivial perfect smooth connected normal k-subgroups of G, so the G_j's pairwise commute and multiplication defines a k-homomorphism $\pi : \prod G_j \to G$ that is surjective with central kernel [CGP, Prop. 3.1.8]. For a maximal k-torus T of G the associated k-subtori $T_j = T \cap G_j \subset G_j$ are maximal [CGP, Cor. A.2.7], and the map $\prod T_j \to T$ is an isogeny that identifies $\Phi(G,T)$ with $\coprod \Phi(G_j, T_j)$ inside $\mathrm{X}(T)_{\mathbf{Q}} = \prod \mathrm{X}(T_j)_{\mathbf{Q}}$ [CGP, Prop. 3.2.10]. Thus, each G_j has a reduced root system.

Since G is of minimal type, so are its normal k-subgroups G_j (Lemma 2.3.10). The central quotient π restricts to an isomorphism between corresponding root groups, and it restricts to an isomorphism between Cartan k-subgroups due to Proposition 4.3.3, so π restricts to an isomorphism between open cells and hence is an isomorphism (as for any birational homomorphism between smooth connected k-groups). In particular, each $(G_j)^{\mathrm{ss}}_{\overline{k}}$ is a direct factor of $G^{\mathrm{ss}}_{\overline{k}}$ and hence is simply connected. Consequently, it suffices to treat each G_j separately, so we may and do assume G is (absolutely) pseudo-simple over $k = k_s$.

The hypotheses on G say that it is non-standard of minimal type with a reduced root system of rank $n \geqslant 1$ and $G^{\mathrm{ss}}_{\overline{k}}$ is simply connected. If $F \subset K$ is the root field of G then Proposition 3.3.6 implies $G = \mathscr{D}(\mathrm{R}_{F/k}(G_F^{\mathrm{prmt}}))$ where G_F^{prmt} over F satisfies the analogous hypotheses as G does over k. Hence, we may replace G and k with G_F^{prmt} and F respectively so that G has root field k.

Let K/k denote the minimal field of definition over k for $\mathscr{R}_u(G_{\overline{k}}) \subset G_{\overline{k}}$, so $G' := G_K/\mathscr{R}_{u,K}(G_K)$ is a simply connected and absolutely simple connected semisimple K-group. Since G is of minimal type, by Proposition 2.3.4 the natural map

$$i_G : G \to \mathrm{R}_{K/k}(G')$$

has trivial kernel.

Since G is not standard, by Theorem 3.4.2 the reduced and irreducible root system Φ of rank $n \geqslant 1$ must be G_2 if $p = 3$ and either B_n $(n \geqslant 1)$, C_n $(n \geqslant 1)$, or F_4 if $p = 2$. If it is G_2 with $p = 3$ or F_4 with $p = 2$ then G is exotic by Proposition 3.4.3, so we may and do assume $p = 2$ and Φ is either B_n or C_n with $n \geqslant 1$. The rank-1 case is settled by Proposition 3.1.8, so we may assume $n \geqslant 2$. Hence, the possibilities for G up to isomorphism are classified by Theorem 3.4.1(iii) for types B and C with L simply connected (having root system Φ).

Consider type B_n with $n \geqslant 3$, so G with root field k is classified by the subfield $K_> = k$ containing kK^2 (so $K^2 \subset k$) and the $K^\times$-homothety class of a nonzero k-subspace $V \subset K$ satisfying $k\langle V\rangle = K$. Since $K^2 \subset k$ and $K \neq k$, the same field-theoretic and linear-algebraic data arises from a type-B generalized basic exotic k-group, so G must be that group by Proposition 8.2.5.

For type C_n with $n \geqslant 3$, G with root field k is classified by the subfield $K_> \subset K$ containing kK^2 and the $K_>^\times$-homothety class of a nonzero kK^2-subspace $V_> \subset K_>$ such that $\{\lambda \in K_> \mid \lambda \cdot V_> \subset V_>\} = k$ (so $K^2 \subset k$) and $k\langle V_>\rangle = K_>$. By Proposition 8.2.5, the invariants $(K/k, V_>)$ arise from a type-C generalized basic exotic k-group.

Finally, consider type $\mathrm{B}_2 = \mathrm{C}_2$, so G with root field k is classified up to k-isomorphism by a triple of invariants $(K_>/k, V, V_>)$ as in Theorem 3.4.1(iii) with $V_>$ satisfying $\{\lambda \in K_> \mid \lambda \cdot V_> \subset V_>\} = k$. In particular, since $V_>$ is a kK^2-subspace of $K_>$, necessarily $K^2 \subset k$. If $V = K$ then the type-C generalized basic exotic construction for $(K/k, V_>)$ recovers G. If $\dim_k(V_>) = 1$ then the type-B generalized basic exotic construction for $(K/k, V)$ recovers G. Thus, we may assume $V \neq K$ and $\dim_k(V_>) > 1$.

Let $F = \{\lambda \in K \mid \lambda \cdot V \subset V\}$. Since $F\langle V\rangle = K$, the hypothesis $V \neq K$ rules out the possibility $[K : F] \leqslant 2$, so $[K : F] \geqslant 4$ (since $K^2 \subset k \subset F$). Likewise, since $k\langle V_>\rangle = K_>$ with k the root field of $V_>$, the hypothesis $\dim_k(V_>) > 1$ ensures that $[K_> : k] \geqslant 4$ and $V_> \neq K_>$. Thus, G is as in Corollary 8.3.7. □

8.5 Structure of Z_G and $Z_{G,C}$

The structure of the center of a basic exotic k-group is given in [CGP, Cor. 7.2.5]. We now give an analogue in the generalized basic exotic and basic exceptional cases for types B and C (types G_2 and F_4 are settled basic exotic cases).

Let k be imperfect with char$(k) = 2$, let G be a generalized basic exotic or rank-2 basic exceptional k-group, and let $T \subset G$ be a maximal k-torus. Assume $\Phi := \Phi(G_{k_s}, T_{k_s})$ equals B_n or C_n with $n \geqslant 1$. Let K/k be the minimal field of definition of $\mathscr{R}_u(G_{\overline{k}}) \subset G_{\overline{k}}$ and let $G' := G_K^{\mathrm{ss}}$, so $K^2 \subset k$ and $i_G : G \to \mathrm{R}_{K/k}(G')$ has trivial kernel.

Proposition 8.5.1. *Using notation as above, $Z_G = G \cap \mathrm{R}_{K/k}(Z_{G'})$. In particular, $G/Z_G \subset \mathrm{R}_{K/k}(G'/Z_{G'})$. More explicitly:*

(i) *If $\Phi = C_n$ with even $n > 2$ then $Z_G = \mathrm{R}_{K/k}(Z_{G'})$.*
(ii) *Assume T is split and Φ equals B_n with $n \geqslant 1$ or C_n with odd $n > 2$. Choose $c \in \Phi_>$ for type C_n and $c \in \Phi_<$ for type B_n (including $n = 2$). For a nonzero k-subspace $V \subset K$ such that $G_c \simeq H_{V,K/k}$, the unique K-isomorphism $Z_{G'} \simeq \mu_2$ identifies Z_G with $V^*_{K/k} \cap \mathrm{R}_{K/k}(\mu_2) = V^*_{K/k}[2]$ inside $\mathrm{R}_{K/k}(\mathrm{GL}_1)$.*

Proof. We may assume $k = k_s$, so G is pseudo-split. Define $T' = T_K \subset G'$ and let Δ be a basis of $\Phi = \Phi(G, T) = \Phi(G', T')$. Since G is a k-subgroup of $\mathrm{R}_{K/k}(G')$, it is obvious that $G \cap \mathrm{R}_{K/k}(Z_{G'}) \subset Z_G$. To prove the reverse inclusion, note that the map $G_K \to G'$ associated to the inclusion $G \hookrightarrow \mathrm{R}_{K/k}(G')$ is a quotient map, so it carries $(Z_G)_K$ into $Z_{G'}$. Hence, $Z_G \subset \mathrm{R}_{K/k}(Z_{G'})$, so $Z_G \subset G \cap \mathrm{R}_{K/k}(Z_{G'})$.

Consider G with root system C_n for $n \geqslant 3$. Suppose n is *even*, so $Z_{G'}$ is contained in the direct product of the coroot groups associated to the short roots in Δ. For each $b \in \Phi_<$ we have $G_b = \mathrm{R}_{K/k}(G'_b)$ by comparing dimensions of root groups (since $G'_b \simeq \mathrm{SL}_2$). Thus, $\mathrm{R}_{K/k}(b^\vee(\mathrm{GL}_1)) \subset G_b$ for all $b \in \Phi_<$, so the description of $Z_{G'}$ using the direct product structure with respect to $\Delta_<$ (as n is even) implies that $\mathrm{R}_{K/k}(Z_{G'}) \subset G$ and hence $Z_G = \mathrm{R}_{K/k}(Z_{G'})$.

Assume Φ is of type C_n with *odd* $n \geqslant 3$. Let a be the unique long root in Δ, so the Dynkin diagram is:

Clearly the center $Z_G = G \cap \mathrm{R}_{K/k}(Z_{G'})$ is equal to $Z_G(T) \cap \mathrm{R}_{K/k}(Z_{G'})$. To compute this intersection, we shall describe $Z_G(T)$ inside $\mathrm{R}_{K/k}(T') = \prod_j \mathrm{R}_{K/k}((a_j)^\vee_K(\mathrm{GL}_1))$. By Proposition 8.2.5,

$$Z_G(T) = \mathrm{R}_{K/k}(a^\vee_K)(V^*_{K/k}) \times \prod_{b \in \Delta_<} \mathrm{R}_{K/k}(b^\vee_K(\mathrm{GL}_1))$$

and $Z_{G'}$ is μ_2 diagonally embedded into T' along the a_j-coroots for odd j. Since $a = a_n$ with n odd, the projection pr_a onto the a-factor of $Z_G(T)$ carries $Z_G = G \cap \mathrm{R}_{K/k}(Z_{G'})$ isomorphically onto $V^*_{K/k} \cap \mathrm{R}_{K/k}(\mu_2)$ (via the inverse of the isomorphism $\mathrm{R}_{K/k}(a^\vee_K) : \mathrm{R}_{K/k}(\mathrm{GL}_1) \simeq \mathrm{R}_{K/k}(T' \cap G'_a)$).

Now consider G with root system B_n ($n \geqslant 1$). For $n = 1$, there is an isomorphism $G \simeq H_{V,K/k}$ carrying T into the diagonal k-torus. It is obvious in this case that the isomorphism $Z_{G'} \simeq \mu_2$ identifies Z_G with $V^*_{K/k} \cap \mathrm{R}_{K/k}(\mu_2) = V^*_{K/k}[2]$, so let us assume $n \geqslant 2$. There is a unique short root b in Δ, and $\Delta^\vee$ is a $\mathbf{Z}$-basis of $\mathrm{X}_*(T') = \mathrm{X}_*(T)$ (since G' is simply connected). The center $Z_{G'}$ is the 2-torsion in the coroot group $b^\vee(\mathrm{GL}_1) \subset G'_b$. If $n \geqslant 3$ then $G_b \simeq H_{V,K/k}$ for a nonzero k-subspace V of K such that $k\langle V \rangle = K$, and $G_a \simeq \mathrm{SL}_2$ for $a \in \Delta_>$. By Proposition 8.2.5 we conclude that

$$Z_G(T) = Z_{G_b}(T_b) \times \prod_{a \in \Delta_>} a^\vee(\mathrm{GL}_1)$$

with $T_b := T \cap G_b$. Hence, if $n \geqslant 3$ then

$$Z_G(T) \cap \mathrm{R}_{K/k}(Z_{G'}) = Z_{G_b}(T_b)[2].$$

Since $\mathrm{R}_{K/k}(b^\vee_K)$ identifies $V^*_{K/k}$ with $Z_{G_b}(T \cap G_b)$, the desired description of Z_G is established for $n \geqslant 3$.

Finally, we address type $\mathrm{B}_2 (= \mathrm{C}_2)$, so $\Delta = \{a, b\}$ with a long and b short.

Applying Corollary 8.3.7, for some nonzero k-subspace $V_>$ of K we have

$$Z_G(T) = \mathrm{R}_{K/k}(a_K^\vee)((V_>)^*_{K/k}) \times \mathrm{R}_{K/k}(b_K^\vee)(V^*_{K/k}).$$

Thus, the k-group $Z_G = Z_G(T) \cap \mathrm{R}_{K/k}(Z_{G'})$ again corresponds to $V^*_{K/k}[2]$ via $\mathrm{R}_{K/k}(b_K^\vee)$. □

Propositions 6.2.2 and 6.2.4 addressed the existence of the affine finite type automorphism scheme $\mathrm{Aut}_{G/k}$ of a pseudo-semisimple group G over any field k, as well as the structure of $(\mathrm{Aut}^{\mathrm{sm}}_{G/k})^0$. By Proposition 6.3.4(i), $\mathrm{Aut}^{\mathrm{sm}}_{G/k}$ is connected if the Dynkin diagram of the root system of G_{k_s} does not admit nontrivial automorphisms. (See Example 6.3.9 and Proposition 6.3.10 for all absolutely pseudo-simple counterexamples to the converse.)

For a Cartan k-subgroup $C \subset G$, let $\mathrm{Aut}_{G,C} \subset \mathrm{Aut}_{G/k}$ be the closed k-subgroup scheme classifying automorphisms of G restricting to the identity on C, so by [CGP, Thm. 2.4.1] and Proposition 6.1.4 the maximal smooth closed k-subgroup $Z_{G,C}$ of $\mathrm{Aut}_{G,C}$ is a commutative pseudo-reductive group. By Proposition 6.2.4, $(\mathrm{Aut}^{\mathrm{sm}}_{G/k})^0 = (G \rtimes Z_{G,C})/C$ (so $\mathscr{D}((\mathrm{Aut}^{\mathrm{sm}}_{G/k})^0) = G/Z_G$) and moreover if the maximal k-torus of C is k-split then

$$(\mathrm{Aut}^{\mathrm{sm}}_{G/k})^0(k) = (G(k) \rtimes Z_{G,C}(k))/C(k);$$

the latter coincides with $\mathrm{Aut}_k(G)$ when $\mathrm{Aut}^{\mathrm{sm}}_{G/k}$ is connected.

Our interest in $Z_{G,C}$ is largely due its role in a generalization of the standard construction from §2.1 that we shall introduce in §9.1. As motivation, note that in the standard construction associated to a triple $(G', k'/k, T')$, the adjoint torus $\overline{T}' = T'/Z_{G'}$ intervenes in (2.1.2.1) due to its identification with $Z_{G',T'}$. In generalizations of the standard construction we will use G' with fibers over $\mathrm{Spec}(k')$ that are pseudo-simple and generalized exotic, and the role of $\overline{T}'$ in the standard case will be replaced by $Z_{G',C'}$ for Cartan k'-subgroups $C' \subset G'$.

Example 8.5.2. If $\mathrm{char}(k) = 2$ and $[k : k^2] = 2$ then *generalized standard* pseudo-reductive k-groups in [CGP, §10.1] rest on a diagram similar to (2.1.2.1) with the second map having the form $C \to \mathrm{R}_{k'/k}(Z_{G',C'})$ where $k' = \prod k'_i$ for fields k'_i finite over k and $G' \to \mathrm{Spec}(k')$ is a group scheme with absolutely pseudo-simple fibers that are either simply connected semisimple or basic exotic.

8.5.3. In view of Lemma 6.1.2(ii) and Proposition 6.1.7, to describe $Z_{G,C}$ in non-standard cases we focus on four classes of absolutely pseudo-simple groups G of minimal type over imperfect fields k of characteristic $p \in \{2, 3\}$: (i) basic exotic groups of type F_4 (with $p = 2$) or G_2 (with $p = 3$), (ii) generalized basic exotic groups of types B or C (with $p = 2$), (iii) rank-2 basic exceptional groups (with $p = 2$), and (iv) absolutely pseudo-simple groups of minimal type with a non-reduced root system over k_s (which only exist when $p = 2$). In case (iv) the scheme-theoretic center is trivial [CGP, Prop. 9.4.9] and $Z_{G,C}$ is explicitly described in [CGP, Prop. 9.8.15]. In rank-1 generalized basic exotic cases the structure of $Z_{G,C}$ is given by Lemma 6.1.3. Thus, we now consider higher-rank cases with a reduced root system (over k_s).

Let G be in cases (i), (ii), or (iii) above with rank $\geqslant 2$ and $p = \mathrm{char}(k) \in \{2,3\}$. Let K/k be the minimal field of definition for $\mathscr{R}_u(G_{\overline{k}}) \subset G_{\overline{k}}$ and define $G' = G_K/\mathscr{R}_{u,K}(G_K)$. Let $\pi : G' \to \overline{G}'$ be the very special isogeny and let f denote $\mathrm{R}_{K/k}(\pi) : \mathrm{R}_{K/k}(G') \to \mathrm{R}_{K/k}(\overline{G}')$. View G inside $\mathrm{R}_{K/k}(G')$ via the homomorphism i_G and let $\overline{G}$ denote $f(G)$.

Choose a Cartan k-subgroup $C \subset G$, and let $T \subset C$ be its unique maximal k-torus (so $C = Z_G(T)$). Let $\Phi = \Phi(G_{k_s}, T_{k_s})$. The minimal field of definition K/k for $\mathscr{R}_u(G_{\overline{k}}) \subset G_{\overline{k}}$ satisfies $K^p \subset k$, and the action of $Z_{G,C}$ on G induces an action of $(Z_{G,C})_K$ on G_K that must preserve $\mathscr{R}_{u,K}(G_K)$ (as may be checked on K_s-points since $(Z_{G,C})_K$ is K-smooth). Hence, this yields an action of $(Z_{G,C})_K$ on $G' = G_K/\mathscr{R}_{u,K}(G_K)$. The action is the identity on $T' = T_K \subset G'$, so it is classified by a K-homomorphism

$$(Z_{G,C})_K \to \mathrm{Aut}_{G',T'} = T'/Z_{G'} =: T'^{\mathrm{ad}}.$$

This map corresponds to a k-homomorphism

$$Z_{G,C} \to \mathrm{R}_{K/k}(T'^{\mathrm{ad}}) = \mathrm{R}_{K/k}(T_K^{\mathrm{ad}}), \tag{8.5.3}$$

where T^{ad} is the unique quotient of T that is a k-descent of the quotient T'^{ad} of T' (or equivalently, T^{ad} is the maximal k-torus of $\mathrm{R}_{K/k}(T'^{\mathrm{ad}})$). Considering (8.5.3) on k_s-points, a k_s-automorphism φ of G_{k_s} restricting to the identity on C_{k_s} is sent to $t' \in T'^{\mathrm{ad}}(K_s)$ that classifies the K_s-automorphism $\varphi'_{K_s} \in \mathrm{Aut}_{G',T'}(K_s)$ induced by φ_{K_s} on G_{K_s}.

We shall prove that (8.5.3) is a closed immersion and identify its image in terms of root fields. The formulation of this result requires notation, as follows. Let $\overline{T}^{\mathrm{ad}}$ be the unique maximal k-torus of $\mathrm{R}_{K/k}(\overline{T}'^{\mathrm{ad}})$ (where $\overline{T}'^{\mathrm{ad}} := \overline{T}'/Z_{\overline{G}'}$ with $\overline{T}' = \pi(T')$), and let $T^{\mathrm{ad}} \to \overline{T}^{\mathrm{ad}}$ be the restriction of $\mathrm{R}_{K/k}(\pi^{\mathrm{ad}})$, where $\pi^{\mathrm{ad}} : G'^{\mathrm{ad}} \to \overline{G}'^{\mathrm{ad}}$ is induced by the very special K-isogeny $\pi : G' \to \overline{G}'$. Equivalently, $T^{\mathrm{ad}} \to \overline{T}^{\mathrm{ad}}$ is the unique k-descent of $T'/Z_{G'} \to \overline{T}'/Z_{\overline{G}'}$.

Proposition 8.5.4. *For $(G, K/k, T, C)$ as above and the respective long and short root fields $F_>, F_< \subset K$ as in Definition* 3.3.7, *the map* (8.5.3) *is a closed immersion that lands inside $\mathrm{R}_{F_</k}(T^{\mathrm{ad}}_{F_<})$ and fits into a fiber product diagram*

$$\begin{array}{ccc} Z_{G,C} & \longrightarrow & \mathrm{R}_{F_</k}(T^{\mathrm{ad}}_{F_<}) \\ \downarrow & & \downarrow \\ \mathrm{R}_{F_>/k}(\overline{T}^{\mathrm{ad}}_{F_>}) & \longrightarrow & \mathrm{R}_{F_</k}(\overline{T}^{\mathrm{ad}}_{F_<}) \end{array}$$

where the bottom map is the natural inclusion (so $\mathrm{R}_{F_>/k}(T^{\mathrm{ad}}_{F_>}) \subset Z_{G,C}$).

If T is split and Δ is a basis of $\Phi = \Phi(G,T)$ with respective subsets of short and long roots denoted $\Delta_<$ and $\Delta_>$ then via the isomorphism $T'^{\mathrm{ad}} \simeq \mathrm{GL}_1^{\Delta}$ induced by t' mod $Z_{G'} \mapsto \prod_{c\in\Delta} c(t')$ we have

$$Z_{G,C} = \prod_{c\in\Delta_<} \mathrm{R}_{F_</k}(\mathrm{GL}_1) \times \prod_{c\in\Delta_>} \mathrm{R}_{F_>/k}(\mathrm{GL}_1) \qquad (8.5.4)$$

inside $\mathrm{R}_{K/k}(T'^{\mathrm{ad}})$.

By Theorem 3.3.8, for types F_4 and G_2 we have $F_< = K$ and $F_> = K_>$.

Proof. The closed immersion property in the basic exotic case (for types B, C, F_4, and G_2) is part of [CGP, Cor. 8.2.7], and the same argument works in general because (as in basic exotic cases) the inclusion $G \hookrightarrow \mathrm{R}_{K/k}(G')$ corresponds to the maximal reductive quotient map $G_K \twoheadrightarrow G'$. For the rest of the argument we may assume $k = k_s$, so T is split. By [CGP, Prop. 7.1.5], the isomorphism $\mathrm{X}(\overline{T}')_{\mathbf{Q}} \simeq \mathrm{X}(T')_{\mathbf{Q}}$ induced by the very special isogeny $(G', T') \to (\overline{G}', \overline{T}')$ identifies $\Phi(\overline{G}', \overline{T}')_<$ with $\Phi_>$ and identifies $\Phi(\overline{G}', \overline{T}')_>$ with $p\Phi_<$. Thus, since $K^p \subset k$ and $\mathrm{X}(T^{\mathrm{ad}})$ has Δ as a $\mathbf{Z}$-basis, the right side of (8.5.4) coincides with the proposed fiber product description of $Z_{G,C}$ and so it suffices to prove (8.5.4).

The natural action of T'^{ad} on G' preserves every root group via a scaling action through the isomorphism $T'^{\mathrm{ad}} \simeq \mathrm{GL}_1^{\Delta}$ and the expansion of elements of Φ as unique $\mathbf{Z}$-linear combinations of elements of Δ. Since $W(\Phi)$ is generated by reflections in elements of Δ and acts transitively on $\Phi_<$ and $\Phi_>$, the root groups $U_{\pm c}$ of (G,T) for $c \in \Delta$ generate G. Thus, since the action of $\mathrm{R}_{K/k}(T'^{\mathrm{ad}})$ on $\mathrm{R}_{K/k}(G')$ preserves all T-root groups for $\mathrm{R}_{K/k}(G')$, the smooth connected right side of (8.5.4) is precisely the k-subgroup scheme of points of $\mathrm{R}_{K/k}(T'^{\mathrm{ad}})$ whose action on $\mathrm{R}_{K/k}(G')$ preserves the k-subgroup G. It follows that the right side of (8.5.4) is the image of $Z_{G,C} \hookrightarrow \mathrm{R}_{F_</k}(T^{\mathrm{ad}}_{F_<})$. $\square$

9

Generalization of the standard construction

9.1 Generalized standard groups

The notion of "generalized standard" pseudo-reductive group over a field k is defined in [CGP, Def. 10.1.9] subject to the requirement $[k : k^2] \leqslant 2$ when $\mathrm{char}(k) = 2$, and in characteristic 2 such k-groups have a reduced root system over k_s by construction. (This coincides with the notion of standard pseudo-reductive group if $\mathrm{char}(k) \neq 2, 3$.) To remove the restriction on $[k : k^2]$ and permit non-reducedness of the root system when $\mathrm{char}(k) = 2$, the definition we shall give in Definition 9.1.7 will involve root fields (due to their role in Proposition 3.3.6, Definitions 8.2.3 and 8.3.6, and Proposition 9.1.3 below).

The notion of root field has only been defined for absolutely pseudo-simple G whose root system over k_s is reduced, so our first step is to remove that restriction. Recall that when G_{k_s} has a reduced root system, the long and short root fields $F_>$ and $F_<$ in Definition 3.3.7 satisfy $F_> \subset F_<$, and $F_>$ is equal to the root field of G. Thus, in the non-reduced case we will focus on the longest (equivalently, divisible) roots when defining the root field of G.

Let G be an absolutely pseudo-simple k-group of *minimal type* for which G_{k_s} has a non-reduced root system. Fix a maximal k-torus $T \subset G$, and let a be a divisible root in $\Phi(G_{k_s}, T_{k_s})$. Since a is not multipliable, the absolutely pseudo-simple k_s-subgroup $(G_{k_s})_a$ generated by the $\pm a$-root groups has as its roots only $\pm a$; i.e., its root system is A_1 (rather than BC_1). By Lemma 2.3.10, $(G_{k_s})_a$ is of minimal type since G_{k_s} is of minimal type. Since $G^{\mathrm{ss}}_{\overline{k}}$ is simply connected (see [CGP, Thm. 2.3.10]), we have $((G_{k_s})_a)^{\mathrm{ss}}_{\overline{k}} \simeq \mathrm{SL}_2$ by 3.2.2. Hence, $(G_{k_s})_a$ is given by the construction in Proposition 3.1.8 over k_s.

Example 9.1.1. Let K/k be a nontrivial purely inseparable finite extension. Pseudo-simple *pseudo-split* k-groups G of minimal type with root system BC_n $(n \geqslant 1)$ and with minimal field of definition K/k for $\mathscr{R}_u(G_{\overline{k}}) \subset G_{\overline{k}}$ are classified in [CGP, Thm. 9.8.6] in terms of linear algebra data $(V^{(2)}, V')$ when $n \neq 2$ and $(V^{(2)}, V', V'')$ when $n = 2$, with V' a nonzero proper kK^2-subspace of K. The description of G in terms of such data implies that for a split maximal k-torus $T \subset G$ and a divisible root $a \in \Phi(G,T)$, we have $G_a \simeq H_{V',K/k}$. Hence, the common root field F of such G_a's is a proper subfield of K containing kK^2. When $n = 2$, V'' is a nonzero F-subspace of K.

The linear algebra data classifying G makes sense relative to the extension K/F instead of K/k. Letting $G_{K/F}$ denote the analogous BC_n-construction made over F via the same linear algebra data, we obtain a non-canonical isomorphism from [CGP, Prop. 9.8.13] (using the pair (F,k) in place of (k,k_0)):

$$G \simeq \mathscr{D}(\mathrm{R}_{F/k}(G_{K/F})). \tag{9.1.1}$$

A canonical formulation of (9.1.1) will be provided in Proposition 9.1.3, but first we use Example 9.1.1 to make a definition beyond the pseudo-split case:

Definition 9.1.2. Let G be an absolutely pseudo-simple k-group of minimal type with a non-reduced root system over k_s. The *root field* F of G is the k_s/k-descent of the common root field of $(G_{k_s})_a$ for divisible $a \in \Phi(G_{k_s}, T)$ with $T \subset G_{k_s}$ any maximal k_s-torus.

Letting K/k be the minimal field of definition for $\mathscr{R}_u(G_{\overline{k}}) \subset G_{\overline{k}}$, applying Example 9.1.1 to G_{k_s} shows that $kK^2 \subset F \subset K$. The root field F as just defined satisfies an analogue of Proposition 3.3.6:

Proposition 9.1.3. *The maximal pseudo-reductive quotient of minimal type G_F^{prmt} of G_F is absolutely pseudo-simple with a non-reduced root system over F_s, its geometric unipotent radical has minimal field of definition K/F, and its root field is F. Moreover, the natural map*

$$j_G : G \to \mathscr{D}(\mathrm{R}_{F/k}(G_F^{\mathrm{prmt}}))$$

is an isomorphism.

Proof. Without loss of generality we may (and do) assume that $k = k_s$, so G is pseudo-split and $F = F_s$. Thus, the non-canonical isomorphism (9.1.1) takes the form of a k-isomorphism $f : G \simeq \mathscr{D}(\mathrm{R}_{F/k}(H))$ for an absolutely pseudo-simple F-group H of minimal type with a non-reduced root system, minimal field of definition K/F for its geometric unipotent radical, and root field F. But Proposition 2.3.13 implies that the natural map $G_F \to H$ arising from f identifies H with G_F^{prmt}, and using the universal property of Weil restriction to proceed in reverse identifies j_G with f. Hence, the canonical j_G is an isomorphism. $\square$

Remark 9.1.4. Assume G as above is pseudo-split. In contrast with the non-reducedness of the root system for G_F^{prmt} (or equivalently for its pseudo-reductive central extension G_F^{pred}), it is a nontrivial condition on a proper subfield $E \subset K$ over kK^2 that G_E^{prmt} has a non-reduced root system. Indeed, if $[K : kK^2] > 2$ then there exists pseudo-split G as above and a proper subextension E of K/kK^2 such that the maximal pseudo-reductive quotient G_E^{pred} has a reduced root system (and hence likewise for the central quotient G_E^{prmt} of G_E^{pred}); this is shown in [CGP, 9.8.17, 9.8.18].

If $[K : kK^2] = 2$ then necessarily $E = kK^2 = F$, so G_E^{prmt} always has a non-reduced root system in such cases. Since $[K : kK^2] = 2$ whenever $[k : k^2] = 2$, the phenomenon of G_E^{prmt} having a reduced root system for some proper subfield $E \subset K$ over kK^2 is only seen when $[k : k^2] > 2$.

In view of Proposition 9.1.3, for absolutely pseudo-simple k-groups of minimal type with a non-reduced root system over k_s the condition that the root field coincides with the ground field (i.e., $F = k$) is a reasonable one to impose for the purpose of a structure theorem. This is used in:

Definition 9.1.5. Let k be a field, k' a nonzero finite reduced k-algebra, and $G' \to \mathrm{Spec}(k')$ a smooth affine group scheme. The pair $(G', k'/k)$ is *primitive* if for each factor field k'_i of k' the k'_i-fiber G'_i is one of the following:

(1) connected semisimple, absolutely simple, and simply connected group,
(2) basic exotic of type F_4 or G_2, generalized basic exotic of types B or C, or rank-2 basic exceptional group,
(3) minimal type absolutely pseudo-simple group with a non-reduced root system over the separable closure of k'_i and root field equal to k'_i.

Groups mentioned in (2) exist only if k is imperfect with characteristic 2 or 3, and groups in (3) exist only if k is imperfect with characteristic 2. Also, for imperfect k, generalized exotic k-groups are precisely k-groups of the form $\mathscr{D}(\mathrm{R}_{k'/k}(G'))$ for primitive pairs $(G', k'/k)$ such that G'_i is as in (2) for all i.

For every i the k'_i-group G'_i is of minimal type, and its maximal geometric semisimple quotient is simply connected. The pseudo-split possibilities for G'_i are parameterized by known invariants: the root system for case (1), the root system along with the minimal field of definition K'_i/k'_i for the geometric unipotent radical and some additional field-theoretic and linear-algebraic data inside K'_i as in Theorem 3.4.1 for case (2) (also see Proposition 8.2.5 and Corollary 8.3.7 for types B and C), and similarly in [CGP, Thm. 9.8.6, Prop. 9.8.9] for case (3).

If $C'_i \subset G'_i$ is a Cartan k'_i-subgroup with maximal k'_i-torus T'_i then the k'_i-group $Z_{G'_i, C'_i}$ is commutative and pseudo-reductive (see Proposition 6.1.4) and its structure can be expressed in terms of T'_i: in case (1) it is $T'_i/Z_{G'_i}$, in case (2) it is described by Lemma 6.1.3 for rank 1 and by Proposition 8.5.4 for rank $\geqslant 2$, and in case (3) it is described by [CGP, Prop. 9.8.15].

Proposition 9.1.6. *Let $(G', k'/k)$ be a primitive pair as above, and define $G = \mathscr{D}(\mathrm{R}_{k'/k}(G'))$. The k-group G is pseudo-semisimple of minimal type, $G^{\mathrm{ss}}_{\overline{k}}$ is simply connected, and G is absolutely pseudo-simple if and only if k' is a field purely inseparable over k. Moreover, the center Z_G is k-tame.*

Proof. We may assume $k = k_s$ (since if k' is not a field purely inseparable over k then $k' \otimes_k k_s$ has at least two factor fields and hence G_{k_s} is not pseudo-simple; note that by Proposition 5.1.2, Z_G is k-tame if and only if $Z_{G_{k_s}} = (Z_G)_{k_s}$ is k_s-tame). Write $k' = \prod k'_i$ for fields k'_i, and let G'_i be the k'_i-fiber of G', so

$$G = \mathscr{D}(\mathrm{R}_{k'/k}(G')) = \prod \mathscr{D}(\mathrm{R}_{k'_i/k}(G'_i)).$$

Proposition 2.3.13 gives that $G^{\mathrm{ss}}_{\overline{k}} = \prod (G'_i \otimes_{k'_i} \overline{k})^{\mathrm{ss}}$, and this is simply connected by inspection. In particular, G is pseudo-simple if and only if k' is a field (see [CGP, Lemma 3.1.2]), and it is of minimal type by Proposition 5.3.5(ii).

To prove that Z_G is k-tame we may clearly assume that k' is a field. Since $\mathrm{R}_{k'/k}(G')_{k'} \to G'$ is surjective [CGP, Prop. A.5.11(1)], so is $G_{k'} \to G'$. Thus, $(Z_G)_{k'}$ is carried into $Z_{G'}$, so $Z_G \subset \mathrm{R}_{k'/k}(Z_{G'})$. Arguing as immediately below (5.1.1), it therefore suffices to show that $Z_{G'}$ is k'-tame, so we may assume $k' = k$. If G has a non-reduced root system then Z_G is trivial by [CGP, Prop. 9.4.9] (using that G is of minimal type). Likewise, Z_G is trivial for types G_2 and F_4 by [CGP, Cor. 7.2.5(2)]. For the remaining cases with type B or C, Proposition 8.5.1 identifies Z_G with a k-subgroup of the Weil restriction to k of finite multiplicative type group over a finite extension of k. Again arguing as immediately below (5.1.1), Z_G is k-tame. □

Definition 9.1.7. Let k be a field. A pseudo-reductive k-group G is *generalized standard* if either it is commutative or there exists a 4-tuple $(G', k'/k, T', C)$ consisting of a nonzero finite reduced k-algebra k', a smooth affine k'-group G' such that $(G', k'/k)$ is a primitive pair, a maximal k'-torus $T' \subset G'$, a commutative pseudo-reductive k-group C, and a factorization diagram

$$\mathscr{C} \xrightarrow{\phi} C \xrightarrow{\psi} Z_{\mathscr{G},\mathscr{C}} = \mathrm{R}_{k'/k}(Z_{G',C'}) \qquad (9.1.7.1)$$

with $\mathscr{G} = \mathscr{D}(\mathrm{R}_{k'/k}(G'))$, $C' = Z_{G'}(T')$, and $\mathscr{C} = \mathscr{G} \cap \mathrm{R}_{k'/k}(C')$ (a Cartan k-subgroup of $\mathscr{G}$) such that there is a k-isomorphism

$$(\mathscr{G} \rtimes C)/\mathscr{C} \simeq G, \tag{9.1.7.2}$$

where $\mathscr{C}$ is anti-diagonally embedded as a central k-subgroup of $\mathscr{G} \rtimes C$. (The maximal k-torus $\mathscr{T}$ in $\mathrm{R}_{k'/k}(T')$ is contained in $\mathscr{C}$, so the Cartan k-subgroup $\mathscr{C}$ of $\mathscr{G}$ coincides with $Z_{\mathscr{G}}(\mathscr{T})$.)

The 4-tuple $(G', k'/k, T', C)$ equipped with the factorization (9.1.7.1) and isomorphism (9.1.7.2) is called a *generalized standard presentation* of G.

Remark 9.1.8. The equality in (9.1.7.1) is an instance of Proposition 6.1.7, and Definition 9.1.7 recovers the definition of the same terminology in [CGP, Def. 10.1.9] assuming $[k : k^2] \leqslant 2$ when $\mathrm{char}(k) = 2$ except for one crucial aspect: when $[k : k^2] = 2$ and $\mathrm{char}(k) = 2$ the definition in [CGP] avoids case (3) of Definition 9.1.5. That avoidance is reasonable when $[k : k^2] = 2$ due to a splitting result in [CGP, Thm. 10.2.1(1)], but such splitting generally fails beyond the perfect minimal-type case whenever $[k : k^2] > 2$ (see Example 6.1.5).

Any k-group G of the form (9.1.7.2) is obviously non-commutative. Let us show that any such G is pseudo-reductive. Since $\mathscr{C}$ is a Cartan k-subgroup of $\mathscr{G}$ and the formation of Cartan k-subgroups of smooth connected affine k-groups is compatible with quotients, the evident inclusion $C \to G$ makes C a Cartan k-subgroup of the central quotient G in (9.1.7.2). The centrality of $\mathscr{C}$ and the evident pseudo-reductivity of $\mathscr{G} \rtimes C$ therefore imply that G is pseudo-reductive, by [CGP, Prop. 1.4.3].

The pseudo-semisimple $\mathscr{D}(G)$ coincides with the central quotient $\mathscr{G}/(\ker\phi)$ since C is commutative, so if $G = \mathscr{D}(G)$ then $C = \phi(\mathscr{C})$. (The k-group $\ker\phi$ is central in $\mathscr{G}$ because $\psi \circ \phi$ encodes the action of $\mathscr{C}$ on $\mathscr{G}$ via conjugation.)

Remark 9.1.9. For $(G', k'/k)$ as in Definition 9.1.7 and $\mathscr{G} := \mathscr{D}(\mathrm{R}_{k'/k}(G'))$, any *pseudo-reductive* central quotient $G = \mathscr{G}/Z$ is generalized standard. To see this, choose a Cartan k-subgroup $\mathscr{C}$ of $\mathscr{G}$, so $C := \mathscr{C}/Z$ is identified with a Cartan k-subgroup of the pseudo-reductive G. Thus, C is pseudo-reductive and the natural action of C on $\mathscr{G}$ (arising from conjugation by $\mathscr{C}$ and the centrality of Z in $\mathscr{G}$) provides a factorization diagram

$$\mathscr{C} \twoheadrightarrow C \to Z_{\mathscr{G},\mathscr{C}}$$

giving a generalized standard presentation $(G', k'/k, T', C)$ for G, where T' is the (unique) maximal k'-torus in G' such that $\mathscr{C} = \mathscr{G} \cap \mathrm{R}_{k'/k}(Z_{G'}(T'))$.

9.1.10. In the setting of Definition 9.1.7 we claim that the k-group $\mathscr{G}$ is determined by the central quotient k-group $\mathscr{G}/(\ker\phi) = \mathscr{D}(G)$ functorially with respect to isomorphisms in this central quotient, so by Proposition 8.4.3 the triple $(G', k'/k, j)$ incorporating the k-homomorphism $j : \mathscr{D}(\mathrm{R}_{k'/k}(G')) \twoheadrightarrow \mathscr{D}(G) \hookrightarrow G$ is determined by the k-group G functorially with respect to isomorphisms in G. In [CGP, Prop. 10.1.12 (1)] this is treated via a splitting result for central extensions by affine k-group schemes Z of finite type such that $Z^0(k_s) = 1$, assuming $[k : k^2] \leqslant 2$ when $\mathrm{char}(k) = 2$. (This splitting result rests on [CGP, Prop. 5.1.3, Ex. 5.1.4, Prop. 8.1.2], and to allow G with a non-reduced root system over k_s with $[k : k^2] = 2$ one can use [CGP, Thm. 10.2.1(1)].)

The preceding method does not apply when $\mathrm{char}(k) = 2$ with $[k : k^2] > 2$ because for every such k and $n \geqslant 1$ there exists a pseudo-split absolutely pseudo-simple k-group H with root system BC_n such that H is not of minimal type, so H is a non-split k-tame central extension of $G := H/\mathscr{C}_H$. Moreover, when $[k : k^2] \geqslant 16$ analogous examples exist with $H/\mathscr{C}_H$ generalized basic exotic of type B or C (with any rank $n \geqslant 1$). Appendix B provides many such H, and proves it is necessary that $[k : k^2] \geqslant 16$ when the root system is reduced. We shall use universal smooth k-tame central extensions to bypass this problem via:

Lemma 9.1.11. *Let $(G', k'/k)$ and $(G'_0, k'_0/k)$ be primitive pairs. Define $\widetilde{G} = \mathscr{D}(\mathrm{R}_{k'/k}(G'))$ and $\widetilde{G}_0 = \mathscr{D}(\mathrm{R}_{k'_0/k}(G'_0))$, and let $Z \subset Z_{\widetilde{G}}$ and $Z_0 \subset Z_{\widetilde{G}_0}$ be closed k-subgroup schemes. Define $G = \widetilde{G}/Z$ and $G_0 = \widetilde{G}_0/Z_0$.*

(i) *For any separable extension field F/k, every F-isomorphism $f : G_F \simeq (G_0)_F$ uniquely lifts to an F-homomorphism $\widetilde{f} : \widetilde{G}_F \to (\widetilde{G}_0)_F$, and $\widetilde{f}$ is an isomorphism. Moreover, if $(G'_0, k'_0/k) = (G', k'/k)$ then $\widetilde{f}$ restricts to the identity on a smooth connected F-subgroup of $\widetilde{G}_F$ if f restricts to the identity on its image in G_F.*

(ii) *Let H be a smooth connected k-group equipped with a left action on G. This lifts uniquely to a left action on $\widetilde{G}$. The lifted action restricts to the trivial action on a smooth connected k-subgroup of $\widetilde{G}$ if H acts trivially on its image in G.*

Proof. The special case when G' and G'_0 have semisimple fibers is [CGP, Prop. 5.1.7]. In the present additional generality, note that the initial setup is compatible with separable extension of the ground field and in all cases the fiber of $Z_{G'}$ over each factor field k'_i of k' is contained in the Weil restriction to k'_i of a finite group scheme over a finite extension field of k'_i (see Proposition 8.5.1 for the generalized basic exotic and rank-2 basic exceptional cases in characteristic 2). Hence, the fibers of $Z_{G'}$ do not admit a nontrivial map of pointed schemes from a smooth connected scheme, so we can argue exactly as in the proof of [CGP, Prop. 5.1.7] to reduce (ii) to (i) and furthermore reduce (i) to the special case $F = k = k_s$. The problem is to uniquely fill in a commutative diagram

$$\begin{array}{ccc} \widetilde{G} & \xrightarrow{?} & \widetilde{G}_0 \\ {\scriptstyle \pi}\downarrow & & \downarrow{\scriptstyle \pi_0} \\ G & \xrightarrow[f]{\simeq} & G_0 \end{array}$$

where π and π_0 are the respective quotient maps modulo Z and Z_0, and to show that the map $\widetilde{G} \to \widetilde{G}_0$ across the top of the diagram is an isomorphism.

The central k-subgroups Z and Z_0 are k-tame since $Z_{\widetilde{G}}$ and $Z_{\widetilde{G}_0}$ are k-tame (Proposition 9.1.6). Thus, via the isomorphism f, the perfect smooth connected k-groups $\widetilde{G}$ and $\widetilde{G}_0$ are k-tame central extensions of G. Each has maximal geometric reductive quotient that is simply connected, and this property uniquely characterizes (up to unique isomorphism) the universal smooth k-tame central extension. Hence, there is a unique k-homomorphism $\widetilde{G} \to \widetilde{G}_0$ over G and it is an isomorphism. □

As an application of Lemma 9.1.11, we now prove a rigidity property of generalized standard presentations. This involves the notion of "pseudo-isogeny" that is defined in §A.1:

Proposition 9.1.12. *Let G be a non-commutative generalized standard k-group arising from a 4-tuple $(G', k'/k, T', C)$ and the factorization diagram* (9.1.7.1).

(i) *The natural map $j : \mathscr{D}(\mathrm{R}_{k'/k}(G')) \to G$ arising from the generalized standard presentation is a pseudo-isogeny onto $\mathscr{D}(G)$ with central kernel.*
(ii) *The triple $(G', k'/k, j)$ is unique up to unique isomorphism in the sense that if $(G'_0, k'_0/k, T'_0, C_0)$ is part of a generalized standard presentation of G then there is a unique isomorphism of primitive pairs $(G', k'/k) \simeq (G'_0, k'_0/k)$ compatible with the quotient maps j and j_0 onto $\mathscr{D}(G)$.*

Proof. Using notation as in Definition 9.1.7, the central quotient presentation $(\mathscr{G} \rtimes C)/\mathscr{C} \simeq G$ with commutative C and perfect $\mathscr{G}$ implies that the natural map $j : \mathscr{G} \to G$ has image $\mathscr{D}(G)$. Moreover, the kernel of $j : \mathscr{G} \twoheadrightarrow \mathscr{D}(G)$ is central because $\mathscr{C}$ is central in $\mathscr{G} \rtimes C$. The uniqueness up to unique isomorphism in (ii) is an immediate consequence of Lemma 9.1.11 applied to $\mathscr{D}(G)$ as a central quotient of $\mathscr{G} := \mathscr{D}(\mathrm{R}_{k'/k}(G'))$.

To show that j is a pseudo-isogeny onto the image $\mathscr{D}(G)$, the problem is to prove that $\ker j$ contains no nontrivial tori. Using the notation from (9.1.7.1), this kernel is identified with

$$\ker(\mathscr{C} \to C) \subset \ker(\mathscr{C} \to Z_{\mathscr{G},\mathscr{C}}).$$

Since $\mathscr{C} \to Z_{\mathscr{G},\mathscr{C}}$ is a pseudo-isogeny by Proposition 6.1.7, we are done. $\square$

By Proposition 9.1.12, if G is a generalized standard k-group then *every* generalized standard presentation of G encodes the same underlying triple $(G', k'/k, j)$ in a manner that is functorial with respect to isomorphisms in G. The following result goes further and makes precise the sense in which a choice of generalized standard presentation of a given G is "the same" as a choice of maximal k-torus of G, or a choice of maximal k'-torus of G':

Proposition 9.1.13. *Let G be a non-commutative generalized standard k-group arising from the 4-tuple $(G', k'/k, T', C)$ and factorization diagram* (9.1.7.1). *Let $\mathscr{G} = \mathscr{D}(\mathrm{R}_{k'/k}(G'))$ and $C' = Z_{G'}(T')$.*

(i) *The maximal k-torus $\mathscr{T}$ in $\mathrm{R}_{k'/k}(T')$ is a maximal k-torus in $\mathscr{G}$, and if T is the maximal k-torus of G that is the almost direct product of $j(\mathscr{T})$ and the maximal central k-torus Z of G then $C = Z_G(T) = Z_G(\mathscr{T})$.*

(ii) *There is a bijection between the set of maximal k'-tori $S' \subset G'$ and the set of maximal k-tori in G via $S' \mapsto j(\mathscr{S}) \cdot Z$ where $\mathscr{S}$ is the maximal k-torus in $\mathrm{R}_{k'/k}(S')$ (necessarily also a maximal k-torus in $\mathscr{G}$).*

(iii) *Let S be a maximal k-torus in G, S' the corresponding maximal k'-torus in G', and $\mathscr{S}$ the maximal k-torus in $\mathrm{R}_{k'/k}(S')$. The conjugation action of $Z_G(S)$ on $\mathscr{D}(G) = j(\mathscr{G})$ uniquely lifts to an action on $\mathscr{G}$ that restricts to the identity on $\mathscr{S}$ (and hence on $Z_{\mathscr{G}}(\mathscr{S})$). Using the resulting factorization diagram*

$$Z_{\mathscr{G}}(\mathscr{S}) \to Z_G(S) \to Z_{\mathscr{G}, Z_{\mathscr{G}}(\mathscr{S})} = \mathrm{R}_{k'/k}(Z_{G',S'})$$

(equality by Proposition 6.1.7*), the natural k-homomorphism*

$$(\mathscr{G} \rtimes Z_G(S))/Z_{\mathscr{G}}(\mathscr{S}) \to G$$

is an isomorphism.

Proof. We saw in the proof of Lemma 9.1.11 that the fibers of $Z_{G'}$ do not admit a nontrivial map of pointed schemes from a smooth connected scheme, so by using Proposition 6.1.7 as a replacement for [CGP, Lemma 10.1.8] and using

Lemma 9.1.11 as a replacement for [CGP, Prop. 10.1.12] we can argue exactly as in the special case when G' has semisimple fibers (see [CGP, Prop. 4.1.4]). □

Corollary 9.1.14. *A pseudo-reductive k-group G is generalized standard if and only if $\mathscr{D}(G)$ is generalized standard.*

Proof. The case of commutative G is trivial, so we may assume G is non-commutative. The analogue for the standardness property is [CGP, Prop. 5.2.1], and by using Lemma 9.1.11(ii) we can easily adapt the same proof to the generalized standard case since (by Proposition 9.1.13) we may use whatever maximal k-torus we wish (in G or in $\mathscr{D}(G)$) to underlie a choice of generalized standard presentation when one exists (for the non-commutative G or $\mathscr{D}(G)$). □

9.2 Structure theorem

Over any imperfect field k (with any positive characteristic) there are standard absolutely pseudo-simple k-groups that are not of minimal type [CGP, Ex. 5.3.7]. There is a weaker condition that is satisfied by all pseudo-reductive k-groups G if $\mathrm{char}(k) \neq 2$ or if $\mathrm{char}(k) = 2$ and $[k : k^2] \leqslant 2$: G is locally of minimal type (see Remark 4.3.2). This condition is also satisfied if $\mathrm{char}(k) = 2$ and $[k : k^2] \leqslant 8$ provided that G_{k_s} has a reduced root system (see Proposition B.3.1).

Theorem 9.2.1. *Let G be a pseudo-reductive group over a field k. Then G is generalized standard if and only if it is locally of minimal type.*

Proof. The implication "⇒" is easy, as follows. We may assume $k = k_s$. By Corollary 9.1.14, we may replace G with $\mathscr{D}(G)$ so that G is perfect. Thus, G is a central quotient of $\mathscr{D}(\mathrm{R}_{k'/k}(G'))$ for a primitive pair $(G', k'/k)$. Identifying the root systems of G and G' relative to compatible maximal tori, and working with G_c and G'_c in place of G and G' respectively for each c in the root system, reduces us to the case where G has rank 1. Now k' is a field, G' has rank 1, and G' is of minimal type over k' with maximal geometric reductive quotient SL_2. By Proposition 5.3.5(ii), the rank-1 pseudo-simple k-group $\mathscr{D}(\mathrm{R}_{k'/k}(G'))$ (that is a central extension of G) is of minimal type over k.

Now we assume G is locally of minimal type and prove that it is generalized standard. By Corollary 9.1.14, we may and do assume G is perfect. The universal smooth k-tame central extension $\widetilde{G}$ of G is of minimal type (Proposition 5.3.3). It suffices to prove that there exists a primitive pair $(G', k'/k)$ such that $\widetilde{G} \simeq \mathscr{D}(\mathrm{R}_{k'/k}(G'))$, as then it would follow from Remark 9.1.9 that G is also generalized standard. Thus, we may replace G with $\widetilde{G}$ to arrange that

G is of minimal type with $G^{\mathrm{ss}}_{\overline{k}}$ simply connected, and we seek a primitive pair $(G', k'/k)$ such that there is a k-isomorphism $f : G \simeq \mathscr{D}(\mathrm{R}_{k'/k}(G'))$.

By Proposition 8.4.3, $(G', k'/k, f)$ is unique up to unique isomorphism if it exists. Galois descent thereby reduces the proof of existence of $(G', k'/k)$ to the case $k = k_s$. By [CGP, Prop. 3.1.8], there are finitely many normal pseudo-simple k-subgroups G_i of G and they pairwise commute and generate G, with multiplication defining a central quotient map

$$\pi : \prod G_i \to G.$$

The groups $(G_i)^{\mathrm{ss}}_{\overline{k}}$ are the simple normal subgroups of the connected semisimple group $G^{\mathrm{ss}}_{\overline{k}}$ that is simply connected (see [CGP, Prop. 3.1.6, Prop. A.4.8]), so each $(G_i)^{\mathrm{ss}}_{\overline{k}}$ is simply connected. Normality of G_i in G implies that G_i is of minimal type (Lemma 2.3.10). If we can handle the pseudo-simple case (applied to G_i for every i) then each Z_{G_i} is k-tame (by Proposition 9.1.6), so $\prod Z_{G_i}$ is also k-tame. Hence, π would be a k-tame central extension of the perfect smooth connected affine k-group G that is its own universal smooth k-tame central extension, so π is an isomorphism and thus we would obtain the desired description of G. We may therefore assume that G is pseudo-simple.

Let T be a maximal k-torus in G. We may assume that G is not standard (as otherwise we are done), so by Theorem 3.4.2 the field k is imperfect with characteristic $p \in \{2, 3\}$ and the root system $\Phi = \Phi(G, T)$ is G_2 if $p = 3$ and is B_n $(n \geqslant 1)$, C_n $(n \geqslant 1)$, F_4, or BC_n $(n \geqslant 1)$ if $p = 2$. We have arranged that G is of minimal type, so if $\Phi = \mathrm{BC}_n$ $(n \geqslant 1)$ then $G = \mathscr{D}(\mathrm{R}_{k'/k}(G'))$ for a primitive pair $(G', k'/k)$ by Proposition 9.1.3. Thus, G is generalized standard in the BC_n-cases, so we may assume Φ is reduced. Now as G is a non-standard absolutely pseudo-simple group of minimal type with a reduced root system over $k = k_s$, and $G^{\mathrm{ss}}_{\overline{k}}$ is simply connected, Theorem 8.4.5 implies that there is a primitive pair $(G', k'/k)$ (with k' a field) such that $G \simeq \mathscr{D}(\mathrm{R}_{k'/k}(G'))$. □

Appendix A
Pseudo-isogenies

A.1 Main result

Observe that over a field k, a surjective k-homomorphism $f : G \to G'$ between connected semisimple k-groups is an isogeny if and only if it restricts to an isogeny between maximal k-tori. More generally:

Definition A.1.1. Let G and G' be perfect smooth connected affine k-groups. A k-homomorphism $f : G \to G'$ is a *pseudo-isogeny* if it is surjective and restricts to an isogeny between maximal k-tori.

A natural example of a pseudo-isogeny that is not an isogeny is

$$\mathrm{R}_{k'/k}(\mathrm{SL}_p) \to \mathrm{R}_{k'/k}(\mathrm{SL}_p)/\mathrm{R}_{k'/k}(\mu_p) = \mathscr{D}(\mathrm{R}_{k'/k}(\mathrm{PGL}_p))$$

for a nontrivial purely inseparable finite extension k'/k in characteristic $p > 0$.

Our formulation of a "Pseudo-Isogeny Theorem" will use the following convenient shorthand: if G is a pseudo-reductive k-group and S is a split maximal k-torus of G then for $a \in \Phi(G, S)$ we define $U_{(a)}$ to be the root group U_a when a is not divisible and to be the root group $U_{a/2}$ when a is divisible. (See [CGP, Def. 2.3.4, 2.3.12–2.3.14] for a broader context.) As usual, G_a denotes $\langle U_a, U_{-a}\rangle$ (see [CGP, Prop. 3.4.1]). For the commutative Cartan k-subgroup $C := Z_G(S)$ and $a \in \Phi$, the intersection $C_a := C \cap G_a$ coincides with the Cartan k-subgroup $Z_{G_a}(G_a \cap S) \subset G_a$. If moreover G is pseudo-semisimple then the product map $\prod_{a\in\Delta} C_a \to C$ is surjective by Proposition 4.3.3(i).

Theorem A.1.2 (Pseudo-Isogeny Theorem). *Let k be a field, and let (G, S) and (G', S') be pseudo-reductive k-groups equipped with split maximal k-tori S and S', with G pseudo-semisimple. Let $C = Z_G(S)$, $C' = Z_{G'}(S')$, and let $\Phi = \Phi(G, S)$ and $\Phi' = \Phi(G', S')$ be the respective root systems. We fix a basis Δ of Φ.*

For $a \in \Phi$ let $C_a := C \cap G_a$ be the Cartan k-subgroup $Z_{G_a}(G_a \cap S)$ of G_a. Let $a \mapsto a'$ be a map $\Delta \to \Phi'$ such that for distinct $a, b \in \Delta$, $U'_{(a')}$ and $U'_{(-b')}$ commute. For each $a \in \Delta$ let $f_a : G_a \to G'$ be a k-homomorphism carrying

$a^\vee(\mathrm{GL}_1)$ into S' such that $f_a(U_{\pm a}) = U'_{(\pm a')}$ and f_a is $b^\vee(\mathrm{GL}_1)$-equivariant with respect to the inclusion $b^\vee(\mathrm{GL}_1) \subset S$ and the map $f_b : b^\vee(\mathrm{GL}_1) \to S'$ for all $b \in \Delta - \{a\}$.

(i) *For all $a \in \Delta$, $f_a(C_a) \subset C'$.*

(ii) *If the k-homomorphism $\prod_{a\in\Delta} C_a \to C'$ defined by $(c_a) \mapsto \prod f_a(c_a)$ factors through a k-homomorphism $f_C : C \to C'$ then there is a unique k-homomorphism $f : G \to G'$ extending f_a for every $a \in \Delta$.*

(iii) *Assume the existence of f_C. If G' is generated by the k-subgroups $U'_{(\pm a')}$ for $a \in \Delta$ then f is surjective. If the k-subgroup scheme* $\ker f_a$ *is central in G_a for all $a \in \Delta$ then* $\ker f$ *is central in G.*

If G' is pseudo-semisimple and the image of Δ in Φ' is a basis then G' is generated by the k-subgroups $U'_{(\pm a')}$ for $a \in \Delta$ due to [CGP, Lemma 3.1.5] (since reflections associated to elements of a basis of Φ' generate $W(G', S')$).

Our proof of the Pseudo-Isogeny Theorem is completely different from the traditional proof of the Isogeny Theorem for connected semisimple groups (see [Spr, Thm. 9.6.5]). It rests on a pseudo-reductive variant of an idea of Steinberg for proving the Isomorphism Theorem in the connected semisimple case by constructing homomorphisms via graphs built as connected semisimple subgroups of a direct product of two connected semisimple groups.

A.2 Proof of Pseudo-Isogeny Theorem

The hypothesis that $U'_{(a')}$ and $U'_{(-b')}$ commute for distinct $a, b \in \Delta$ implies that b' cannot be a positive rational multiple of a', so in particular $a \mapsto a'$ is an injective map of Δ into Φ'. Let $G'_{(a')} := \langle U'_{(a')}, U'_{(-a')}\rangle$, so $f_a(G_a) = G'_{(a')}$. By the functoriality of dynamic constructions with 1-parameter subgroups (e.g., $U_G(\lambda)$) given in [CGP, §2.1], f_a carries $U_a = U_{G_a}(a^\vee)$ into $U_{G'}(f_a \circ a^\vee)$, so on the nonzero $\mathrm{Lie}(f_a(U_a))$ all weights for the GL_1-action through $\lambda_{a'} := f_a \circ a^\vee$ are positive. But all weights for the S'-action on $\mathrm{Lie}(U_{(a')})$ are positive rational multiples of a', so $\langle a', \lambda_{a'}\rangle > 0$ for all $a \in \Delta$.

The codimension-1 subtorus $S'_{a'} = (\ker a')^0_{\mathrm{red}}$ in S' centralizes $G'_{(a')}$, and the cocharacter $\lambda_{a'} : \mathrm{GL}_1 \to S'$ has image that is an isogeny complement to $S'_{a'}$ since $\langle a', \lambda_{a'}\rangle \neq 0$. The image $f_a(C_a) \subset G'_{(a')}$ commutes with the subtori $\lambda_{a'}(\mathrm{GL}_1)$ and $S'_{a'}$ that generate S', so $f_a(C_a) \subset C'$. This establishes (i), and for the rest of the argument we shall assume the existence of f_C as in (ii).

Since $f_a(G_a) \subset \mathscr{D}(G')$ and (by [CGP, Lemma 1.2.5(ii),(iii)]) $C' \cap \mathscr{D}(G') = Z_{\mathscr{D}(G')}(S' \cap \mathscr{D}(G'))$ with $S' \cap \mathscr{D}(G')$ a split maximal k-torus in $\mathscr{D}(G')$, we may replace G' with $\mathscr{D}(G')$ so that G' is also pseudo-semisimple (in addition to G being pseudo-semisimple, by hypothesis). For $a \in \Delta$ the pseudo-split pseudo-reductive k-subgroup $H_a := G_a \cdot C$ has the presentation

$$(G_a \rtimes C)/C_a \simeq H_a,$$

and by construction f_C agrees with f_a on C_a, so there exists a (visibly unique) k-homomorphism $\phi_a : H_a \to G'$ that extends f_a and f_C *provided* that $f_a : G_a \to G'$ is C-equivariant when C acts on G' through f_C.

To establish equivariance of f_a with respect to the quotient C of $\prod_{b\in\Delta} C_b$, it is equivalent to prove C_b-equivariance for all $b \in \Delta$. The case $b = a$ is a tautology since f_a is a k-homomorphism, so we may assume $b \in \Delta - \{a\}$. Since $f_a(G_a) = G'_{(a')}$, to construct ϕ_a it suffices to show that $G_a \twoheadrightarrow G'_{(a')}$ is C_b-equivariant. By hypothesis, equivariance holds for the action of the maximal k-torus $T_b := b^\vee(\mathrm{GL}_1)$ in C_b. To prove the C_b-equivariance we may assume $k = k_s$, so then it suffices to show that for each $z \in C_b(k)$ the map $f_a : G_a \twoheadrightarrow G'_{(a')}$ agrees with $x \mapsto f_b(z)^{-1} f_a(zxz^{-1}) f_b(z)$. This is a comparison of two surjective homomorphisms $G_a \rightrightarrows G'_{(a')}$ between *pseudo-reductive groups*, so by [CGP, Prop. 1.2.2] it is equivalent to check equality for the induced maps between maximal geometric reductive quotients.

By 3.2.2, these quotients over $\overline{k}$ respectively coincide with the subgroups of $G^{\mathrm{ss}}_{\overline{k}}$ and $G'^{\mathrm{ss}}_{\overline{k}}$ generated by the evident opposite root groups. Thus, in view of the T_b-equivariance, we may conclude by noting that the composite maps

$$C_{\overline{k}} \to G^{\mathrm{ss}}_{\overline{k}}, \ \ C'_{\overline{k}} \to G'^{\mathrm{ss}}_{\overline{k}}$$

have images respectively coinciding with the images of the maximal tori $S_{\overline{k}}$ and $S'_{\overline{k}}$. This completes the construction of the k-homomorphism $\phi_a : H_a \to G'$ extending f_a and f_C for all $a \in \Delta$.

Since C' is commutative, f_C must carry S into the unique maximal k-torus S' in C'. Let $f_S : S \to S'$ be this restricted map. This map carries the maximal central torus $S_a := (\ker a)^0_{\mathrm{red}}$ of H_a into the maximal central torus $S'_{a'} = (\ker a')^0_{\mathrm{red}}$ of $G'_{(a')} \cdot C'$ since f_S extends to the map ϕ_a whose image contains the group $G'_{(a')} = f_a(G_a)$. Thus, $a' \circ f_S$ kills S_a, so $a' \circ f_S$ is a rational multiple of a. But

$$\langle a' \circ f_S, a^\vee \rangle = \langle a', f_S \circ a^\vee \rangle = \langle a', f_a \circ a^\vee \rangle = \langle a', \lambda_{a'} \rangle > 0,$$

so $a' \circ f_S$ is a positive rational multiple of a.

Let $\mathscr{G} = G \times G'$, and let $\mathscr{S} \subset \mathscr{G}$ be the graph of f_S, so $\mathscr{S}$ is a k-torus in $\mathscr{G}$ such that the projection map $\mathrm{pr}_1 : \mathscr{S} \to S$ is an isomorphism. This identifies Δ with a linearly independent subset of $\mathrm{X}(\mathscr{S})$. The composite map

$$\mathrm{X}(S') \stackrel{\mathrm{X}(\mathrm{pr}_2)}{\longrightarrow} \mathrm{X}(\mathscr{S}) \simeq \mathrm{X}(S)$$

is $\mathrm{X}(f_S)$, so it carries a' to a positive rational multiple of a.

Let $\mathscr{H}_a \subset \mathscr{G}$ be the graph of ϕ_a, and let $\mathscr{C} \subset \mathscr{H}_a$ be the graph of f_C. Since $\mathrm{pr}_1 : \mathscr{H}_a \to H_a$ is an isomorphism carrying $\mathscr{S}$ onto S, $\mathscr{H}_a$ is pseudo-reductive and its Cartan k-subgroup $Z_{\mathscr{H}_a}(\mathscr{S})$ is identified with $Z_{H_a}(S) = C$. More precisely, $Z_{\mathscr{H}_a}(\mathscr{S}) = \mathscr{C}$ inside $\mathscr{G}$. The set of nontrivial $\mathscr{S}$-weights on $\mathscr{H}_a$ coincides with the set of nontrivial S-weights on H_a via the identification $\mathrm{pr}_1 : \mathscr{S} \simeq S$, so this set of weights contains $\pm a$ and is contained inside $\mathbf{Z} \cdot a$. The a-root group $\mathscr{U}_a$ of the pair $(\mathscr{H}_a, \mathscr{S})$ is contained inside the direct product $U_a \times f_a(U_a)$, and by hypothesis $f_a(U_{\pm a}) = U'_{(\pm a')}$. Thus, for any distinct $a, b \in \Delta$, $\mathscr{U}_a$ commutes with $\mathscr{U}_{-b}$ because U_a commutes with U_{-b} and also $U'_{(a')}$ commutes with $U'_{(-b')}$ (the latter by hypothesis).

We conclude via [CGP, Thm. C.2.29] that the k-groups $\{\mathscr{H}_a\}_{a \in \Delta}$ generate a pseudo-split pseudo-reductive k-subgroup $\mathscr{H}$ of $G \times G'$ such that: $\mathscr{S}$ is a maximal k-split torus of $\mathscr{H}$, $Z_{\mathscr{H}}(\mathscr{S}) = \mathscr{C}$, Δ is a basis of $\Phi(\mathscr{H}, \mathscr{S})$, and for $a \in \Delta$ the $\pm a$-root groups for the pair $(\mathscr{H}, \mathscr{S})$ coincide with the $\pm a$-root groups $\mathscr{U}_{\pm a}$ for the pair $(\mathscr{H}_a, \mathscr{S})$; these latter root groups are the graphs of f_a on $U_{\pm a}$, and the coroot associated to a is the 1-parameter subgroup

$$(a^\vee, f_a \circ a^\vee) : \mathrm{GL}_1 \to \mathscr{S} \subset S \times S'.$$

Pseudo-semisimplicity of G implies that it is generated by the k-subgroups $\{G_a\}_{a \in \Delta}$, so the projection $\mathscr{H} \to G$ is surjective. Hence, $\mathscr{D}(\mathscr{H}) \to G$ is also surjective, so the map $\mathscr{C} \cap \mathscr{D}(\mathscr{H}) \to C$ between Cartan k-subgroups is surjective. However, $\mathrm{pr}_1 : \mathscr{C} \to C$ is an isomorphism by design, so $\mathscr{C} = \mathscr{C} \cap \mathscr{D}(\mathscr{H}) \subset \mathscr{D}(\mathscr{H})$. Since $\mathscr{H} = \mathscr{C} \cdot \mathscr{D}(\mathscr{H})$, it follows that $\mathscr{H} = \mathscr{D}(\mathscr{H})$; i.e., $\mathscr{H}$ is pseudo-semisimple.

We claim that $\mathrm{pr}_1 : \mathscr{H} \to G$ is an isomorphism. Since pr_1 restricts to isomorphisms $\mathscr{C} \simeq C$ and $\mathscr{H}_a \simeq H_a$ for all $a \in \Delta$, pr_1 is surjective and the normal closed k-subgroup scheme $\ker \mathrm{pr}_1 \subset \mathscr{H}$ has trivial intersection with the root group of the pair $(\mathscr{H}, \mathscr{S})$ for every root in Δ. Every $N_{\mathscr{H}}(\mathscr{S})(k)$-orbit of a non-divisible root in $\Phi(\mathscr{H}, \mathscr{S})$ meets Δ because $N_{\mathscr{H}}(\mathscr{S})(k)$ maps onto the Weyl group of $\Phi(\mathscr{H}, \mathscr{S})$, so $(\ker \mathrm{pr}_1) \cap \mathscr{U}_a$ is trivial for all non-divisible $a \in \Phi(\mathscr{H}, \mathscr{S})$. Passing to Lie algebras, the $\mathscr{S}$-action on $\mathrm{Lie}(\ker \mathrm{pr}_1)$ has no nontrivial weight and hence is the trivial action. Thus, by [CGP, Cor. A.8.11], $\mathscr{S}$ centralizes $(\ker \mathrm{pr}_1)^0$. But $\mathscr{H}_{\overline{k}}$ is generated by tori since $\mathscr{H}$ is perfect, so $(\ker \mathrm{pr}_1)^0$ is central. The normal $\ker \mathrm{pr}_1$ is therefore also central since $\mathscr{H}$ is perfect (see [CGP, Lemma 5.3.2]), so $\ker \mathrm{pr}_1 \subset \mathscr{C}$. But $\ker(\mathrm{pr}_1|_{\mathscr{C}}) = 1$, so pr_1 is an isomorphism as desired. Composing the inverse of this isomorphism with $\mathrm{pr}_2 : \mathscr{H} \to G'$ defines a k-homomorphism $f : G \to G'$. It is clear that f extends f_C and f_a for all $a \in \Delta$, so (ii) is proved.

The first part of (iii) is obvious since $U'_{(\pm a')} = f(U_{\pm a})$ for all $a \in \Delta$. Now assume $\ker f_a$ is central in G_a for all $a \in \Delta$; we aim to show that $\ker f$ is central in G. As with the study of $\ker \mathrm{pr}_1$ in the proof of (ii), since G is perfect it suffices to show that $U_a \cap \ker f$ is trivial for all $a \in \Delta$. For such a we know that $f|_{G_a} = f_a$, and the centrality hypothesis for $\ker f_a$ in G_a implies that $\ker f_a$ is contained in the Cartan k-subgroup $Z_{G_a}(S \cap G_a)$. Thus, $U_a \cap \ker f_a$ is trivial as a k-subgroup scheme of G_a. The Pseudo-Isogeny Theorem is proved.

A.3 Relation with the semisimple case

The Pseudo-Isogeny Theorem implies the Isogeny Theorem for split connected semisimple groups G and G' as stated in [CGP, Thm. A.4.10] for a p-morphism $(f, b, q) : R(G', S') \to R(G, S)$ between root data (where $p \geqslant 1$ is the characteristic exponent of k and $q(a) = p^{e(a)}$ for some $e(a) \geqslant 0$ for each $a \in \Phi(G, S)$); note the "contravariant" convention in the source and target for this terminology. In [CGP] we gave a proof of the Isogeny Theorem that reduced to the case of an algebraically closed ground field, which in turn is proved in [Spr, Thm. 9.6.5]. Now we sketch a proof of the Isogeny Theorem directly over k via the Pseudo-Isogeny Theorem without reducing to work over $\overline{k}$.

For each $a \in \Delta$ we may identify G_a with SL_2 or PGL_2 by carrying $a^\vee$ to $t \mapsto \mathrm{diag}(t, 1/t)$ or $t \mapsto \mathrm{diag}(t, 1)$ respectively and carrying the a-root group onto the upper triangular unipotent subgroup. For the split connected semisimple target group G' and $a' \in \Delta'$ we likewise identify $G'_{a'}$ with SL_2 or PGL_2.

Consider roots $a \in \Delta$ and $a' \in \Delta'$ that correspond to each other under the given p-morphism between the root datum. The given p-morphism restricts to a p-morphism from the root datum of $(G'_{a'}, a'^\vee(\mathrm{GL}_1))$ to that of $(G_a, a^\vee(\mathrm{GL}_1))$, but there is no p-morphism from the root datum of SL_2 to that of PGL_2 except if $p = 2$, in which case necessary $q(a) \neq 1$. Thus, if $p \neq 2$ then the case $G_a \simeq \mathrm{PGL}_2$ and $G'_{a'} \simeq \mathrm{SL}_2$ cannot occur.

If $p = 1$ then we define $f_a : G_a \to G'_{a'}$ to be the identity endomorphism of SL_2 or PGL_2 or the central quotient map $\mathrm{SL}_2 \to \mathrm{PGL}_2$, and if $p > 2$ then we define f_a to be the composition of that map with the $q(a)$-Frobenius endomorphism of the source. If $p = 2$ then we use the same procedure as if $p > 2$ provided that we are not in the case $G_a \simeq \mathrm{PGL}_2$ and $G'_{a'} \simeq \mathrm{SL}_2$. In this latter case we have seen that necessarily $q(a) \neq 1$, so we define f_a to be the classical (non-central) unipotent isogeny $\mathrm{PGL}_2 \to \mathrm{SL}_2$ if $q(a) = 2$ and its composition with the $q(a)/2$-Frobenius endomorphism of PGL_2 if $q(a) > 2$. Note that in all cases with $p > 1$, the effect of f_a between root groups is a $k^\times$-multiple of the $q(a)$-Frobenius endomorphism of $\mathbf{G}_\mathrm{a}$.

The GL_1-equivariance requirement on f_a with respect to $b^\vee$ for distinct $a, b \in \Delta$ is deduced from the purely formal [Spr, Lemma 9.6.4(i)]. Moreover, the hypothesis concerning existence f_C in (ii) and (iii) is immediate via the given p-morphism of root datum (we can take f_C to correspond to the given map of character lattices). Hence, the Pseudo-Isogeny Theorem provides the desired isogeny, as well as the desired combinatorial characterization of when it is central, and it reduces uniqueness (up to the $(S'/Z_{G'})(k)$-action on G') to the rank-1 case. Consideration of open cells for SL_2 and PGL_2 settles uniqueness for rank 1.

Appendix B

Clifford constructions

Throughout this monograph we have seen the utility of passing to pseudo-reductive k-groups of minimal type when proving general theorems. However, over every imperfect field there exist standard absolutely pseudo-simple groups that are not of minimal type (see [CGP, Ex. 5.3.7]). The concept "locally of minimal type" introduced in Definition 4.3.1 is more robust: in the pseudo-semisimple case it is equivalent to the universal smooth k-tame central extension being of minimal type (Proposition 5.3.3), and in general it characterizes the output of the "generalized standard" construction (Theorem 9.2.1). Hence, it is natural to seek examples of absolutely pseudo-simple k-groups that are not locally of minimal type.

Let k be a field of characteristic 2. If $[k : k^2] \geqslant 16$ then an explicit construction in 4.2.2 produces non-standard pseudo-split absolutely pseudo-simple k-groups G with root system A_1 such that G is not locally of minimal type. The construction rests on a *commutative Clifford algebra*: for a subfield $K \subset k^{1/2}$ of degree 16 over k and 6-dimensional k-subspace $V \subset K$ as in (4.2.2), such G are built by using the Clifford algebra C associated to the squaring map $q : V \to k$ viewed as a quadratic form over k. (The k-algebra C is a non-reduced finite local commutative k-algebra with residue field K.)

Our aim in this appendix is to show that constructions based on commutative Clifford algebras are not limited to the rank-1 case: whenever $[k : k^2] \geqslant 16$ we use such triples $(K/k, V, C)$ to construct pseudo-split absolutely pseudo-simple k-groups not locally of minimal type, with root system B_n or C_n for any $n \geqslant 2$. These k-groups are built as smooth connected central extensions of pseudo-split generalized basic exotic pseudo-reductive k-groups that are *not* basic exotic. (The necessity of avoiding the basic exotic case is due to [CGP, Prop. 8.1.2].)

In §B.3 we show the bound $[k : k^2] \geqslant 16$ is optimal: if k is imperfect of characteristic 2 and $[k : k^2] < 16$ then every pseudo-reductive k-group with a *reduced* root system over k_s is locally of minimal type. The reducedness hypothesis is essential since if $[k : k^2] > 2$ then another technique in §B.4 (not involving Clifford algebras) yields pseudo-split absolutely pseudo-simple k-groups with root system BC_n ($n \geqslant 1$) that are not locally of minimal type.

B.1 Type B

Let $\mathscr{G}$ be a split connected semisimple k-group that is absolutely simple and simply connected with root system B_n for $n \geqslant 2$, and let $\mathscr{T} \subset \mathscr{G}$ be a split maximal k-torus. Let Δ be a basis of $\Phi = \Phi(\mathscr{G}, \mathscr{T})$, so $\Delta^\vee$ is a $\mathbf{Z}$-basis of $\mathrm{X}_*(\mathscr{T})$. For all $c \in \Phi$, $\mathscr{T}_c := c^\vee(\mathrm{GL}_1)$ is a maximal k-torus in $\mathscr{G}_c \simeq \mathrm{SL}_2$. Let $(K/k, V, C)$ be as in 4.2.2.

Clearly $\mathscr{T}$ is a maximal k-torus in $\mathrm{R}_{C/k}(\mathscr{G}_C)$. For the unique short root $b \in \Delta$, choose a k-isomorphism $\mathscr{G}_b \simeq \mathrm{SL}_2$ carrying the maximal k-torus $\mathscr{T}_b$ onto the diagonal k-torus. Define the pseudo-semisimple k-group $H := H_{V,C/k} \subset \mathrm{R}_{C/k}(\mathrm{SL}_2) = \mathrm{R}_{C/k}((\mathscr{G}_b)_C)$ as in 4.2.2. The commutative k-group

$$\mathscr{C} := Z_H(\mathscr{T}_b) \times \prod_{a \in \Delta - \{b\}} \mathscr{T}_a \subset \mathrm{R}_{C/k}(\mathscr{T}_C) \tag{B.1.0}$$

is pseudo-reductive since H is pseudo-reductive. By definition $\mathscr{C}$ normalizes H, and $H \cdot \mathscr{C} = H \rtimes \prod_{a \in \Delta - \{b\}} \mathscr{T}_a$, so $H \cdot \mathscr{C}$ is pseudo-reductive and contains $\mathscr{T}$ with $Z_{H \cdot \mathscr{C}}(\mathscr{T}) = \mathscr{C}$.

We conclude via [CGP, Thm. C.2.29] that the smooth connected k-subgroup $G \subset \mathrm{R}_{C/k}(\mathscr{G}_C)$ generated by H and $\{\mathscr{G}_a\}_{a \in \Delta - \{b\}}$ is pseudo-reductive with maximal k-torus $\mathscr{T}$ and satisfies the following additional properties: $Z_G(\mathscr{T}) = \mathscr{C}$, $\Phi(G, \mathscr{T})$ has basis Δ, and for $c \in \Delta$ the $\pm c$-root groups of $(G, \mathscr{T})$ coincide with those of $(H, \mathscr{T}_b)$ if $c = b$ and coincide with those of $(\mathscr{G}_c, \mathscr{T}_c)$ if $c \neq b$. In particular, by pseudo-semisimplicity of H we see that G is generated by its $\mathscr{T}$-root groups, so G is pseudo-semisimple. Moreover, for all $c \in \Delta$ the coroot $c^\vee$ associated to $(G, \mathscr{T})$ is the same as for $(\mathscr{G}, \mathscr{T})$, so $\Phi(G, \mathscr{T}) = \Phi$ inside $\mathrm{X}(\mathscr{T})$.

The minimal field of definition over k for the geometric unipotent radical of G_c is k if c is long and is K if c is short, as we can compute these extensions of k using $c \in \Delta$. Hence, by Proposition 3.2.5, the minimal field of definition over k for $\mathscr{R}_u(G_{\overline{k}}) \subset G_{\overline{k}}$ is K.

Lemma B.1.1. *The pseudo-reductive k-group G is not locally of minimal type.*

Proof. By construction, G_b is k-isomorphic to $H_{V,C/k}$, so $(G_b)^{\mathrm{ss}}_{\overline{k}} \simeq \mathrm{SL}_2$. Thus, according to Proposition 4.1.6, G_b does not admit a central extension that is absolutely pseudo-simple of minimal type. □

The natural map $\mathrm{R}_{C/k}(\mathscr{G}_C) \twoheadrightarrow \mathrm{R}_{K/k}(\mathscr{G}_K)$ carries G into a type-B generalized basic exotic k-group G of minimal type as in Proposition 8.2.5 using $V \subset K$. Let's show that for the k-group G that is not locally of minimal type, the natural map $f : G \to \mathsf{G}$ is surjective with central kernel isomorphic to α_2 if $v = e_1e_2 + e_3e_4$ and isomorphic to $\mathbf{Z}/(2)$ if $v = e_1e_2e_3e_4$ (with v as in 4.2.2).

By construction, f restricts to the identity map between the natural copies of $\mathscr{T}$ as a maximal k-torus of G and G, so it carries G_c into G_c for all $c \in \Phi$. The resulting map $f_c : G_c \to \mathsf{G}_c$ recovers the identity map on $\mathscr{G}_c$ for $c \in \Delta_>$, and since the composite map $V \hookrightarrow C \twoheadrightarrow K$ is the canonical inclusion it follows that f_b is identified with the natural map $H_{V,C/k} \to H_{V,K/k}$ that (as we reviewed in 4.2.2) is a central quotient map having kernel α_2 if $v = e_1e_2 + e_3e_4$ and kernel $\mathbf{Z}/(2)$ if $v = e_1e_2e_3e_4$. In view of the structure of $\mathscr{C} = Z_G(\mathscr{T})$ as a direct product, the Pseudo-Isogeny Theorem in §A.1 implies that f is a central quotient map with central kernel equal to $\ker f_b$.

B.2 Type C

Now let $\mathscr{G}$ be a split connected semisimple k-group that is absolutely simple and simply connected with root system C_n ($n \geqslant 2$), and let $\mathscr{T}$ be a split maximal k-torus. Let $\pi : \mathscr{G} \to \overline{\mathscr{G}}$ be the very special isogeny. The image $\overline{\mathscr{T}} = \pi(\mathscr{T})$ is a split maximal k-torus in $\overline{\mathscr{G}}$.

Let $(K/k, V, C)$ be as above. Applying the preceding B_n-construction to $(\overline{\mathscr{G}}, \overline{\mathscr{T}})$ yields a pseudo-split and absolutely pseudo-simple k-subgroup $\overline{G} \subset \mathrm{R}_{C/k}(\overline{\mathscr{G}}_C)$ with maximal k-torus $\overline{\mathscr{T}} \subset \mathrm{R}_{C/k}(\overline{\mathscr{T}}_C)$ such that $\overline{\Phi} := \Phi(\overline{G}, \overline{\mathscr{T}})$ is identified with $\Phi(\overline{\mathscr{G}}, \overline{\mathscr{T}}) = \mathrm{B}_n$ and if $c \in \overline{\Phi}_<$ then $\overline{G}_c \simeq H_{V,C/k}$ (so $\overline{G}$ is not locally of minimal type). We showed that the image of $\overline{G}$ under the quotient map $\mathrm{R}_{C/k}(\overline{\mathscr{G}}_C) \twoheadrightarrow \mathrm{R}_{K/k}(\overline{\mathscr{G}}_K)$ is a generalized basic exotic k-group $\overline{\mathsf{G}}$, and that the quotient map $\overline{f} : \overline{G} \twoheadrightarrow \overline{\mathsf{G}}$ has central kernel isomorphic to α_2 or $\mathbf{Z}/(2)$ (depending on V).

We are *not* going to search for the desired group with root system C_n by looking inside $\mathrm{R}_{C/k}(\mathscr{G}_C)$, but rather we will build it from $\overline{G}$ using a fiber product involving Weil restrictions from K rather than from C:

Proposition B.2.1. *The k-group G defined by the fiber square*

$$\begin{array}{ccc} G & \longrightarrow & \mathrm{R}_{K/k}(\mathscr{G}_K) \\ \big\downarrow & & \big\downarrow{\scriptstyle \mathrm{R}_{K/k}(\pi_K)} \\ \overline{G} & \longrightarrow & \mathrm{R}_{K/k}(\overline{\mathscr{G}}_K) \end{array}$$

is pseudo-split and absolutely pseudo-simple with root system C_n. It is not locally of minimal type.

Proof. Choose a basis Δ of the root system

$$\Phi := \Phi(\mathscr{G}, \mathscr{T}) = \Phi(\mathrm{R}_{K/k}(\mathscr{G}_K), \mathscr{T}) = \mathrm{C}_n,$$

and let a be the unique long root in Δ. Since $K^2 \subset k$, it follows from [CGP, Prop. 7.1.5(1)] that the very special isogeny $\pi : (\mathscr{G}, \mathscr{T}) \to (\overline{\mathscr{G}}, \overline{\mathscr{T}})$ carries $\mathscr{G}_a$ isomorphically onto $\overline{\mathscr{G}}_{\overline{a}}$, so $\mathrm{R}_{K/k}(\pi_K)$ carries $\mathrm{R}_{K/k}((\mathscr{G}_a)_K)$ isomorphically onto $\mathrm{R}_{K/k}((\overline{\mathscr{G}}_{\overline{a}})_K)$. Thus, the k-subgroup $\overline{\mathsf{G}}_{\overline{a}} \simeq H_{V,K/k}$ of $\mathrm{R}_{K/k}((\overline{\mathscr{G}}_{\overline{a}})_K)$ is identified with its inverse image in $\mathrm{R}_{K/k}((\mathscr{G}_a)_K)$ that we define to be G_a. By (the proof of) Proposition 8.2.5, inside $\mathrm{R}_{K/k}(\mathscr{G}_K)$ the k-subgroup G_a and the k-subgroups $\mathrm{R}_{K/k}((\mathscr{G}_c)_K)$ for $c \in \Phi_<$ generate a type-C generalized basic exotic k-subgroup $\mathsf{G} \subset \mathrm{R}_{K/k}(\mathscr{G}_K)$ containing $\mathscr{T}$.

By [CGP, Prop. 7.1.5(2)], $\mathrm{R}_{K/k}(\pi_K)(\mathsf{G}) = \overline{\mathsf{G}}$. The kernel $\ker \mathrm{R}_{K/k}(\pi_K) = \mathrm{R}_{K/k}(\ker \pi_K)$ is directly spanned by closed k-subgroup schemes of the form $\mathrm{R}_{K/k}(\mu_2)$ and $\mathrm{R}_{K/k}(\alpha_2)$ inside the k-groups $\mathrm{R}_{K/k}((\mathscr{G}_c)_K)$ for $c \in \Phi_<$ due to the description of $\mathrm{Lie}(\ker \pi_K)$ given in [CGP, Lemma 7.1.2], so $\ker \mathrm{R}_{K/k}(\pi_K)$ is connected (as $K^2 \subset k$) and is contained inside G. Thus, $\mathsf{G} = \mathrm{R}_{K/k}(\pi_K)^{-1}(\overline{\mathsf{G}})$ and $G = \overline{G} \times_{\overline{\mathsf{G}}} \mathsf{G}$, so $G \to \mathsf{G}$ is faithfully flat with central kernel α_2 or $\mathbf{Z}/(2)$ (depending on V) and $G \to \overline{G}$ is faithfully flat with connected kernel $\mathrm{R}_{K/k}(\ker \pi_K)$. The k-group G therefore inherits connectedness from $\overline{G}$.

An inspection of compatible open cells for the pairs

$$(\overline{G}, \overline{\mathscr{T}}),\ (\overline{\mathsf{G}}, \overline{\mathscr{T}}),\ (\mathsf{G}, \mathscr{T})$$

shows that the fiber product G is *smooth*, so G inherits pseudo-reductivity from its isogenous central quotient G. By centrality of the finite kernel of $G \twoheadrightarrow \mathsf{G}$, we see that G is absolutely pseudo-simple with $\Phi(G, \mathscr{T})$ equal to $\Phi(\mathsf{G}, \mathscr{T}) = \Phi = \mathrm{C}_n$. For any $c \in \Phi_>$ and the corresponding $\overline{c} \in \overline{\Phi}_<$, $\pi : \mathscr{G}_c \to \overline{\mathscr{G}}_{\overline{c}}$ is an isomorphism by [CGP, Prop. 7.1.5(1)]. Thus, $\mathrm{R}_{K/k}(\pi_K)$ carries G_c isomorphically onto $\overline{G}_{\overline{c}} \simeq H_{V,C/k}$, so G is not locally of minimal type (via the same reasoning as we used for the type-B analogue). □

B.3 Cases with $[k : k^2] \leqslant 8$

The following result was brought to our attention by Gabber.

Proposition B.3.1. *Let k be imperfect of characteristic 2 such that $[k : k^2] \leqslant 8$. If G is a pseudo-reductive k-group whose root system over k_s is reduced then G is locally of minimal type.*

The reducedness hypothesis is necessary; see Examples B.4.1 and B.4.3.

Proof. We may assume $k = k_s$ and G is absolutely pseudo-simple of rank 1 with root system A_1. By Proposition 5.3.3 we may replace G with $\widetilde{G}$ so that $G^{ss}_{\overline{k}} = SL_2$. Let K/k be the minimal field of definition for the geometric unipotent radical of G, so the canonical minimal type central quotient $G/\mathscr{C}_G$ is of the form $H_{V,K/k}$ for a nonzero kK^2-subspace $V \subset K$ containing 1 and satisfying $k[V] = K$. It suffices to show that no such $H_{V,K/k}$ has a nontrivial central extension by an affine finite type k-group scheme Z with no nontrivial smooth connected k-subgroup (e.g., $Z = \mathscr{C}_G$).

We shall use the splitting criterion in [CGP, Prop. 5.1.3]: for the diagonal k-torus $D \subset H_{V,K/k}$ and its centralizer $V^*_{K/k}$, it suffices to express $V^*_{K/k}$ rationally in terms of the D-root groups $U^{\pm}_V$ of $H_{V,K/k}$. That is, we seek an ordered finite sequence of rational maps $h_j : V^*_{K/k} \dashrightarrow U^{\pm}_V$ such that the product $\prod h_j : V^*_{K/k} \dashrightarrow H_{V,K/k}$ coincides (as a rational map) with the canonical inclusion.

Let $F = \{\lambda \in K \mid \lambda \cdot V \subset V\}$, so $kK^2 \subset F$ and Example 3.3.5 gives that $H_{V,K/k} = \mathscr{D}(R_{F/k}(H_{V,K/F}))$. Note that $[F:F^2] = [k:k^2]$ (see Remark 8.3.1 for a general such degree identity in any positive characteristic). If $\dim_F(V) \leqslant 2$ then $V = F[V] = k[V] = K$ and $G = R_{K/k}(SL_2)$, in which case h_j's are provided by the classical formula

$$\mathrm{diag}(t, 1/t) = u_+(t)u_-(-1/t)u_+(t-1)u_-(1)u_+(-1) \qquad \text{(B.3.1)}$$

(with the standard parameterizations $u_{\pm} : \mathbf{G}_a \simeq U^{\pm}$ of the standard root groups of SL_2) expressing a diagonal point of SL_2 universally as a finite product of points in the standard root groups. Hence, we may assume $\dim_F(V) > 2$, so $[K:F] \geqslant 4$.

It is a standard fact that if E'/E is a finite extension of fields with characteristic $p > 0$ then $[E' : EE'^p] \leqslant [E : E^p]$, so $[K : kK^2] \leqslant [k : k^2] \leqslant 8$. For the convenience of the reader, we briefly recall the proof. Since $\Omega^1_{E/\mathbf{F}_p} = \Omega^1_{E/E^p}$ and $\Omega^1_{E'/E} = \Omega^1_{E'/EE'^p}$, by the relation between p-bases and differential bases [Mat, Thm. 26.5] it is equivalent to show that

$$\dim_{E'} \Omega^1_{E'/E} \leqslant \dim_{E'}(E' \otimes_E \Omega^1_{E/\mathbf{F}_p}).$$

This inequality follows from a formula of Cartier [Mat, Thm. 26.10], which gives that

$$\dim_{E'} \Omega^1_{E'/E} = \dim_{E'} \ker(E' \otimes_E \Omega^1_{E/\mathbf{F}_p} \to \Omega^1_{E'/\mathbf{F}_p}).$$

Now we may assume $[K:F]=4,8$ with $2<\dim_F(V)<[K:F]$. Suppose $[K:F]=4$, so $\dim_F(V)=3$. Since $1\in V$, clearly V has an F-basis of the form $\{1,e,e'\}$ with ee' extending this to an F-basis of K (because $F(e)=F\oplus Fe$ and $[K:F(e)]=2$). For $z,z'\in \mathrm{R}_{F[e]/k}(\mathbf{G}_{\mathrm{a}})$ such that $z\in \mathrm{R}_{F[e]/k}(\mathrm{GL}_1)$, the point $z+z'e'$ in $\mathrm{R}_{K/k}(\mathbf{G}_{\mathrm{a}})$ has the form $z(1+ye')$ for $y:=z'/z\in \mathrm{R}_{F[e]/k}(\mathbf{G}_{\mathrm{a}})$. But $1+ye'=((1/e'^2)e'+y)e'$ and the D-root groups of $H_{V,K/k}$ correspond to $V=\mathrm{span}_F(1,e,e')\subset K$ inside the D-root groups of $\mathrm{R}_{K/k}(\mathrm{SL}_2)$, so we get the required rational maps h_j via (B.3.1). This also shows that $V^*_{K/k}=\mathrm{R}_{K/k}(\mathrm{GL}_1)$ when $[K:F]=4$.

Suppose $[K:F]=8$, so $F=kK^2$. We note that $\dim_{kK^2}(V)\geqslant 4$, for otherwise $[K:kK^2]=\dim_{kK^2}(k[V])\leqslant 4$. As $1\in V$ and $k[V]=K$, V must contain a 2-basis $\{e,e',e''\}$ of K/k. Let V_0 be the kK^2-span of $\{1,e,e',e''\}$, so $\dim(V_0)^*_{K/k}\leqslant \dim \mathrm{R}_{kK^2/k}((V_0)^*_{K/kK^2})\leqslant 7[kK^2:k]$ by [CGP, Prop. 9.1.9].

Lemma B.3.2. *The k-group $(V_0)^*_{K/k}$ has dimension $7[kK^2:k]$ (so $(V_0)^*_{K/k}=\mathrm{R}_{kK^2/k}((V_0)^*_{K/kK^2})$) and $(V_0)^*_{K/k}$ is expressed rationally in terms of the root groups of $H_{V_0,K/k}$ relative to the diagonal k-torus.*

Proof. Consider the map of k-schemes

$$f:\mathrm{R}_{kK^2/k}(\mathrm{GL}_1)\times \mathrm{R}_{kK^2/k}(\mathbf{G}_{\mathrm{a}})^6\to \mathrm{R}_{K/k}(\mathbf{G}_{\mathrm{a}})$$

that carries $(c,x_0,y_0,y_1,z_0,z_1,z_2)$ to

$$c(1+x_0e)(1+(y_0+y_1e)e')(1+(z_0+z_1e+z_2e')e'').$$

Let $\Omega\subset \mathrm{R}_{kK^2/k}(\mathbf{G}_{\mathrm{a}})^6$ be the Zariski-dense open locus defined by

$$1+x_0e,\,1+(y_0+y_1e)e',\,1+(z_0+z_1e+z_2e')e''\in \mathrm{R}_{K/k}(\mathrm{GL}_1),$$

so f carries $U:=\mathrm{R}_{kK^2/k}(\mathrm{GL}_1)\times\Omega$ into $(V_0)^*_{K/k}$ because the factors

$$\begin{aligned}1+(y_0+y_1e)e'&=e'(y_0+y_1e+e'^{-2}\cdot e'),\\ 1+(z_0+z_1e+z_2e')e''&=e''(z_0+z_1e+z_2e'+e''^{-2}e'')\end{aligned}\qquad\text{(B.3.2)}$$

lie in $e'\underline{V}_0$ and $e''\underline{V}_0$ respectively. Note that $\dim U=7[kK^2:k]$.

We claim that $f|_U$ is a monomorphism. Granting this, let us see how to conclude. By monicity applied to dual-number points, $f:U\to(V_0)^*_{K/k}$ is injective on tangent spaces at $\overline{k}$-points. Thus, by smoothness and dimension considerations, $\dim(V_0)^*_{K/k}=7[kK^2:k]$ and f is étale on U. But an étale monomorphism is an open immersion [EGA, IV$_4$, 17.9.1], so $f|_U$ is an open immersion and hence (via (B.3.1) and (B.3.2)) $(V_0)^*_{K/k}$ is expressed rationally in terms of root groups of $H_{V_0,K/k}$ relative to the diagonal k-torus. It therefore remains to prove monicity on U.

Note that $K = kK^2[e] \oplus kK^2[e]e' \oplus kK^2[e,e']e''$. Consider the k-scheme map

$$f' : \mathrm{R}_{kK^2/k}(\mathrm{GL}_1 \times \mathbf{G}_\mathrm{a}) \times \mathrm{R}_{kK^2[e]/k}(\mathbf{G}_\mathrm{a}) \times \mathrm{R}_{kK^2[e,e']/k}(\mathbf{G}_\mathrm{a}) \to \mathrm{R}_{K/k}(\mathbf{G}_\mathrm{a})$$

defined by $f'((u,x),y,z) = u(1+xe)(1+ye')(1+ze'')$. Let Ω' be the Zariski-open locus of points (x,y,z) such that $1+xe, 1+ye', 1+ze'' \in \mathrm{R}_{K/k}(\mathrm{GL}_1)$. We can write any point of $\mathrm{R}_{K/k}(\mathbf{G}_\mathrm{a})$ in the unique form $a_0 + a_1e + be' + ce''$ where $a_0, a_1 \in \mathrm{R}_{kK^2/k}(\mathbf{G}_\mathrm{a})$, $b \in \mathrm{R}_{kK^2[e]/k}(\mathbf{G}_\mathrm{a})$, and $c \in \mathrm{R}_{kK^2[e,e']/k}(\mathbf{G}_\mathrm{a})$. The combined conditions

$$a_0, a_0 + a_1e, a_0 + a_1e + be', a_0 + a_1e + be' + ce'' \in \mathrm{R}_{K/k}(\mathrm{GL}_1)$$

define a dense open subscheme of $\mathrm{R}_{K/k}(\mathbf{G}_\mathrm{a})$, and f' carries $\mathrm{R}_{kK^2/k}(\mathrm{GL}_1) \times \Omega'$ isomorphically onto that open subscheme. Since $\Omega \subset \Omega'$, it follows that $f|_U$ is monic. (This argument adapts in an evident manner to more general towers of field extensions in place of $K/kK^2/k$.) □

Now we may suppose $V \neq V_0$, so $\dim_{kK^2} V \geqslant 5$ and hence we can apply:

Lemma B.3.3. *Let F be a field of characteristic 2 and K/F an extension of degree 8 with $K^2 \subset F$. Let V be an F-subspace of K containing F such that $\dim_F(V) \geqslant 5$.*

If $\dim_F(V) \geqslant 6$ then V contains a 6-dimensional F-subspace V' such that F is strictly contained in $\{\lambda \in K \mid \lambda \cdot V' \subset V'\}$, and if $\dim_F(V) = 5$ then there is a 2-basis $\{e, e', e''\}$ of K over F such that $V = \mathrm{span}_F\{1, e, e', e'', v\}$ where $v = ee'$ or $v = ee'e''$. If $v = ee'e''$ then $e''V$ is the F-span of $\{1, \varepsilon, \varepsilon', \varepsilon'', \varepsilon\varepsilon'\}$ for the 2-basis $\varepsilon = ee'$, $\varepsilon' = e'e''$, $\varepsilon'' = e''$ of K over F.

Proof. First assume $\dim_F(V) = 5$. Clearly $F[V] = K$, so V certainly contains a 2-basis $\{e_1, e_2, e_3\}$ of K over F. Thus, V has an F-basis of the form $\{1, e_1, e_2, e_3, v\}$ where v involves some e_ie_j $(i < j)$ or $e_1e_2e_3$. Suppose v does not involve $e_1e_2e_3$, so we may take it to be

$$v = e_1e_2 + ae_2e_3 + be_1e_3$$

for some $a, b \in F$ that we may assume are not both zero. Hence, we can assume $b = 1$, so

$$v = e_1(e_2 + e_3) + ae_2e_3 = (e_1 + ae_3)(e_2 + e_3) + ae_3^2.$$

Since $e_3^2 \in F$, we can use the 2-basis $\{e_1 + ae_3, e_2 + e_3, e_3\}$ to arrive at the case $v = ee'$.

Suppose v involves $e_1e_2e_3$. By replacing each e_j with $e_j - c_j$ for some $c_j \in F$ we can eliminate all e_ie_j's $(i < j)$ that appear in v, so we arrive at the case $v = ee'e''$. It is trivial to check that in this case $\{ee', e'e'', e''\}$ is a 2-basis of K over F that presents $e''V$ as being in the preceding case.

If $\dim_F(V) \geqslant 6$ then by the preceding considerations we can choose the 2-basis so that V contains ee'. Hence, either V contains $F(e, e') \oplus F(ee')e''$ or (after change in the 2-basis) V contains $F(e, e') \oplus F(e)e''$. These 6-dimensional F-subspaces V' are subspaces over a strictly larger subfield of K than F. □

If $\dim_{kK^2}(V) \geqslant 6$ then Lemma B.3.3 provides a kK^2-subspace $V' \subset V$ of dimension 6 such that $F' := \{\lambda \in K \mid \lambda \cdot V' \subset V'\}$ strictly contains kK^2, so $[K : F'] \leqslant 4$. Our earlier arguments imply that $V'^*_{K/k} = \mathrm{R}_{K/k}(\mathrm{GL}_1)$ and that this is expressed rationally in terms of the standard root groups of $H_{V',K/k} \subset H_{V,K/k}$. We may therefore assume $\dim_{kK^2}(V) = 5$. Since V only matters up to $K^\times$-multiple, by Lemma B.3.3 we may replace it with a suitable such multiple so that V is the kK^2-span of $\{1, e, e', e'', v = ee'\}$ for some 2-basis $\{e, e', e''\}$ of K/k (equivalently, of K/kK^2).

The proof of Lemma B.3.2 provides a dense open $\Omega' \subset \mathrm{R}_{kK^2/k}(\mathbf{G}_\mathrm{a})^7$ and an open immersion

$$f' : \mathrm{R}_{kK^2/k}(\mathrm{GL}_1) \times \Omega' \to \mathrm{R}_{K/k}(\mathrm{GL}_1)$$

that lands inside $V^*_{K/k}$ because $f'|_{\{1\}\times\Omega'}$ is a product of maps valued in $\underline{V}$, $e'\underline{V}$, $e''\underline{V}$ respectively. Thus, $V^*_{K/k} = \mathrm{R}_{K/k}(\mathrm{GL}_1)$ and $V^*_{K/k}$ is rationally expressed in terms of standard root groups of $H_{V,K/k}$. This completes the proof of Proposition B.3.1. □

Proposition B.3.4. *Let k be imperfect of characteristic 2 such that $[k : k^2] \leqslant 8$. For any generalized exotic pseudo-semisimple k-group G and commutative affine finite type k-group scheme Z that does not contain a nontrivial smooth connected k-subgroup, every central extension of G by Z uniquely splits.*

The interest in this proposition is due to the role of central extensions (1.4.1.2) in [CGP] when $\mathrm{char}(k) \neq 2$ or $\mathrm{char}(k) = 2$ with $[k:k^2] = 2$.

Proof. We shall apply the splitting criterion in [CGP, Prop. 5.1.3]. It is sufficient to verify this criterion when $k = k_s$ (as we now assume). Let T be a (split) maximal k-torus, and Δ a basis of $\Phi(G,T)$. In view of the direct product decomposition in Proposition 4.3.3, it suffices to show that for each $a \in \Delta$ and the split maximal k-torus $T_a := T \cap G_a$ in G_a, the Cartan k-subgroup $Z_{G_a}(T_a)$ of G_a is expressed rationally in terms of the root groups for (G_a, T_a). This property of the minimal type rank-1 pseudo-simple groups G_a is exactly what was established in the proof of Proposition B.3.1 (since $[k:k^2] \leqslant 8$). □

As an application of Proposition B.3.1 and Proposition 3.1.9, we now remove the "minimal type" hypothesis from the maximality property of (2.3.13), subject to some restrictions when $\mathrm{char}(k) = 2$. This is of interest even in the standard case since over every imperfect field there exist standard absolutely pseudo-simple groups that are not of minimal type [CGP, Ex. 5.3.7].

Proposition B.3.5. *Let k be an arbitrary field, and if* $\mathrm{char}(k) = 2$ *then assume* $[k:k^2] \leqslant 8$. *Let K/k be a purely inseparable finite extension and let G' be a pseudo-semisimple K-group. Assume the root system of G'_{K_s} is reduced (automatic when* $\mathrm{char}(k) \neq 2$).

Via the natural quotient map $q : \mathscr{D}(\mathrm{R}_{K/k}(G'))_K \twoheadrightarrow G'$, the induced map $\mathscr{D}(\mathrm{R}_{K/k}(G'))_K^{\mathrm{pred}} \twoheadrightarrow G'$ is an isomorphism. In particular, G' is canonically determined by K/k and the k-group $\mathscr{D}(\mathrm{R}_{K/k}(G'))$.

In §B.4 we will see that the reducedness hypothesis on the root system cannot be dropped when $\mathrm{char}(k) = 2$ and $[k:k^2] > 2$.

Proof. We claim that for any pseudo-semisimple K-group H' such that H'_{K_s} has a *reduced* root system, if $(H' \otimes_K \overline{k})^{\mathrm{ss}}$ is simply connected then H' is of minimal type. To prove this, by Proposition 4.3.3 it is equivalent to show that H' is locally of minimal type. If $\mathrm{char}(k) \neq 2$ then by Proposition 3.1.9 (applied over K_s) every pseudo-reductive K-group is locally of minimal type. The same holds if $\mathrm{char}(k) = 2$ by Proposition B.3.1 since $[K:K^2] = [k:k^2] \leqslant 8$ (degree equality proved in Remark 8.3.1).

Assume $(G' \otimes_K \overline{k})^{\mathrm{ss}}$ is simply connected, so G' is of minimal type. For the pseudo-semisimple $\mathscr{G}' := \mathscr{D}(\mathrm{R}_{K/k}(G'))_K^{\mathrm{pred}}$ and the unique factorization

$$\mathscr{D}(\mathrm{R}_{K/k}(G'))_K \twoheadrightarrow \mathscr{G}' \twoheadrightarrow G'$$

of q, we shall prove that $\mathscr{G}' \twoheadrightarrow G'$ is an isomorphism. Since $\mathscr{G}' \twoheadrightarrow G'$ is a quotient of minimal type, it factors through $\mathscr{G}'/\mathscr{C}_{\mathscr{G}'}$; i.e., $\mathscr{G}'/\mathscr{C}_{\mathscr{G}'}$ dominates G' as pseudo-reductive quotients of $\mathscr{D}(\mathrm{R}_{K/k}(G'))_K$ of minimal type. The maximality of G' as such a quotient follows from Proposition 2.3.13, so $\mathscr{G}'/\mathscr{C}_{\mathscr{G}'} = G'$. Hence, $\mathscr{G}'_{K_s}$ and G'_{K_s} have the same root data, so $(\mathscr{G}' \otimes_K \overline{k})^{\mathrm{ss}}$ is simply connected (as we are temporarily assuming $(G' \otimes_K \overline{k})^{\mathrm{ss}}$ is simply connected) and $\mathscr{G}'_{K_s}$ has a *reduced* root system. Thus, $\mathscr{G}'$ is of minimal type and so $\mathscr{G}' = G'$ as desired.

Now consider the general case, and again let $\mathscr{G}' = \mathscr{D}(\mathrm{R}_{K/k}(G'))_K^{\mathrm{pred}}$. Let $f : \mathscr{D}(\mathrm{R}_{K/k}(G'))_K \to \mathscr{G}'$ be the natural surjective K-homomorphism. (We do *not* assume that the root system of $\mathscr{G}'_{K_s}$ is reduced.) To prove the proposition in general, it suffices to show that f factors uniquely through q. Uniqueness follows from the surjectivity of q. Letting $\pi : \widetilde{G}' \to G'$ be the universal smooth K-tame central extension, by centrality $\widetilde{G}'$ has the same associated field K/k (see Proposition 3.2.6) and π_{K_s} induces an isomorphism between root systems and an isomorphism between root groups for corresponding roots over K_s. Hence, $\widetilde{G}'_{K_s}$ has a reduced root system, so the desired result has been settled for $\widetilde{G}'$, and consideration of root groups over K_s shows that $\mathscr{D}(\mathrm{R}_{K/k}(\pi))_K$ is surjective. Thus, via f we may view $\mathscr{G}'$ as a pseudo-reductive quotient of $\mathscr{D}(\mathrm{R}_{K/k}(\widetilde{G}'))_K$ and so (since the original problem is settled for $\widetilde{G}'$) we get a unique factorization

$$\mathscr{D}(\mathrm{R}_{K/k}(\widetilde{G}'))_K \xrightarrow{\widetilde{q}} \widetilde{G}' \xrightarrow{\varphi} \mathscr{G}'$$

of $f \circ \mathscr{D}(\mathrm{R}_{K/k}(\pi))_K$. We claim that φ factors through π and that the resulting map $G' \to \mathscr{G}'$ provides a factorization of f through q.

In the commutative diagram

$$\begin{array}{ccccc}
\widetilde{G}' & \xrightarrow{\widetilde{\iota}} & \mathscr{D}(\mathrm{R}_{K/k}(\widetilde{G}'))_K & \xrightarrow{\widetilde{q}} & \widetilde{G}' \\
{\scriptstyle \pi}\downarrow & & \downarrow & & \downarrow{\scriptstyle \pi} \\
G' & \xrightarrow{\iota} & \mathscr{D}(\mathrm{R}_{K/k}(G'))_K & \xrightarrow[q]{} & G'
\end{array}$$

(with canonical inclusions $\widetilde{\iota}$ and ι), the horizontal compositions are the identity maps. Thus, the existence of a factorization of φ through π may be checked after composing it with $\widetilde{q} \circ \widetilde{\iota}$. Since $\varphi \circ \widetilde{q} = f \circ \mathscr{D}(\mathrm{R}_{K/k}(\pi))_K$, the commutativity of the left square does the job. It is clear from the surjectivity of the vertical maps in the diagram that the map $G' \to \mathscr{G}'$ through which φ uniquely factors has composition with q that recovers f. $\square$

B.4 Type BC

If $[k:k^2]=2$ then every pseudo-reductive k-group is locally of minimal type (see Remark 4.3.2). But when $[k:k^2]>2$, we shall construct pseudo-split pseudo-simple k-groups G with a non-reduced root system such that G is not of minimal type. As $G^{\rm ss}_{\overline{k}}$ is simply connected, by Proposition 4.3.3 such G cannot be locally of minimal type. The following example in rank 1 was suggested by Gabber. For examples of such G with rank >1, see Example B.4.3.

Example B.4.1. Let k be a field of characteristic 2 such that $[k:k^2]>2$, so we can choose a subfield $K\subset k^{1/2}$ with degree 4 over k. Explicitly, $K=k(u_1,u_2)$ where $u_1^2=e$, $u_2^2=e'$ for $\{e,e'\}$ part of a 2-basis of k/k^2. Let $K'=k(\sqrt{u_1},\sqrt{u_2})\subset K^{1/2}$, so $[K':K]=4$, and define $V\subset K'$ to be the K-span of $\{1,\sqrt{u_1},\sqrt{u_2}\}$. Let $H=H_{V,K'/K}\subset \mathrm{R}_{K'/K}(\mathrm{SL}_2)$.

The k-group $\mathrm{R}_{K/k}(H)$ has derived group $H_{V,K'/k}$ by (3.1.6). For the k-subspace $V\subset K'$ we have $K=\{\lambda\in K'\,|\,\lambda\cdot V\subset V\}$ with $[K':K]=4$ and $\dim_K V=3$, so the arguments for the degree-4 case in the proof of Proposition B.3.1 imply $V^*_{K'/k}=\mathrm{R}_{K'/k}(\mathrm{GL}_1)$ with dimension 16. Hence, the inclusion $H_{V,K'/k}\subset \mathrm{R}_{K/k}(H)$ inside $\mathrm{R}_{K/k}(\mathrm{SL}_2)$ is an equality by comparison of Cartan k-subgroups and the associated root groups, so $\mathrm{R}_{K/k}(H)$ is pseudo-simple and the minimal field of definition over k for its geometric unipotent radical is $k\langle V\rangle=K'$.

For $\mathbf{1}:=\mathrm{diag}(1,-1)$, let $\mathfrak{n}=\mathfrak{sl}_2(K)+V\cdot\mathbf{1}$ inside $\mathrm{Lie}(H)\subset\mathfrak{sl}_2(K')$ (so $\mathfrak{n}$ is the direct sum of $\mathfrak{sl}_2(K)$ and any K-plane in $V\cdot\mathbf{1}$ complementary to $K\cdot\mathbf{1}$). This is a p-Lie subalgebra (with $p=2$), so by [CGP, Prop. A.7.14, Ex. A.7.16] it is the Lie algebra of a unique K-subgroup scheme $N\subset\ker F_{H/K}$ of the Frobenius kernel of H. Since H is generated by its root groups relative to the diagonal K-torus, by computing the adjoint action of such root groups we see that the K-subspace $\mathfrak{n}\subset\mathrm{Lie}(H)$ is Ad_H-stable. Hence, N is normal in H [CGP, Ex. A.7.16], so $\mathrm{R}_{K/k}(N)$ is a closed normal k-subgroup scheme of $\mathrm{R}_{K/k}(H)$.

Consider the k-group

$$G:=\mathrm{R}_{K/k}(H)/\mathrm{R}_{K/k}(N),$$

so $\mathscr{R}_u(G_{\overline{k}})=\mathscr{R}_{u,K'}(G_{K'})_{\overline{k}}$. Clearly

$$\dim \mathrm{R}_{K/k}(H)=[K:k]([K':K]+2\dim_K(V))=10[K:k],$$

and if we ignore the group scheme structure then $N=\alpha_2^5$, so $\mathrm{R}_{K/k}(N)$ has dimension $5([K:k]-1)$ since $\mathrm{R}_{K/k}(\alpha_2)$ has dimension $[K:k]-1$ (as $K^2\subset k$).

Thus, $\dim G = 5[K:k]+5 = 25$. We will show that G is pseudo-semisimple with root system BC_1 and that it is *not* of minimal type.

Step 1. Let D be the diagonal maximal k-torus in $\mathrm{R}_{K/k}(H) = H_{V,K'/k}$, with image T in G. The k-group $Z_G(T)$ is the image of the Cartan k-subgroup $\mathrm{R}_{K'/k}(\mathrm{GL}_1) = Z_{\mathrm{R}_{K/k}(H)}(D)$ of $\mathrm{R}_{K/k}(H)$, so

$$Z_G(T) = \mathrm{R}_{K'/k}(\mathrm{GL}_1)/\mathrm{R}_{K/k}(N_0)$$

where $N_0 := N \cap \mathrm{R}_{K'/K}(\mathrm{GL}_1)$ is the K-subgroup scheme killed by Frobenius for which $\mathrm{Lie}(N_0) = V \subset K' = \mathrm{Lie}(\mathrm{R}_{K'/k}(\mathrm{GL}_1))$. Clearly $\dim \mathrm{R}_{K/k}(N_0) = 3([K:k]-1) = 9$, so $\dim Z_G(T) = 16-9 = 7$.

Now we prove $Z_G(T)$ is pseudo-reductive. The inclusion

$$N_0 \subset \mathrm{R}_{K'/K}(\mathrm{GL}_1)[2] = \mathrm{R}_{K'/K}(\mu_2)$$

yields a map

$$Z_G(T) \twoheadrightarrow \mathrm{R}_{K'/k}(\mathrm{GL}_1)/\mathrm{R}_{K'/k}(\mu_2) = [2](\mathrm{R}_{K'/k}(\mathrm{GL}_1)) \subset \mathrm{R}_{K'/k}(\mathrm{GL}_1),$$

so since $\mathrm{R}_{K'/k}(\mathrm{GL}_1)$ is commutative and pseudo-reductive we see that

$$\mathscr{R}_{u,k}(Z_G(T)) \subset \mathrm{R}_{K'/k}(\mu_2)/\mathrm{R}_{K/k}(N_0) = \mathrm{R}_{K/k}(\mathrm{R}_{K'/K}(\mu_2))/\mathrm{R}_{K/k}(N_0).$$

We shall prove this latter quotient has no nontrivial k_s-points, so it would follow that $\mathscr{R}_{u,k}(Z_G(T))$ is trivial.

Since $\mathrm{R}_{K'/K}(\mu_2)/\mu_2 = \mathrm{R}_{K'/K}(\mathrm{GL}_1)/\mathrm{GL}_1$ is a K-form of $\mathbf{G}_{\mathrm{a}}^3$ for the fppf topology, its Frobenius kernel is an fppf form of α_2^3 and thus is K-isomorphic to α_2^3 (because α_p^r has no nontrivial fppf forms over any field of characteristic p, as its automorphism functor is represented by GL_r). The Frobenius kernel M of $\mathrm{R}_{K'/K}(\mu_2)$ has order 2^4 since it coincides with the Frobenius kernel of the smooth 4-dimensional $\mathrm{R}_{K'/K}(\mathrm{GL}_1)$, so M is a commutative extension of α_2^3 by μ_2. Hence, via the inclusion $\mu_2 \subset N_0$ inside $\mathrm{R}_{K'/K}(\mu_2)$ we see that $M/N_0 \simeq \alpha_2$. The underlying K-schemes of M and N_0 (ignoring their K-group structure) are α_2^4 and α_2^3 respectively, so $\mathrm{R}_{K/k}(M)$ and $\mathrm{R}_{K/k}(N_0)$ have respective dimensions $4([K:k]-1) = 12$ and $3([K:k]-1) = 9$. Hence, the closed immersion $\mathrm{R}_{K/k}(M)/\mathrm{R}_{K/k}(N_0) \hookrightarrow \mathrm{R}_{K/k}(M/N_0) = \mathrm{R}_{K/k}(\alpha_2)$ is an isomorphism since $\mathrm{R}_{K/k}(\alpha_2)$ is reduced and irreducible of dimension 3 (defined by $x^2+ey^2+e'z^2+ee'w^2 = 0$).

The k-group $\mathrm{R}_{K'/k}(\mu_2)$ is defined by the system of four equations

$$x_j^2 + ey_j^2 + e'z_j^2 + ee'w_j^2 = c_j$$

for $0 \leqslant j \leqslant 3$ with $c_0 = 1$ and $c_1, c_2, c_3 = 0$, so it is *irreducible* of dimension 12. Thus, the inclusion

$$\mathrm{R}_{K/k}(\alpha_2) = \mathrm{R}_{K/k}(M)/\mathrm{R}_{K/k}(N_0) \hookrightarrow \mathrm{R}_{K'/k}(\mu_2)/\mathrm{R}_{K/k}(N_0)$$

is between irreducible k-groups of dimension 3, so it is an equality of underlying reduced schemes and hence the target has no nontrivial k_s-points (since $\mathrm{R}_{K/k}(\alpha_2)(k_s) = \alpha_2(K_s) = 0$). This concludes the proof that $Z_G(T)$ is pseudo-reductive.

Step 2. To prove that $\mathscr{R}_{u,k}(G)$ is trivial, we first note that any T-weight occurring on $\mathrm{Lie}(\mathscr{R}_{u,k}(G))$ must be nontrivial since $Z_G(T)$ is commutative and pseudo-reductive. Hence, it suffices to rule out the occurrence of nontrivial T-weights on this Lie algebra.

The diagonal k-torus $D \subset H_{V,K'/k} = \mathrm{R}_{K/k}(H)$ meets $\mathrm{R}_{K/k}(N)$ in $D[2]$, so $T = D/D[2]$ in G. Let $\lambda : \mathrm{GL}_1 \simeq T$ be the isomorphism such that 2λ is induced by composing the quotient map $D \twoheadrightarrow D/D[2] = T$ with the isomorphism $\mathrm{GL}_1 \simeq D$ defined by $t \mapsto \mathrm{diag}(t, 1/t)$. Let $a : T \simeq \mathrm{GL}_1$ be the inverse isomorphism. Define $U_\pm = U_{\mathscr{R}_{u,k}(G)}(\pm\lambda) = U_{\langle\pm a\rangle}(\mathscr{R}_{u,k}(G))$, so $\mathrm{Lie}(U_\pm)$ is the span of the T-weight spaces in $\mathrm{Lie}(\mathscr{R}_{u,k}(G))$ for the T-weights having a fixed sign (relative to a).

The K-subgroups N and SL_2 in $H \subset \mathrm{R}_{K'/K}(\mathrm{SL}_2)$ have intersection equal to $\ker F_{\mathrm{SL}_2/K}$, so $\mathrm{R}_{K/k}(N)$ meets the evident Levi k-subgroup $L := \mathrm{SL}_2 \subset \mathrm{R}_{K/k}(H)$ in $\ker F_{L/k}$ (as we check by computing with points valued in k-algebras). Hence, the isomorphism $\mathrm{R}_{K/k}(H)^{\mathrm{red}}_{K'} \simeq L_{K'}$ carries $\mathrm{R}_{K/k}(N)_{K'}$ onto $(\ker F_{L/k})_{K'}$, so $G^{\mathrm{red}}_{K'} \simeq L^{(2)}_{K'} \simeq \mathrm{SL}_2$. This provides a canonical k-homomorphism

$$f : G \to \mathrm{R}_{K'/k}(G^{\mathrm{red}}_{K'}) = \mathrm{R}_{K'/k}(\mathrm{SL}_2)$$

carrying $U_G(\pm\lambda)$ into the root groups relative to the diagonal k-torus, and $U_\pm \subset U_G(\pm\lambda) \cap \ker f$ since $\mathscr{R}_{u,k}(G) \subset \ker f$. We will prove that $U_G(\pm\lambda) \cap \ker f \simeq \mathrm{R}_{K/k}(\alpha_2 \otimes_K (V/K))$. This has no nontrivial k_s-points, so it forces $U_\pm = 1$ and hence $\mathscr{R}_{u,k}(G)$ is trivial.

The quotient map $q : \mathrm{R}_{K/k}(H) \to G$ carries the D-root groups $\mathrm{R}_{K/k}(\underline{V})$ onto $U_G(\pm\lambda)$, and via the mapping property of $\mathrm{R}_{K'/k}(\cdot)$ and computations with

k_s-points of root groups we see that $\widetilde{f} := f \circ q$ is equal to the composite map

$$\mathrm{R}_{K/k}(H) = H_{V,K'/k} \longrightarrow \mathrm{R}_{K'/k}(\mathrm{SL}_2) \xrightarrow{\mathrm{R}_{K'/k}(F_{\mathrm{SL}_2/K'})} \mathrm{R}_{K'/k}(\mathrm{SL}_2)$$

carrying the D-root groups $\mathrm{R}_{K/k}(\underline{V})$ into the root groups $\mathrm{R}_{K'/k}(\mathbf{G}_{\mathrm{a}})$ relative to the diagonal k-torus. These maps $\widetilde{f}_{\pm} : \mathrm{R}_{K/k}(\underline{V}) \rightrightarrows \mathrm{R}_{K'/k}(\mathbf{G}_{\mathrm{a}})$ are given on points in a k-algebra A by the common map $V \otimes_k A \to K' \otimes_k A$ defined by the inclusion $V \hookrightarrow K'$ and squaring on $K' \otimes_k A$. This squaring is valued in $K \otimes_k A$, and relative to the K-basis $\{1, \sqrt{u_1}, \sqrt{v_2}\}$ of V and the k-basis $\{1, u_1, u_2, u_1u_2\}$ of K the resulting map $V \otimes_k A \to K \otimes_k A$ is given by

$$x_0 + x_1\sqrt{u_1} + x_2\sqrt{u_2} \mapsto x_0^2 + u_1x_1^2 + u_2x_2^2$$

for $x_0, x_1, x_2 \in K \otimes_k A$ since $x_j^2 \in A$ for all j. Clearly $x_0^2 + u_1x_1^2 + u_2x_2^2 = 0$ if and only if each x_j^2 vanishes; i.e., $x_0, x_1, x_2 \in \mathrm{R}_{K/k}(\alpha_2)$ inside $\mathrm{R}_{K/k}(\mathbf{G}_{\mathrm{a}})$. Hence, $\ker \widetilde{f}_{\pm} = \mathrm{R}_{K/k}(\alpha_2 \otimes_K V)$. Since the map $q_{\pm} : \mathrm{R}_{K/k}(\underline{V}) \to U_G(\pm\lambda)$ is faithfully flat with kernel $\mathrm{R}_{K/k}(\alpha_2) \subset \mathrm{R}_{K/k}(\alpha_2 \otimes_K V) = \ker \widetilde{f}_{\pm}$ corresponding to $j : K \hookrightarrow V$, yet this kernel coincides with the intersection of $\ker \widetilde{f}_{\pm}$ and $\ker q = \mathrm{R}_{K/k}(N)$, $U_G(\pm\lambda) \cap \ker f \simeq \mathrm{R}_{K/k}(\alpha_2 \otimes_K (V/K))$ since j has a K-linear section.

Step 3. Now we compute the root system $\Phi(G,T)$ for the pseudo-split pseudo-semisimple k-group G. The positive and negative T-weight spaces in $\mathrm{Lie}(G)$ are swapped under the adjoint action of the image $w \in G(k)$ of the standard Weyl element in the Levi k-subgroup $\mathrm{SL}_2 \subset H_{V,K'/k} = \mathrm{R}_{K/k}(H)$, so these subspaces of $\mathrm{Lie}(G)$ each have dimension $(1/2)(25-7) = 9$.

The D-root spaces in $\mathrm{Lie}(H) = \mathrm{Lie}(\mathrm{R}_{K/k}(H))$ are V with weights ± 2, so on their images V/K in $\mathrm{Lie}(G)$ the action of $D/D[2] = T \simeq \mathrm{GL}_1$ has weights ± 1. Note that $\dim_k(V/K) = 2[K:k] = 8$. Since $\mathrm{R}_{K/k}(N) \cap L = \ker F_{L/k}$, we naturally identify $L/\ker F_{L/k} = L^{(p)} \simeq \mathrm{SL}_2$ with a k-subgroup of G containing T as its diagonal k-torus, so the Lie algebra of this k-subgroup provides 1-dimensional T-weight spaces with weights ± 2 in $\mathrm{Lie}(G)$. Thus, $\Phi(G,T) = \mathrm{BC}_1$.

Step 4. Finally, we prove that G is not of minimal type. Any absolutely pseudo-simple k-group of minimal type has trivial center if its root system over k_s is non-reduced [CGP, Prop. 9.4.9], so it suffices to prove Z_G is nontrivial.

By considering the images V/K in $\mathrm{Lie}(G)$ of the D-root spaces in the Lie algebra $\mathrm{Lie}(\mathrm{R}_{K/k}(H))$, we see that the center

$$Z_G \subset Z_G(T) = \mathrm{R}_{K'/k}(\mathrm{GL}_1)/\mathrm{R}_{K/k}(N_0)$$

is contained in the image in G of the visibly central k-subgroup $\mathrm{R}_{K'/k}(\mu_2) \subset \mathrm{R}_{K/k}(H)$; the reverse containment is obvious. Since $N_0 \subset \mathrm{R}_{K'/K}(\mathrm{GL}_1)[2] = \mathrm{R}_{K'/K}(\mu_2)$, we conclude that

$$Z_G = \mathrm{R}_{K'/k}(\mu_2)/\mathrm{R}_{K/k}(N_0).$$

But $\dim \mathrm{R}_{K/k}(N_0) = 9$ and the dimension of

$$\mathrm{R}_{K'/k}(\mu_2) = \ker(\mathrm{R}_{K'/k}(\mathrm{GL}_1) \stackrel{t^2}{\twoheadrightarrow} \mathrm{R}_{K/k}(\mathrm{GL}_1))$$

is $[K':k]-[K:k] = 12$, so Z_G is nontrivial. (Computing as in Step 1, $Z_G \simeq \mathrm{R}_{K/k}(\alpha_2)$ as k-groups.)

Remark B.4.2. For G as in Example B.4.1, the absolutely pseudo-simple minimal type central quotient $G/\mathscr{C}_G$ with a non-reduced root system must have trivial center [CGP, Prop. 9.4.9], so $\mathscr{C}_G = Z_G$. In terms of the classification of minimal type pseudo-split absolutely pseudo-simple k-groups with a non-reduced root system (see [CGP, Thm. 9.8.6, Prop. 9.8.9(iii)]), G/Z_G is classified by the pair $(V^{(2)}, V') = (K^2 u_1 + K^2 u_2, k)$.

Example B.4.3. The construction in Example B.4.1 admits a higher-rank analogue (with root system BC_n, $n \geqslant 2$), as follows. Let $(K/k, K', V)$ be as in that rank-1 construction, choose $n \geqslant 2$, and let $\mathscr{G}$ denote the K-group Sp_{2n}. Choose a split maximal K-torus $\mathscr{T}$ in $\mathscr{G}$, and let $\mathscr{G}' = \mathscr{G}_{K'}$, $\mathscr{T}' = \mathscr{T}_{K'}$. Let Δ be a basis of $\Phi := \Phi(\mathscr{G}, \mathscr{T})$ with associated positive system of roots Φ^+ and fix a pinning over K; this identifies $\mathscr{G}_c$ with SL_2 for each $c \in \Delta$. Let $\Phi^+_< = \Phi^+ \cap \Phi_<$ and $\Phi^+_> = \Phi^+ \cap \Phi_>$.

Step 1. By [CGP, Thm. C.2.29], there exists a unique K-subgroup $H \subset \mathrm{R}_{K'/K}(\mathscr{G}')$ containing $\mathrm{R}_{K'/K}(\mathscr{T}')$ that is pseudo-semisimple and satisfies:

- $\Phi(H, \mathscr{T}) = \Phi$,
- $H_a = H_{V,K'/K}$ inside $\mathrm{R}_{K'/K}(\mathscr{G}'_a) = \mathrm{R}_{K'/K}(\mathrm{SL}_2)$ for the unique long root $a \in \Delta$ (and $V^*_{K'/K} = \mathrm{R}_{K'/K}(\mathrm{GL}_1)$ since $\dim_K V = 3$),
- $H_b = \mathrm{R}_{K'/K}(\mathscr{G}'_b)$ for every short root $b \in \Delta$,

so $\{H_c\}_{c\in\Delta}$ generates H. Clearly the Levi K-subgroup $\mathscr{G} \subset \mathrm{R}_{K'/K}(\mathscr{G}')$ lies inside H, so $\mathscr{G}$ is a Levi K-subgroup of H by Proposition 2.1.2(i). In particular, the subgroup $\mathscr{R}_u(H_{\overline{K}}) \subset H_{\overline{K}}$ admits K' as a field of definition over K, with $H^{\text{ss}}_{K'} = \mathscr{G}'$, and $\mathrm{R}_{K'/K}(\mathscr{T}') \subset H$ (so this K-subgroup of H is $Z_H(\mathscr{T})$).

Inside $\mathrm{Lie}(H)$ define the K-subspace

$$\mathfrak{n} = (\sum_{a \in \Phi^+_>} \mathfrak{n}_a) + (\sum_{b \in \Phi^+_<} \mathfrak{n}_b),$$

where $\mathfrak{n}_b := \mathrm{Lie}(H_b) = \mathrm{Lie}(\mathscr{G}'_b)$ and $\mathfrak{n}_a := \mathrm{Lie}(\mathscr{G}_a) + V \cdot \mathrm{Lie}(a^\vee)(t\partial_t|_{t=1}) \subset \mathrm{Lie}(\mathscr{G}'_a)$. We claim that $\mathfrak{n}$ is Ad_H-stable. It suffices to show that $\mathrm{Ad}_H(H_c)$ carries $\mathfrak{n}_{c'}$ into $\mathfrak{n}$ for all $c, c' \in \Phi$ satisfying $\langle c', c^\vee \rangle \neq 0$ (as the adjoint action of H_c on $\mathfrak{n}_{c'}$ is trivial if $\langle c', c^\vee \rangle = 0$).

The computations early in Example B.4.1 show that $\mathfrak{n}_a$ is $\mathrm{Ad}_H(H_a)$-stable for every $a \in \Phi^+_>$, and clearly $\mathfrak{n}_b$ is $\mathrm{Ad}_H(H_b)$-stable for all $b \in \Phi^+_<$. For distinct $b, b' \in \Phi^+_<$ such that $\langle b', b^\vee \rangle \neq 0$ the K'-groups $\mathscr{G}'_b$ and $\mathscr{G}'_{b'}$ generate a K'-group $\mathscr{G}'_{b,b'}$ of type A_2, so $\mathrm{Lie}(\mathscr{G}'_{b,b'}) \subset \sum_{c \in \Phi^+_<} \mathfrak{n}_c \subset \mathfrak{n}$. Hence, $\mathrm{Ad}(H_b)$ carries $\mathfrak{n}_{b'}$ into $\mathfrak{n}$ for such b, b'. Since the k-groups $\mathscr{G}_a$ for $a \in \Phi^+_>$ pairwise commute (as $\Phi_> = \mathrm{A}_1^n$ for $\Phi = \mathrm{C}_n$), to verify that $\mathfrak{n}$ is Ad_H-stable it remains to check that for $a \in \Phi^+_>$ and $b \in \Phi^+_<$ such that $\langle a, b^\vee \rangle \neq 0$, the adjoint actions of H_b on $\mathfrak{n}_a$ and of H_a on $\mathfrak{n}_b$ are valued in $\mathfrak{n}$. We may replace $\mathscr{G}$ with the derived group Sp_4 of the centralizer of the codimension-2 torus $((\ker a) \cap (\ker b))^0_{\mathrm{red}}$ to reduce to the case $n = 2$. By negating b if necessary, we arrive at the situation that $\{a, b\}$ is a basis of Φ.

Step 2. The chosen pinning of $(\mathscr{G}, \mathscr{T})$ over K provides root group parameterizations $u_c : \mathbf{G}_\mathrm{a} \simeq \mathscr{U}_c$ for $c \in \Phi$ such that the Chevalley commutation relations [Hum, §33.4] are satisfied. (There are no sign ambiguities since $\mathrm{char}(K) = 2$.) Let $X_c := \mathrm{Lie}(u_c)(\partial_x|_{x=0})$ and $Z_c = \mathrm{Lie}(c^\vee)(t\partial_t|_{t=1})$.

Consider the H_b-action on $\mathfrak{n}_a$. Clearly H_b is generated by the K-subgroups $\mathrm{R}_{K'/K}((\mathscr{U}_{\pm b})_{K'})$, so it suffices to check that $\mathrm{Ad}_{\mathscr{G}}(u_{\pm b}(x'))(\mathfrak{n}_a) \subset \mathfrak{n}$ for all $x' \in K'$. Since $\mathscr{U}_{-b}$ commutes with $\mathscr{U}_a$, and $\mathscr{U}_b$ commutes with $\mathscr{U}_{-a}$, it suffices to show that for $v \in V \subset K'$,

$$\mathrm{Ad}_{\mathscr{G}}(u_{\pm b}(x'))(X_{\pm a}),\ \mathrm{Ad}_{\mathscr{G}}(u_{\pm b}(x'))(vZ_a) \in \mathfrak{n}.$$

The commutation relations give

$$\begin{aligned} u_{\pm b}(x') u_{\pm a}(y) u_{\pm b}(x')^{-1} &= (u_{\pm b}(x'), u_{\pm a}(y)) u_{\pm a}(y) \\ &= u_{\pm(a+2b)}(x'^2 y) u_{\pm(a+b)}(x'y) u_{\pm a}(y), \end{aligned}$$

so differentiating at $y = 0$ gives

$$\mathrm{Ad}_{\mathscr{G}}(u_{\pm b}(x'))(X_{\pm a}) = x'^2 X_{\pm(a+2b)} + x' X_{\pm(a+b)} + X_{\pm a} \in \mathfrak{n}$$

since the roots $\pm(a+b)$ are short and $x'^2 \in K$. Likewise, since $\langle b, a^\vee \rangle = -1$, the vector

$$\mathrm{Ad}_{\mathscr{G}}(u_{\pm b}(x'))(vZ_a) = v\mathrm{Ad}_{\mathscr{G}}(u_{\pm b}(x'))(Z_a)$$

is equal to $x'vX_{\pm b} + vZ_a$, and this lies in $\mathfrak{n}$ since b is short.

Step 3. Next, consider the H_a-action on $\mathfrak{n}_b$. Since $H_a = H_{V,K'/K}$ is generated by its root groups relative to the diagonal K-torus, it suffices to show that $\mathrm{Ad}_{\mathscr{G}}(u_{\pm a}(v))(\mathfrak{n}_b) \subset \mathfrak{n}$ for all $v \in V$. The K'-vector space $\mathfrak{n}_b$ is spanned by $\{X_b, X_{-b}, Z_b\}$, and $\mathscr{U}_{-a}$ (resp. $\mathscr{U}_a$) commutes with $\mathscr{U}_b$ (resp. $\mathscr{U}_{-b}$), so it suffices to show

$$K' \cdot \mathrm{Ad}_{\mathscr{G}}(u_{\pm a}(v))(X_{\pm b}), \;\; K' \cdot \mathrm{Ad}_{\mathscr{G}}(u_{\pm a}(v))(Z_b) \subset \mathfrak{n}$$

for all $v \in V \subset K'$. For $y' \in K'$ we have

$$u_{\pm a}(v)u_{\pm b}(y')u_{\pm a}(v)^{-1} = u_{\pm(a+b)}(y'v)u_{\pm(a+2b)}(y'^2 v)u_{\pm b}(y'),$$

and differentiating at $y' = 0$ gives

$$\mathrm{Ad}_{\mathscr{G}}(u_{\pm a}(v))(X_{\pm b}) = vX_{\pm(a+b)} + X_{\pm b}$$

since $\mathrm{char}(K) = 2$; its K'-span lies in $\mathfrak{n}$ because the roots $\pm(a+b)$ are short. As $\langle a, b^\vee \rangle = -2$ and $\mathrm{char}(K) = 2$, similarly $\mathrm{Ad}_{\mathscr{G}}(u_{\pm a}(v))(Z_b) = Z_b$.

This proves that $\mathfrak{n}$ is Ad_H-stable in $\mathrm{Lie}(H)$, so it is a Lie ideal in $\mathrm{Lie}(H)$ and thus is a Lie subalgebra. For $p = 2$, Frobenius-semilinearity of the p-operation on $\mathrm{Lie}(H)$ implies that $\mathfrak{n}$ is a p-Lie subalgebra of $\mathrm{Lie}(H)$ since $v^2 \in K$ for all $v \in V$. Hence, as in Example B.4.1, $\mathfrak{n} = \mathrm{Lie}(N)$ for a unique closed K-subgroup $N \subset \ker F_{H/K}$ that is moreover a normal K-subgroup scheme of H.

Step 4. Consideration of Lie algebras shows that $\mathscr{T} \cap N = \mathscr{T}[2]$. The k-subgroup scheme $\mathrm{R}_{K/k}(N) \subset \mathrm{R}_{K/k}(H)$ is normal, and the calculations in the rank-1 case in Example B.4.1 immediately imply (by consideration of open cells) that the pseudo-reductive k-group $\mathrm{R}_{K/k}(H)$ is perfect and that the split maximal k-torus S in $\mathrm{R}_{K/k}(\mathscr{T})$ is a maximal k-torus of $\mathrm{R}_{K/k}(H)$. In particular, $S \cap \mathrm{R}_{K/k}(N) = S \cap \mathrm{R}_{K/k}(\mathscr{T}[2]) = S[2]$. The perfect smooth connected affine k-group

$$G := \mathrm{R}_{K/k}(H)/\mathrm{R}_{K/k}(N)$$

contains the split maximal k-torus $T := S/S[2]$. We claim that G is pseudo-semisimple with $\Phi(G,T) = \mathrm{BC}_n$ and that G is *not* locally of minimal type (hence also not of minimal type).

For each $c \in \Phi = \Phi(H, \mathscr{T}) = \Phi(\mathrm{R}_{K/k}(H), S)$, let $S_c \subset S$ and $T_c \subset T = S/S[2]$ be the codimension-1 subtori corresponding to $(\ker c)^0_{\mathrm{red}} \subset S$, and let $N_c \subset \ker F_{H_c/K}$ correspond to $\mathfrak{n}_c \subset \mathrm{Lie}(H_c)$. Lie algebra considerations with [CGP, Lemma 7.1.1(1)] (and its proof) imply that $H_c \cap N = N_c$ for all $c \in \Phi$, so clearly $\mathrm{R}_{K/k}(H_c) \cap \mathrm{R}_{K/k}(N) = \mathrm{R}_{K/k}(N_c)$. Thus, since $Z_{\mathrm{R}_{K/k}(H)}(S_c) \to Z_G(T_c)$ is surjective and $\mathrm{R}_{K/k}(H_c)$ is perfect for all $c \in \Phi$ (clear for short c, and established early in Example B.4.1 for long c), it follows that

$$\mathscr{D}(Z_G(T_c)) = \mathrm{R}_{K/k}(H_c)/\mathrm{R}_{K/k}(N_c)$$

for all $c \in \Phi$, with maximal k-torus given by the image of $c^\vee \in \mathrm{X}_*(\mathscr{T}) = \mathrm{X}_*(S) = 2\mathrm{X}_*(T)$.

Step 5. We shall now prove that $\mathscr{D}(Z_G(T_c))$ is pseudo-simple for every c. (This will be used in our proof of pseudo-reductivity of G in Step 6.) By the rank-1 case treated in Example B.4.1, if $a \in \Phi_>$ then $\mathscr{D}(Z_G(T_a))$ is pseudo-simple with set of T-weights in $\mathrm{X}^*(T) = 2\mathrm{X}^*(\mathscr{T})$ given by $\{\pm a, \pm 2a\}$. Likewise, if $b \in \Phi_<$ then $\mathscr{D}(Z_G(T_b))$ is identified with

$$Q := \mathrm{R}_{K'/k}(\mathrm{SL}_2)/\mathrm{R}_{K/k}(\ker F_{\mathrm{R}_{K'/K}(\mathrm{SL}_2)/K}),$$

and Q is naturally a k-subgroup of $\mathrm{R}_{K/k}(\mathrm{R}_{K'/K}(\mathrm{SL}_2)^{(p)})$ (with $p = 2$) due to left exactness of $\mathrm{R}_{K/k}$. We claim that $Q \simeq \mathrm{R}_{K/k}(\mathrm{SL}_2)$.

Letting $\varphi : K \hookrightarrow K$ be the p-power inclusion (with $p = 2$),

$$\begin{aligned}\mathrm{R}_{K'/K}(\mathrm{SL}_2)^{(p)} &:= \mathrm{R}_{K'/K}(\mathrm{SL}_2) \otimes_{K,\varphi} K \\ &= \mathrm{R}_{(K' \otimes_{K,\varphi} K)/K}(\mathrm{SL}_2) \\ &= \mathrm{R}_{(K'^p \otimes_{K^p} K)/K}(\mathrm{SL}_2),\end{aligned}$$

so applying $\mathrm{R}_{K/k}$ to this yields $\mathrm{R}_{(K'^p \otimes_{K^p} K)/k}(\mathrm{SL}_2)$.

The composite map $\mathrm{R}_{K'/k}(\mathrm{SL}_2) \twoheadrightarrow Q \hookrightarrow \mathrm{R}_{(K'^p \otimes_{K^p} K)/k}(\mathrm{SL}_2)$ evaluated on the Zariski-dense set of k-points is the map $\mathrm{SL}_2(K') \to \mathrm{SL}_2(K'^p \otimes_{K^p} K)$ given by $x \mapsto x^p \otimes 1$ on matrix entries, so this lands inside $\mathrm{SL}_2(K'^p \otimes_{K^p} k)$. Thus, $Q \subset \mathrm{R}_{(K'^p \otimes_{K^p} k)/k}(\mathrm{SL}_2)$. But the quotient map $K'^2 \otimes_{K^2} k \twoheadrightarrow k K'^2 = K$ is between k-algebras with the same k-dimension, so it is an *isomorphism*. Hence, we have a closed immersion of k-group schemes $j : Q \hookrightarrow \mathrm{R}_{K/k}(\mathrm{SL}_2)$. Since $\mathrm{R}_{K/k}(\mathrm{SL}_2)$ is smooth and connected of dimension 12, j is an isomorphism if $\dim Q = 12$. The Frobenius kernel of $\mathrm{R}_{K'/K}(\mathrm{SL}_2)$ is $(\alpha_2)^{12}$ as a K-scheme (not as a K-group scheme), so

$$\begin{aligned}\dim Q &= [K':k]\dim \mathrm{SL}_2 - \dim \mathrm{R}_{K/k}((\alpha_2)^{12}) \\ &= 16\cdot 3 - 12\cdot \dim \mathrm{R}_{K/k}(\alpha_2) \\ &= 16\cdot 3 - 12\cdot 3 = 12.\end{aligned}$$

Step 6. We next establish that the perfect smooth connected k-group G is pseudo-reductive. Note that the k-group $Z_G(T) = \mathrm{R}_{K'/k}(\mathscr{T}')/\mathrm{R}_{K/k}(\mathscr{T}[2])$ is commutative, and $Z_G(T)$ is a direct product of copies of the k-group

$$\mathrm{R}_{K'/k}(\mathrm{GL}_1)/\mathrm{R}_{K/k}(\mu_2)$$

that is pseudo-reductive (as this quotient is an extension of $[2]\mathrm{R}_{K'/k}(\mathrm{GL}_1) = \mathrm{R}_{K/k}(\mathrm{GL}_1)$ by the k-group scheme $\mathrm{R}_{K'/k}(\mu_2)/\mathrm{R}_{K/k}(\mu_2)$ that has no nontrivial k_s-points by [CGP, Ex. 5.3.7]). Thus, it suffices to show that $\mathscr{R}_{u,k}(G) \subset Z_G(T)$, or equivalently that no nontrivial T-weight occurs on $\mathrm{Lie}(\mathscr{R}_{u,k}(G))$.

By [CGP, Rem. 2.3.6] applied to the quotient map $\mathrm{R}_{K/k}(H) \twoheadrightarrow G$, any nontrivial T-weight that occurs on $\mathrm{Lie}(G)$ is $\mathbf{Q}$-linearly dependent on an element of Φ. Suppose such a nontrivial T-weight exists on $\mathrm{Lie}(\mathscr{R}_{u,k}(G))$, so it lies in $\mathbf{Q}c$ for some $c \in \Phi$. By [CGP, Thm. 3.3.11] there exists a T-stable nontrivial smooth connected k-subgroup $U \subset \mathscr{R}_{u,k}(G)$ such that the T-weights occurring on $\mathrm{Lie}(U)$ lie in $\mathbf{Q}c - \{0\}$. Clearly $U \subset Z_G(T_c)$, so $U \subset \mathscr{D}(Z_G(T_c))$ since T-conjugation on the commutative quotient $Z_G(T_c)/\mathscr{D}(Z_G(T_c))$ is trivial. But

$$\mathscr{R}_u(\mathscr{D}(Z_G(T_c))_{\overline{k}}) = \mathscr{D}(Z_G(T_c))_{\overline{k}} \cap \mathscr{R}_u(G_{\overline{k}})$$

by [CGP, Prop. A.4.8] (applied over $\overline{k}$), so $U_{\overline{k}} \subset \mathscr{R}_u(\mathscr{D}(Z_G(T_c))_{\overline{k}})$. By [CGP, Lemma 1.2.1] the pseudo-reductivity of $\mathscr{D}(Z_G(T_c))$ thereby forces $U = 1$, a contradiction. It follows that no such weight c exists, as desired.

Step 7. We have shown that G is pseudo-semisimple. The root system $\Phi(G,T)$ that is non-reduced must be irreducible because the connected semisimple $G_{\overline{k}}^{\mathrm{ss}}$ is clearly simple of rank n. Thus, $\Phi(G,T) = \mathrm{BC}_n$. For any long $a \in \Phi$, the rank-1 pseudo-semisimple k-subgroup $\mathscr{D}(Z_G(T_a))$ is as in Example B.4.1 and hence is not of minimal type. This k-subgroup is visibly generated by the $\pm a$-root groups (via the identification of the divisible root $a \in 2\mathrm{X}^*(\mathscr{T})$ as an indivisible character of T), so G is not locally of minimal type and thus is not of minimal type.

Appendix C
Pseudo-split and quasi-split forms

Let G be a pseudo-reductive group over an arbitrary field k. Motivated by the case of connected reductive groups, it is natural to wonder if G admits a pseudo-split k_s/k-form; i.e., does there exist a pseudo-reductive k-group H such that $G_{k_s} \simeq H_{k_s}$ and H admits a split maximal k-torus? In Proposition C.1.1 we show that if such an H exists then it is unique up to k-isomorphism. The existence problem for pseudo-split forms seems to be hopeless for commutative G in general, so for the study of existence we will focus on pseudo-semisimple G.

In contrast with the reductive case, in Example C.1.2 we give (in every positive characteristic) many pseudo-semisimple G without a pseudo-split form due to an elementary field-theoretic obstruction that only arises for G that are not absolutely pseudo-simple. We also provide more subtle examples without a pseudo-split form in the *absolutely pseudo-simple* case in characteristic 2 and show that no such examples exist away from characteristic 2.

For any pseudo-reductive k-group G, there is a finite Galois extension k'/k such that $G_{k'}$ is pseudo-split. Put another way, for a given finite Galois extension k'/k and a *pseudo-split* pseudo-reductive k'-group G', we may consider k-descents G of G' (i.e., k-groups G equipped with a k'-isomorphism $G_{k'} \simeq G'$). The specification of such a G in terms of G' is an instance of Galois descent: a $\mathrm{Gal}(k'/k)$-indexed collection of k'-isomorphisms $f_\gamma : {}^\gamma G' \simeq G'$ satisfying the cocycle condition $f_\gamma \circ {}^\gamma(f_\delta) = f_{\gamma\delta}$ for all $\gamma, \delta \in \mathrm{Gal}(k'/k)$. (See [BLR, §6.1, §6.2 Ex. B] for a general discussion of this formalism.) If H is a pseudo-split k_s/k-form of G then $H_{k'}$ is a pseudo-split k'_s/k'-form of the pseudo-split G' and hence $H_{k'} \simeq G'$ due to the uniqueness of pseudo-split forms (Proposition C.1.1). Thus, exhibiting examples without a pseudo-split form is the same as giving a finite Galois extension k'/k and a pseudo-split pseudo-reductive k'-group G' that admits a descent to a k-group yet has no pseudo-split k-descent.

C.1 General characteristic

We begin our discussion of k_s/k-forms by settling the uniqueness problem for pseudo-split forms over a general field k.

Proposition C.1.1. *If G and H are pseudo-split pseudo-reductive k-groups such that $G_{k_s} \simeq H_{k_s}$ then $G \simeq H$. In particular, a pseudo-reductive k-group has at most one pseudo-split k_s/k-form.*

Proof. The case of pseudo-semisimple groups is an immediate consequence of Theorem 6.3.11, so $\mathscr{D}(G) \simeq \mathscr{D}(H)$. Thus, the main work is to incorporate a (commutative) Cartan subgroup. Choose split maximal k-tori $S \subset G$ and $T \subset H$, and a k_s-isomorphism $f : G_{k_s} \simeq H_{k_s}$ carrying S_{k_s} onto T_{k_s} (found by composing an initial choice with some $H(k_s)$-conjugation). For $C := Z_G(S)$ and $\mathscr{C} := Z_H(T)$, the isomorphism $\iota : C_{k_s} = Z_{G_{k_s}}(S_{k_s}) \simeq Z_{H_{k_s}}(T_{k_s}) = \mathscr{C}_{k_s}$ induced by f descends to a k-isomorphism $C \simeq \mathscr{C}$ since the $\mathrm{Gal}(k_s/k)$-equivariance of ι may be checked between the maximal tori (due to [CGP, Prop. 1.2.2]).

By [CGP, Lemma 1.2.5, Prop. 1.2.6], $G = C \cdot \mathscr{D}(G)$ and $C' := C \cap \mathscr{D}(G) = Z_{\mathscr{D}(G)}(S')$ for the split maximal k-torus $S' = S \cap \mathscr{D}(G)$ of $\mathscr{D}(G)$. Note that $G = (\mathscr{D}(G) \rtimes C)/C'$, and likewise $H = (\mathscr{D}(H) \rtimes \mathscr{C})/\mathscr{C}'$ for the Cartan k-subgroup $\mathscr{C}' = Z_{\mathscr{D}(H)}(T')$ of $\mathscr{D}(H)$ with $T' := T \cap \mathscr{D}(H)$. The isomorphism $C_{k_s} \simeq \mathscr{C}_{k_s}$ induced by f clearly carries C'_{k_s} onto $\mathscr{C}'_{k_s}$, and so the k-descent $C \simeq \mathscr{C}$ carries C' onto $\mathscr{C}'$. This latter isomorphism is uniquely determined by its restriction between the split maximal k-tori S' and T'.

As we have noted at the beginning, there exists a k-isomorphism $\theta : \mathscr{D}(G) \simeq \mathscr{D}(H)$. By k-rational conjugacy of maximal split k-tori [CGP, Thm. C.2.3], θ can be chosen to carry S' onto T' and hence C' onto $\mathscr{C}'$. It suffices to arrange that this isomorphism $C' \simeq \mathscr{C}'$ agrees with the restriction of the k-isomorphism $C \simeq \mathscr{C}$ built from f, as then this latter isomorphism can be glued with θ to build the desired k-isomorphism $G \simeq H$. Consider the composition of θ_{k_s} with the inverse of the restriction f' of f between derived groups. This is an automorphism of $\mathscr{D}(G)_{k_s}$ that preserves S'_{k_s}, and by precomposing θ with conjugation by a suitable element of $N_{\mathscr{D}(G)}(S')(k)$ we can ensure that this automorphism preserves a chosen positive system of roots Φ^+ in $\Phi := \Phi(\mathscr{D}(G), S')$.

By Propositions 6.3.4(i) and 6.3.7, we can precompose θ with a suitable k-automorphism of $(\mathscr{D}(G), S', \Phi^+)$ so that $f'^{-1} \circ \theta_{k_s}$ induces the identity automorphism of the Dynkin diagram and thus is the identity on S'_{k_s}. Hence, $f'^{-1} \circ \theta_{k_s}$ restricts to the identity on the commutative pseudo-reductive C'_{k_s}, so f and θ now yield the same k-isomorphism $C' \simeq \mathscr{C}'$, as required. □

In every positive characteristic there are pseudo-reductive groups that do not admit a pseudo-split form, due to elementary field-theoretic obstructions:

Example C.1.2. Let F/k be an extension of degree p^2 with $p = \mathrm{char}(k)$ such that $[F : k]_s = p$ and there is no degree-p subextension of F/k that is inseparable over k (i.e., the inseparable part of F/k cannot be put "in the bottom"). For example, such an F exists when $k = \mathbf{F}_p(x, y)$ (but not when $[k : k^p] = p$, nor when $k = k_s$). Let G' be a connected semisimple F-group that is absolutely simple and simply connected. The pseudo-reductive k-group $G = \mathrm{R}_{F/k}(G')$ is perfect. We claim that G has no pseudo-split k_s/k-form.

By [CGP, Prop. A.5.14], the k_s/k-forms of G are the k-groups $\mathrm{R}_{E/k}(H')$, where E is a k_s/k-form of F (i.e., $E \otimes_k k_s \simeq F \otimes_k k_s$ as k_s-algebras) and H' is an E-group whose fibers over the factor fields of E are absolutely simple, simply connected, and of the same type as G'. The maximal k-tori in $\mathrm{R}_{E/k}(H')$ are in natural bijection with the maximal E-tori T' in H' by assigning to each T' the maximal k-torus T in $\mathrm{R}_{E/k}(T')$ [CGP, Prop. A.5.15(2)]. Such T cannot be k-split if some factor field of E is not purely inseparable over k, so if $\mathrm{R}_{E/k}(H)$ is pseudo-split then the factor fields of E are purely inseparable over k.

Now assume $\mathrm{R}_{E/k}(H')$ is pseudo-split, so since E is a k_s/k-form of F we see that each factor field of E has degree p over k (as the inseparable degree is unaffected by passing to factor fields after scalar extension by $k \to k_s$). Hence, $E = \prod_{i=1}^{p} E_i$ for purely inseparable degree-p extensions E_i of k. A k_s-isomorphism $F \otimes_k k_s \simeq E \otimes_k k_s$ provides a k-embedding of F into $E \otimes_k k_s$, and hence into a separable closure $(E_1)_s$ of E_1 over k. Under this k-embedding, F is linearly disjoint from E_1 over k since we assume F/k contains no degree-p subextension purely inseparable over k. Thus, the subfield $F \otimes_k E_1 = FE_1 \subset (E_1)_s$ has inseparable degree p^2 over k, contradicting that $(E_1)_s$ has inseparable degree p over k_s. Hence, G has no pseudo-split k_s/k-form.

In the absolutely pseudo-simple case there is an affirmative result:

Proposition C.1.3. *Let G be an absolutely pseudo-simple k-group, with K/k the minimal field of definition for $\mathscr{R}_u(G_{\overline{k}}) \subset G_{\overline{k}}$. Then G admits a pseudo-split k_s/k-form in each of the following cases:*

(i) char$(k) \neq 2$,

(ii) char$(k) = 2$ *and G is standard except possibly when the following all hold: G_K^{ss} is of outer type D_{2n} ($n \geqslant 2$), Z_G is nontrivial, and G is not its own universal smooth k-tame central extension.*

(iii) char$(k) = 2$ *and G is either non-standard with root system over k_s equal to F_4 or has universal smooth k-tame central extension that is exotic with root system over k_s of type B_n or C_n for some $n \geqslant 2$.*

A refinement if char$(k) = 2$ with $[k : k^2] \leqslant 4$ is given in Corollary C.2.12.

Proof. By Theorem 3.4.2 and Proposition 3.4.3, in case (i) either G is standard or char$(k) = 3$ with $G = \mathrm{R}_{K_>/k}(\mathscr{G})$ for a purely inseparable finite extension

$K_>/k$ and a basic exotic $K_>$-group $\mathscr{G}$ of type G_2. In the latter case, by applying Theorem 3.3.8(ii) to G_{k_s} we see that the minimal field of definition over $K_>$ for the geometric unipotent radical of $\mathscr{G}$ is K (so $kK^3 \subset K_> \subsetneq K$). Likewise, in case (iii) we have $G \simeq \mathrm{R}_{K_>/k}(\mathscr{G})/Z$ where $K_>$ is a proper subfield of K containing kK^2, $\mathscr{G}$ is a basic exotic $K_>$-group, Z is a closed k-subgroup scheme of $\mathrm{R}_{K_>/k}(Z_{\mathscr{G}})$, and the minimal field of definition over $K_>$ for the geometric unipotent radical of $\mathscr{G}$ is K (see Proposition 3.2.6). Let $\mathscr{G}_0$ be the pseudo-split $K_{>,s}/K_>$-form of $\mathscr{G}$, so $\mathrm{R}_{K_>/k}(\mathscr{G}_0)$ is a pseudo-split k_s/k-form of $\mathrm{R}_{K_>/k}(\mathscr{G})$.

Consider types B and C in (iii). The $K_>$-group $\mathscr{G}$ is built from $\mathscr{G}_0$ through Galois-twisting against a continuous 1-cocycle on $\mathrm{Gal}(K_{>,s}/K_>)$ valued in $\mathrm{Aut}_{\mathscr{G}_0/K_>}(K_{>,s})$. By Proposition 6.3.4(i), the $K_>$-group $\mathrm{Aut}^{\mathrm{sm}}_{\mathscr{G}_0/K_>}$ is connected since there are no nontrivial automorphisms of Dynkin diagrams of type B or C. Hence, $\mathrm{Aut}^{\mathrm{sm}}_{\mathscr{G}_0/K_>}$ is as described as in Proposition 6.2.4, so its action on $Z_{\mathscr{G}_0}$ is trivial. Consequently, the center $\mathrm{R}_{K_>/k}(Z_{\mathscr{G}})$ of $\mathrm{R}_{K_>/k}(\mathscr{G})$ is naturally identified with $\mathrm{R}_{K_>/k}(Z_{\mathscr{G}_0})$, so Z corresponds to a closed k-subgroup scheme $Z_0 \subset \mathrm{R}_{K_>/k}(Z_{\mathscr{G}_0})$. Clearly $\mathrm{R}_{K_>/k}(\mathscr{G}_0)/Z_0$ is a pseudo-split k_s/k-form of G.

Now we may assume G is standard with $\mathrm{char}(k)$ arbitrary. By standardness and Proposition 3.2.6, $G = \mathrm{R}_{K/k}(G')/Z$ for an absolutely simple and simply connected K-group G' and a closed k-subgroup scheme Z of $\mathrm{R}_{K/k}(Z_{G'})$ such that the Cartan k-subgroup

$$\mathrm{R}_{K/k}(T')/Z$$

is pseudo-reductive for some (equivalently, any) maximal K-torus T' of G'; this latter condition encodes that G is pseudo-reductive (see [CGP, Lemma 9.4.1]).

Let G'_0 be the split K_s/K-form of G', and let $f : (G'_0)_{K_s} \simeq G'_{K_s}$ be a K_s-isomorphism. It suffices to construct a k-subgroup Z_0 of the central k-subgroup $\mathrm{R}_{K/k}(Z_{G'_0})$ of $\mathrm{R}_{K/k}(G'_0)$ such that $(Z_0)_{k_s}$ is carried over to Z_{k_s} under $\mathrm{R}_{K_s/k_s}(f)$, as then $G_0 := \mathrm{R}_{K/k}(G'_0)/Z_0$ is a pseudo-split k_s/k-form of G. The case $Z = 1$ is trivial, as is the case $Z = \mathrm{R}_{K/k}(Z_{G'})$ (corresponding to when $Z_G = 1$ by Lemma 4.1.1 since $\mathrm{R}_{K/k}(Z_{G'})$ is the center of $\mathrm{R}_{K/k}(G')$ by [CGP, Prop. A.5.15(1)]), so we may assume G does not coincide with its universal smooth k-tame central extension $\mathrm{R}_{K/k}(G')$ and that Z_G is nontrivial.

We aim to construct a pseudo-split k_s/k-form provided that $\mathrm{char}(k) \neq 2$ when G' is of outer type D_{2n} with $n \geqslant 2$. The K-group G' is obtained from the split G'_0 through $\mathrm{Gal}(K_s/K)$-twisting against automorphisms of $(G'_0)_{K_s}$. If the root system of G'_0 is not of type D_{2n} with $n \geqslant 2$ then any automorphism of $(G'_0)_{K_s}$ acts on $(Z_{G'_0})_{K_s}$ through either the identity or inversion. Since inversion preserves all subgroup schemes of a group scheme, it follows that away from outer type D_{2n} ($n \geqslant 2$) every k-subgroup scheme Z of $\mathrm{R}_{K/k}(Z_{G'})$ has base change over k_s that descends to a k-subgroup scheme Z_0 of $\mathrm{R}_{K/k}(Z_{G'_0})$. This settles cases where G' is not of outer type D_{2n} ($n \geqslant 2$), and so settles (ii).

Suppose G' is of outer type D_{2n} ($n \geqslant 2$), so we may assume $\mathrm{char}(k) \neq 2$ since (ii) is settled. Since $Z_{G'_0} = \mu_2 \times \mu_2$, so $Z_{G'}$ is étale, we have $Z = \mathrm{R}_{K/k}(Z')$ for a closed K-subgroup Z' of $Z_{G'}$; see [CGP, Prop. A.5.13]. Hence, $G = \mathrm{R}_{K/k}(G'/Z')$, so a pseudo-split k_s/k-form of G is obtained by applying $\mathrm{R}_{K/k}$ to the split K_s/K-form of G'/Z'. □

To explain the sense in which Proposition C.1.3 is optimal (apart from the case $[k : k^2] \leqslant 4$ in characteristic 2), we consider absolutely pseudo-simple groups G over imperfect fields k with characteristic 2 such that G is not covered by Proposition C.1.3(ii),(iii). If G is standard then $G_{\overline{K}}^{\mathrm{ss}}$ must be of outer type D_{2n} for some $n \geqslant 2$, and we will construct many such G without a pseudo-split k_s/k-form in Proposition C.1.4. Suppose G is not standard, so its root system over k_s is of type B, C, or BC with rank $n \geqslant 1$. If $[k : k^2] \leqslant 4$ then in all non-standard cases there is a pseudo-split form; see Corollary C.2.12. If $[k : k^2] \geqslant 8$ then the *generalized* exotic cases over k provide many examples without a pseudo-split k_s/k-form, as we discuss in §C.3–§C.4. (Such generalized exotic cases for types B_n or C_n with $n \geqslant 2$ must lie beyond the exotic case, by Proposition C.1.3(iii).)

Now consider standard absolutely pseudo-simple groups G over imperfect fields k with $\mathrm{char}(k) = 2$, so $G = \mathrm{R}_{K/k}(G')/Z$ for $(G', K/k, Z)$ as in the proof of Proposition C.1.3. Assume G' is outer type D_{2n} ($n \geqslant 2$), so $\mathrm{Gal}(K_s/K)$ maps onto a nontrivial subgroup of the diagram automorphism group of D_{2n} ($n \geqslant 2$).

The diagram automorphism group for D_{2n} is $\mathbf{Z}/2\mathbf{Z}$ when $n > 2$ and is S_3 when $n = 2$. Thus, a necessary condition for the existence of such G which does not admit a pseudo-split k_s/k-form is that K admits a quadratic Galois extension when $n > 2$ and admits a quadratic or cubic Galois extension when $n = 2$. Since K/k is purely inseparable, it is equivalent to say that k must admit a quadratic Galois extension when $n > 2$ and admit a quadratic or cubic Galois extension when $n = 2$. Over any such k admitting a quadratic Galois extension, examples without a pseudo-split form exist for any $n \geqslant 2$ (and Remark C.1.5 addresses k with a cubic Galois extension when $n = 2$):

Proposition C.1.4. *Assume k is imperfect with* $\mathrm{char}(k) = 2$ *and that k admits a quadratic Galois extension k'. Let $K = k(a^{1/4})$ for $a \in k - k^2$. For $n \geqslant 2$, let $\mathscr{G}$ be the unique (up to isomorphism) absolutely simple and simply connected non-split quasi-split k-group of type* D_{2n} *that splits over k'.*

There is a closed k-subgroup Z of the center of $\mathrm{R}_{K/k}(\mathscr{G}_K)$ *such that $G :=$* $\mathrm{R}_{K/k}(\mathscr{G}_K)/Z$ *is absolutely pseudo-simple with maximal geometric semisimple quotient $G_{\overline{k}}^{\mathrm{ss}}$ of adjoint type and G does not admit a pseudo-split k_s/k-form.*

Proof. Let $\mathcal{G}$ be the absolutely simple and simply connected k-*split* group of type D_{2n}, so $\mathcal{G}_{k'} \simeq \mathscr{G}_{k'}$ since $\mathscr{G}_{k'}$ is split. Identify these k'-groups via a fixed k'-isomorphism $f : \mathcal{G}_{k'} \to \mathscr{G}_{k'}$. For the nontrivial $\sigma \in \mathrm{Gal}(k'/k)$, $\varphi := f^{-1} \circ {}^{\sigma}f$ is an outer automorphism of $\mathcal{G}_{k'}$ (equivalently, it induces a nontrivial automorphism of the Dynkin diagram) and the k-descent of $\mathcal{G}_{k'}$ via φ is the quasi-split k-group $\mathscr{G}$. The effect of φ on the center $Z_{\mathcal{G}_{k'}}$ is an involution, and we can choose an identification $Z_{\mathcal{G}} \simeq \mu_2 \times \mu_2$ over k such that this involution on $(Z_{\mathcal{G}})_{k'} = Z_{\mathcal{G}_{k'}}$ swaps the two μ_2-factors. Let μ denote $Z_{\mathscr{G}}$.

Let $K' = k' \otimes_k K$ and let σ also denote the nontrivial K-automorphism of K'. The center $\mathrm{R}_{K'/k'}(\mu_{K'}) = \mathrm{R}_{K'/k'}(Z_{\mathcal{G}_{K'}}) = \mathrm{R}_{K'/k'}(\mu_2) \times \mathrm{R}_{K'/k'}(\mu_2)$ of $\mathrm{R}_{K'/k'}(\mathscr{G}_{K'}) = \mathrm{R}_{K'/k'}(\mathcal{G}_{K'})$ is identified with $\mathrm{R}_{K/k}(\mu)_{k'}$, and a closed k'-subgroup scheme Z' of $\mathrm{R}_{K'/k'}(\mu_2 \times \mu_2)$ descends to a closed k-subgroup of $\mathrm{R}_{K/k}(\mu_K)$ if and only if the K'/K-twist ${}^{\sigma}Z'$ is obtained from Z' by swapping of the two $\mathrm{R}_{K'/k'}(\mu_2)$-factors. Likewise, Z' descends to a closed k-subgroup of $\mathrm{R}_{K/k}(Z_{\mathcal{G}_K}) = \mathrm{R}_{K/k}(\mu_2 \times \mu_2)$ if and only if ${}^{\sigma}Z' = Z'$.

We shall construct an intermediate closed k-subgroup scheme

$$\mu \subset Z \subset \mathrm{R}_{K/k}(\mu_K) = Z_{\mathrm{R}_{K/k}(\mathscr{G}_K)}$$

such that:

(i) $\mathrm{R}_{K/k}(\mu_K)/Z$ has no nontrivial smooth connected k-subgroup,

(ii) the k'-subgroup $Z_{k'} \subset \mathrm{R}_{K/k}(\mu_K)_{k'} = \mathrm{R}_{K/k}(Z_{\mathscr{G}_K})_{k'} = \mathrm{R}_{K/k}(Z_{\mathcal{G}_K})_{k'}$ (with the latter identification defined by $\mathrm{R}_{K'/k'}(f_{K'})^{-1}$) does not descend to a closed k-subgroup scheme of $\mathrm{R}_{K/k}(Z_{\mathcal{G}_K})$.

Assume such a Z exists. By Corollary 4.1.4, the perfect central quotient

$$G = \mathrm{R}_{K/k}(\mathscr{G}_K)/Z$$

is pseudo-reductive in view of (i), so it is absolutely pseudo-simple. Since $\mu \subset Z$, the connected semisimple $G^{\mathrm{ss}}_{\overline{k}}$ is of adjoint type.

As G is standard, any k_s/k-form of G is standard by [CGP, Cor. 5.2.3]. Thus, by [CGP, Prop. A.5.14] and Proposition 5.3.1(i) a pseudo-split k_s/k-form of G is isomorphic to $\mathrm{R}_{K/k}(\mathcal{G}_K)/\mathcal{Z}$ for a closed k-subgroup scheme $\mathcal{Z}$ of the center $\mathrm{R}_{K/k}(Z_{\mathcal{G}_K})$ of $\mathrm{R}_{K/k}(\mathcal{G}_K)$. Assuming such a k_s/k-form exists, we shall deduce a contradiction via condition (ii). Since the pseudo-split k'-groups $G_{k'} = \mathrm{R}_{K/k}(\mathscr{G}_K)_{k'}/Z_{k'}$ and $(\mathrm{R}_{K/k}(\mathcal{G}_K)/\mathcal{Z})_{k'} = \mathrm{R}_{K'/k'}(\mathcal{G}_{K'})/\mathcal{Z}_{k'}$ are isomorphic over k_s, they must be k'-isomorphic (by Proposition C.1.1). Such a k'-isomorphism arises from a unique k'-isomorphism $\phi : \mathrm{R}_{K/k}(\mathcal{G}_K)_{k'} \to$

$\mathrm{R}_{K/k}(\mathscr{G}_K)_{k'}$ that carries $\mathcal{Z}_{k'}$ onto $Z_{k'}$. Since $\mathscr{G}$ is an outer form of $\mathcal{G}$ that splits over k', $\phi^{-1} \circ {}^{\sigma}\phi$ is an outer automorphism of $\mathrm{R}_{K/k}(\mathcal{G}_K)_{k'}$.

Consider the k'-automorphism $\psi = (\mathrm{R}_{K'/k'}(f_{K'}))^{-1} \circ \phi$ of $\mathrm{R}_{K'/k'}(\mathcal{G}_{K'})$. We claim that the restriction of ψ to the center $\mathrm{R}_{K'/k'}(Z_{\mathcal{G}_{K'}})$ descends to a k-automorphism θ of $\mathrm{R}_{K/k}(Z_{\mathcal{G}_K})$. The k'-automorphism ψ of $\mathrm{R}_{K'/k'}(\mathcal{G}_{K'})$ arises as the Weil restriction of a unique K'-automorphism α of $\mathcal{G}_{K'}$ by [CGP, Prop. A.5.14], so it suffices to show that the action of α on $Z_{\mathcal{G}_{K'}} = (Z_{\mathcal{G}_K})_{K'}$ is defined over K. The effect of α on the center only depends on the associated K'-point of $\pi_0(\mathrm{Aut}_{\mathcal{G}_{K'}/K'}) = \pi_0(\mathrm{Aut}_{\mathcal{G}_K/K})_{K'}$. But $\pi_0(\mathrm{Aut}_{\mathcal{G}_K/K})$ is a constant K-group, so its K'-points are K-points and thus the desired k-descent θ of ψ on the center is obtained.

It is harmless to replace $\mathcal{Z}$ with the closed central k-subgroup scheme $\theta(\mathcal{Z})$, so the isomorphism $\mathrm{R}_{K'/k'}(f_{K'}) : \mathrm{R}_{K/k}(\mathcal{G}_K)_{k'} \to \mathrm{R}_{K/k}(\mathscr{G}_K)_{k'}$ carries $\mathcal{Z}_{k'}$ onto $Z_{k'}$. But $\mathcal{Z}_{k'}$ descends to a closed k-subgroup scheme $\mathcal{Z}$ of $\mathrm{R}_{K/k}(Z_{\mathcal{G}_K})$, contradicting (ii). Thus, there is no pseudo-split k_s/k-form of G when Z exists as above. We will find Z as an intermediate closed k-subgroup scheme

$$\mathrm{R}_{k(\sqrt{a})/k}(\mu_{k(\sqrt{a})}) \subset Z \subset \mathrm{R}_{K/k}(\mu_K). \tag{C.1.4}$$

Letting $U = \mathrm{R}_{K/k}(\mu_2)/\mathrm{R}_{k(\sqrt{a})/k}(\mu_2)$, we can express our problem in terms of a k'-subgroup scheme

$$(Z/\mathrm{R}_{k(\sqrt{a})/k}(\mu_{k(\sqrt{a})}))_{k'} \subset (\mathrm{R}_{K'/k'}(\mu_2)/\mathrm{R}_{k'(\sqrt{a})/k'}(\mu_2))^2 = U_{k'} \times U_{k'}$$

as follows. Letting σ be the nontrivial automorphism of k'/k, we seek a closed k'-subgroup $Z' \subset U_{k'} \times U_{k'}$ satisfying three conditions:

(0) $\sigma^*(Z')$ is obtained from Z' via the k'-automorphism of $U_{k'} \times U_{k'}$ that swaps the factors,
(1) $(U_{k'} \times U_{k'})/Z'$ contains no nontrivial smooth connected k'-subgroup,
(2) $\sigma^*(Z') \neq Z'$ inside $U_{k'} \times U_{k'}$.

Indeed, (0) implies that Z' arises from a k-subgroup scheme $Z \subset \mathrm{R}_{K/k}(\mu_K)$ as in (C.1.4), and (1) and (2) say that Z satisfies the above desired properties (i) and (ii) respectively.

To construct Z', observe that $\mathrm{R}_{k(\sqrt{a})/k}(\alpha_2) \simeq U$ via $x \mapsto 1 + a^{1/4}x$. We can choose $t \in k'^{\times}$ such that $\sigma(t) \neq t$, so $c := \sigma(t)/t \in k'^{\times}$ satisfies $\mathrm{N}_{k'/k}(c) = 1$ (i.e., $\sigma(c) = 1/c$) and $c \notin k$. Let $Z' \subset U_{k'} \times U_{k'}$ be the graph of the c-multiplication automorphism of $U_{k'} = \mathrm{R}_{k'(\sqrt{a})/k'}(\alpha_2)$. The k'-subgroup Z' satisfies (0) above because $(x, \sigma(c)x) = (x, x/c) = (cy, y)$ for $y = x/c$, and it satisfies (2) because $c \notin k$. Finally, since Z' is the graph of a k'-group automorphism of $U_{k'}$, the quotient $(U_{k'} \times U_{k'})/Z'$ is identified as a k'-group with $U_{k'} \simeq \mathrm{R}_{k'(\sqrt{a})/k'}(\alpha_2)$, so (1) holds. □

Remark C.1.5. Suppose $n = 2$ and k is imperfect of characteristic 2, admitting a cubic Galois extension k'. Let ι be an order-3 diagram automorphism of D_4, and choose a generator $\sigma \in \mathrm{Gal}(k'/k)$. Let $\mathscr{G}$ be the unique (up to isomorphism) non-split quasi-split simply connected k-group of type D_4 that splits over k' and let $\mu = Z_{\mathscr{G}}$. We may choose an identification $\mu_{k'} = \mu_2 \times \mu_2$ so that the k'-isomorphism ${}^{\sigma}(\mu_{k'}) \simeq \mu_{k'}$ encoding the k-structure μ is obtained from the order-3 automorphism $\left(\begin{smallmatrix} 1 & 1 \\ 1 & 0 \end{smallmatrix}\right) \in \mathrm{Aut}(\mu_2 \times \mu_2) = \mathrm{GL}_2(\mathbf{F}_2)$. Let K/k be as in Proposition C.1.4. For the graph Z' of multiplication by $c \in k'^{\times}$ on $\mathrm{R}_{k'(\sqrt{a})/k'}(\alpha_2)$ and the corresponding k-subgroup scheme $Z \subset \mathrm{R}_{K/k}(\mu_K)$ as in the proof of Proposition C.1.4, the k-group $G = \mathrm{R}_{K/k}(\mathscr{G}_K)/Z$ has exactly the same properties as in Proposition C.1.4 (with $n = 2$) provided that $c \notin k^{\times}$ and $\sigma(c) + 1 = 1/c$. We may use $c = \sigma(b)/b$ for $b \in k'^{\times}$ such that $\sigma(b) \notin kb$ and $\mathrm{Tr}_{k'/k}(b) = 0$. To find such b, note that the differences $\sigma(t) - t$ for $t \in k'$ constitute a k-plane in k', and many b in this plane avoid eigenlines of σ over k.

The field-theoretic obstruction in Example C.1.2 to relaxing the absolutely pseudo-simple hypothesis in Proposition C.1.3 cannot arise when $[k : k^p] = p$. The following alternative example applies without restriction on $[k : k^p] > 1$.

Example C.1.6. Let k be imperfect of characteristic $p > 0$ with a quadratic Galois extension k'/k. Let $K = k(a^{1/p^2})$ for $a \in k - k^p$, $K' = k' \otimes_k K$, $\mathscr{G} = \mathrm{R}_{K'/K}(\mathrm{SL}_p)$, and $\mu = \mathrm{R}_{k'/k}(\mu_p)$. For $U := \mathrm{R}_{K/k}(\mu_p)/\mathrm{R}_{k(a^{1/p})/k}(\mu_p)$ we claim that $U \simeq \mathrm{R}_{k(a^{1/p})/k}(\alpha_p)^{p-1}$ as k-groups. Once this is shown, the construction in the proof of Proposition C.1.4 adapts without difficulty to build a central quotient $G = \mathrm{R}_{K/k}(\mathscr{G})/Z$ with no pseudo-split k_s/k-form. (This G is pseudo-simple but not absolutely pseudo-simple.)

The $\mathrm{R}_{k(a^{1/p})/k}(\mu_p)$-equivariant k-scheme isomorphism

$$\mathrm{R}_{k(a^{1/p})/k}(\mu_p) \times \mathrm{R}_{k(a^{1/p})/k}(\alpha_p)^{p-1} \simeq \mathrm{R}_{K/k}(\mu_p)$$

defined by

$$(z, x_1, \ldots, x_{p-1}) \mapsto z(1 + x_1 a^{1/p^2} + \cdots + x_{p-1} a^{(p-1)/p^2})$$

implies $U \simeq \mathrm{R}_{k(a^{1/p})/k}(\alpha_p)^{p-1}$ as k-schemes. Thus, any k-homomorphism $h : \mathrm{R}_{k(a^{1/p})/k}(\alpha_p)^{p-1} \to U$ with trivial kernel may be viewed as a monic endomorphism of a finite type k-scheme, so h is an isomorphism by [EGA, IV_4, 17.9.6]. It therefore suffices to construct such an h.

Let $F := k(a^{1/p})$, so $\mathrm{R}_{K/F}(\mu_p)/\mu_p = \mathrm{R}_{K/F}(\mathrm{GL}_1)/\mathrm{GL}_1$ is an fppf form of $\mathbf{G}_{\mathrm{a}}^{p-1}$ and hence has Frobenius kernel that is an fppf-form of α_p^{p-1}. This kernel is therefore F-isomorphic to α_p^{p-1} (as α_p^r has no nontrivial fppf forms over fields, since its automorphism functor is represented by GL_r). Fix an F-group inclusion $\theta : \alpha_p^{p-1} \hookrightarrow \mathrm{R}_{K/F}(\mu_p)/\mu_p$ onto the Frobenius kernel. We claim that this lifts to an F-scheme morphism $\widetilde{\theta} : \alpha_p^{p-1} \to \mathrm{R}_{K/F}(\mu_p)$ (not an F-homomorphism!).

More generally, any F-scheme map $f : \alpha_p^r \to \mathrm{R}_{K/F}(\mu_p)/\mu_p$ carrying 0 to 1 lifts to an F-scheme map $\alpha_p^r \to \mathrm{R}_{K/F}(\mu_p)$. Indeed, when viewed as a map $\alpha_p^r \to \mathrm{R}_{K/F}(\mathrm{GL}_1)/\mathrm{GL}_1$, f lifts to an F-scheme map $\widetilde{f} : \alpha_p^r \to \mathrm{R}_{K/F}(\mathrm{GL}_1)$ since there is no nontrivial GL_1-torsor over α_p^r. Multiplying $\widetilde{f}$ by $1/\widetilde{f}(0)$ arranges that $\widetilde{f}(0) = 1$, so $\widetilde{f}$ factors through a unique F-scheme morphism $\alpha_p^r \to \mathrm{R}_{K/F}(\mu_p)$ by the functorial meaning of $\mathrm{R}_{K/F}$.

The map $\mathrm{R}_{F/k}(\widetilde{\theta}) : \mathrm{R}_{F/k}(\alpha_p)^{p-1} \to \mathrm{R}_{K/k}(\mu_p)$ yields a composite map

$$\mathrm{R}_{F/k}(\alpha_p)^{p-1} \xrightarrow{h} \mathrm{R}_{K/k}(\mu_p)/\mathrm{R}_{F/k}(\mu_p) =: U \hookrightarrow \mathrm{R}_{F/k}(\mathrm{R}_{K/F}(\mu_p)/\mu_p)$$

that is equal to the k-homomorphism $\mathrm{R}_{F/k}(\theta)$ whose kernel is trivial (since $\ker\theta = 1$), so h is a *k-homomorphism* with trivial kernel.

C.2 Quasi-split forms

Let G be a pseudo-reductive group over a field k. Its minimal pseudo-parabolic k-subgroups constitute a $G(k)$-conjugacy class [CGP, Thm. C.2.5]. Such a k-subgroup P is called a *pseudo-Borel* k-subgroup if P_{k_s} is minimal among the pseudo-parabolic k_s-subgroups of G_{k_s}, and we say G is *quasi-split* if it admits a pseudo-Borel k-subgroup.

Example C.2.1. If $G = \mathrm{R}_{k'/k}(G')$ for a nonzero finite reduced k-algebra k' and smooth affine k'-group G' with connected reductive fibers then $P' \mapsto \mathrm{R}_{k'/k}(P')$ is an inclusion-preserving bijection between the sets of parabolic k'-subgroups of G' and pseudo-parabolic k-subgroups of G [CGP, Prop. 2.2.13], so by extending scalars to k_s we see that $\mathrm{R}_{k'/k}(P')$ is a pseudo-Borel k-subgroup of G if and only if P' is a Borel k'-subgroup of G'.

By [CGP, Prop. 3.5.1(4)] a pseudo-parabolic k-subgroup P of G is a pseudo-Borel k-subgroup if and only if the pseudo-reductive quotient $P/\mathscr{R}_{u,k}(P)$ is commutative, or equivalently if and only if P is solvable. Thus, in the notation of the Tits classification in Theorem 6.3.11, if G is pseudo-semisimple then it is quasi-split if and only if M is commutative, or equivalently the anisotropic kernel $\mathscr{D}(M)$ is trivial. The triviality of the pseudo-semisimple $\mathscr{D}(M)$ is equivalent to that of its maximal central quotient $\mathscr{D}(M/Z_M)$, as well as to its canonical diagram $\mathrm{Dyn}(\mathscr{D}(M/Z_M))$ being empty. The k-isomorphism class of a quasi-split *pseudo-semisimple* k-group G is therefore determined by the isomorphism class of the pair $(G_{k_s}, \mathrm{Dyn}(G))$ consisting of the k_s-group G_{k_s} and its canonical diagram $\mathrm{Dyn}(G_{k_s})$ equipped with the $*$-action of $\mathrm{Gal}(k_s/k)$ arising from G.

If B is a pseudo-Borel k-subgroup of G then for a maximal k-torus T in B the set $\Phi(B_{k_s}, T_{k_s})$ is a $\mathrm{Gal}(k_s/k)$-stable positive system of roots in $\Phi(G_{k_s}, T_{k_s})$ [CGP, Prop. 3.5.1]. Hence, the natural $\mathrm{Gal}(k_s/k)$-action on $\mathrm{X}(T_{k_s})$ preserves the basis Δ associated to B_{k_s}, so this action coincides with $*$-action of $\mathrm{Gal}(k_s/k)$ on Δ through diagram automorphisms (see Definition 6.3.3).

Lemma C.2.2. *Let G be a pseudo-semisimple k-group, and assume that the Dynkin diagram of the root system of G_{k_s} does not admit nontrivial automorphisms. Any quasi-split k_s/k-form of G is pseudo-split.*

Proof. We may and do assume G is quasi-split and aim to prove that it is pseudo-split. Using notation as above, the absence of nontrivial diagram automorphisms implies that the Galois action on $\mathrm{X}(T_{k_s})$ leaves the roots in Δ invariant. But Δ spans $\mathrm{X}(T_{k_s})_{\mathbf{Q}}$ since G is pseudo-semisimple, so T is k-split. □

The reader is referred to 4.1.2 and Proposition 6.1.4 for the definition and basic properties of the k-group $Z_{G,C}$ associated to a pseudo-reductive k-group G and Cartan k-subgroup C.

Lemma C.2.3. *For a pseudo-reductive k-group G, the $(\mathrm{Aut}^{\mathrm{sm}}_{\mathscr{D}(G)/k})^0$-action on $\mathscr{D}(G)$ uniquely extends to an action on G.*

Proof. The maxmal k-torus T in a Cartan k-subgroup C of G is an almost direct product of the maximal k-torus $\mathscr{T} := T \cap \mathscr{D}(G)$ of $\mathscr{D}(G)$ and the maximal central k-torus Z in G [CGP, Lemma 1.2.5(ii)]. Thus, $\mathscr{C} := C \cap \mathscr{D}(G) = Z_{\mathscr{D}(G)}(\mathscr{T})$ is a Cartan subgroup of $\mathscr{D}(G)$. Since $G = (\mathscr{D}(G) \rtimes C)/\mathscr{C}$ and $(\mathrm{Aut}^{\mathrm{sm}}_{\mathscr{D}(G)/k})^0 = (\mathscr{D}(G) \rtimes Z_{\mathscr{D}(G),\mathscr{C}})/\mathscr{C}$ (see Proposition 6.2.4), the action of $(\mathrm{Aut}^{\mathrm{sm}}_{\mathscr{D}(G)/k})^0$ on $\mathscr{D}(G)$ extends to an action on G that is the identity on the maximal central k-torus Z in G.

Any k-automorphism of G that restricts to the identity on Z is uniquely determined by its restriction to $\mathscr{D}(G)$ due to [CGP, Prop. 1.2.2], so it remains to show that any action on G by a smooth connected k-group H *must* restrict to the identity on Z. We may and do assume $k = k_s$, so considering k-points shows via smoothness of H that the H-action on G restricts to an action on Z. But H is connected and Z is a torus, so such an action on Z is trivial. $\square$

Let G be a pseudo-reductive group over a field k. Lemma C.2.3 provides a $\mathrm{Gal}(k_s/k)$-equivariant inclusion $(\mathrm{Aut}^{\mathrm{sm}}_{\mathscr{D}(G)/k})^0(k_s) \hookrightarrow \mathrm{Aut}(G_{k_s})$. This subgroup meets $\mathrm{Aut}(G)$ in the group

$$\mathrm{PsInn}(G) := (\mathrm{Aut}^{\mathrm{sm}}_{\mathscr{D}(G)/k})^0(k) \tag{C.2.3}$$

whose elements are called *pseudo-inner* automorphisms of G (over k). In general $(G/Z_G)(k) \subset \mathrm{PsInn}(G)$. This inclusion is an equality if G is reductive but generally not otherwise, and $(\mathrm{Aut}^{\mathrm{sm}}_{\mathscr{D}(G)/k})^0$ can fail to be perfect but its derived group is $\mathscr{D}(G)/Z_{\mathscr{D}(G)}$ by Proposition 6.2.4, with $\mathscr{D}(G)/Z_{\mathscr{D}(G)} = \mathscr{D}(G/Z_G)$ as quotients of $\mathscr{D}(G)$. (Note that $Z_{\mathscr{D}(G)}$ is central in G. To see this, fix a Cartan subgroup C of G. Then $C \cap \mathscr{D}(G)$ is a Cartan subgroup of $\mathscr{D}(G)$ and so $Z_{\mathscr{D}(G)}$ is contained in $C \cap \mathscr{D}(G) \subset C$. Now as C is commutative and $G = C \cdot \mathscr{D}(G)$, it is obvious that $Z_{\mathscr{D}(G)}$ is central in G.)

Definition C.2.4. A *pseudo-inner form* of a pseudo-reductive k-group G is a k_s/k-form obtained by twisting against a class in the image of the natural map $\mathrm{H}^1(k_s/k, \mathrm{PsInn}(G_{k_s})) \to \mathrm{H}^1(k_s/k, \mathrm{Aut}(G_{k_s}))$.

For reductive G this recovers the notion of inner form, but for non-reductive G it is more general than twisting against a class in the image of $\mathrm{H}^1(k, G/Z_G)$. Any reductive G admits a unique quasi-split inner form. In the pseudo-reductive case all k_s/k-forms of G are pseudo-inner when G is pseudo-semisimple and $\mathrm{Aut}^{\mathrm{sm}}_{G/k}$ is connected (as occurs when the Dynkin diagram of G_{k_s} has no non-trivial automorphisms, by Proposition 6.3.4(i)), and we will prove uniqueness of quasi-split pseudo-inner forms in general in Proposition C.2.8. The existence of quasi-split pseudo-inner forms will be proved provided that if $\mathrm{char}(k) = 2$ then either G is standard or $[k : k^2] \leqslant 4$ (see Theorem C.2.10).

Remark C.2.5. The restriction $[k : k^2] \leqslant 4$ for existence of a quasi-split pseudo-inner form in the non-standard case when $\mathrm{char}(k) = 2$ is necessary since the examples with no pseudo-split k_s/k-form in §C.3–§C.4 over some k satisfying $[k : k^2] \geqslant 8$ do not admit a quasi-split k_s/k-form, due to Lemma C.2.2.

In every positive characteristic, the pseudo-semisimple groups in Example C.1.6 without a pseudo-split k_s/k-form are quasi-split, and the ones in Example C.1.2 without a pseudo-split k_s/k-form admit a quasi-split pseudo-inner form (replace G' there with a quasi-split inner F_s/F-form); Lemma C.2.2 is not applicable to those groups because the Dynkin diagrams of their root systems over k_s admit a nontrivial automorphism. Proposition C.1.4 provides absolutely pseudo-simple groups without a pseudo-split k_s/k-form over any imperfect field k of characteristic 2 admitting a quadratic Galois extension, and those groups are quasi-split since $\mathscr{G}$ there is quasi-split (and Lemma C.2.2 is not applicable since the Dynkin diagram of D_{2n} has nontrivial automorphisms).

Lemma C.2.6. *Let G be a quasi-split pseudo-reductive k-group and C a Cartan k-subgroup of G contained in a pseudo-Borel k-subgroup B. The group of k-automorphisms of G that arise from* $\mathrm{PsInn}(G)$ *and preserve (C, B) is $Z_{G,C}(k)$.*

Proof. By Galois descent, without loss of generality we may assume that k is separably closed. Hence, by [CGP, Prop. 3.5.1] an automorphism of G carrying the unique maximal k-torus T of C into itself carries B into itself if and only if it carries the associated subset $\Phi(B,T) \subset \Phi(G,T)$ into itself. Every automorphism of G restricting to the identity on C induces the identity automorphism of $X(T) \supset \Phi(G,T)$ and so preserves B. Moreover, $Z_{G,C}(k) \subset \mathrm{PsInn}(G)$ by Lemma 6.1.2(i) (since $Z_{G,C}$ is connected, by Proposition 6.1.4).

It remains to show that if a pseudo-inner automorphism $f \in \mathrm{PsInn}(G)$ preserves (C, B) then $f|_C$ is the identity on C. Since C is commutative and pseudo-reductive, by [CGP, Prop. 1.2.2] it is equivalent to show that the automorphism $f|_T$ of T is the identity. Since B is solvable, the image of $B_{\overline{k}}$ in $G_{\overline{k}}^{\mathrm{red}}$ is parabolic and solvable, hence a Borel subgroup with maximal torus $T_{\overline{k}}$. Lemma C.2.3 provides an action of $(\mathrm{Aut}^{\mathrm{sm}}_{\mathscr{D}(G)/k})^0$ on G uniquely extending its natural action on $\mathscr{D}(G)$, so now we may assume $k = \overline{k}$, G is reductive, and f arises from $\mathrm{Aut}^0_{G/k}(k) = (G/Z_G)(k)$. Preservation of T implies $f \in N_{G/Z_G}(T/Z_G)$. Since f preserves the Borel subgroup B, its image in $W(G/Z_G, T/Z_G)$ is trivial (as this Weyl group acts simply transitively on the set of positive systems of roots). Thus, $f \in T/Z_G$, so f acts as the identity on T. □

Lemma C.2.7. *Let G be a quasi-split pseudo-reductive k-group. As S varies through the maximal k-split tori of G, the k-subgroups $Z_G(S)$ constitute a $G(k)$-conjugacy class of Cartan k-subgroups of G and are precisely the Cartan k-subgroups of the pseudo-Borel k-subgroups of G.*

If C is a member of this distinguished $G(k)$-conjugacy class of Cartan k-subgroups then for any separable extension field k'/k the Cartan k'-subgroup $C_{k'}$ is a member of the analogous distinguished $G(k')$-conjugacy class.

Proof. Maximal k-split tori of G constitute a single $G(k)$-conjugacy class, and likewise for pseudo-Borel k-subgroups of G [CGP, Thm. C.2.3, Thm. C.2.5].

By [CGP, Prop. C.2.4] (and the discussion immediately after its proof), some pseudo-Borel k-subgroup B contains S and moreover $B = Z_G(S) \ltimes \mathscr{R}_{u,k}(B)$. Since $B/\mathscr{R}_{u,k}(B)$ is commutative, so $Z_G(S)$ is commutative, there is a unique maximal k-torus T contained in $Z_G(S)$ and it is certainly maximal in G. Clearly T contains S and $Z_G(T) = Z_G(S)$, so $Z_G(S)$ is a Cartan k-subgroup of G that is visibly also a Cartan k-subgroup of B.

Since any maximal k-torus of B maps isomorphically onto a maximal k-torus in $B/\mathscr{R}_{u,k}(B) \simeq Z_G(S)$, *every* such k-torus contains a maximal split k-torus of B. The $B(k)$-conjugacy of the maximal split k-tori of B then implies that the Cartan k-subgroups of B are a single $B(k)$-conjugacy class.

The desired compatibility with separable extension k'/k is now immediate from the observation that $B_{k'}$ is a pseudo-Borel k'-subgroup of $G_{k'}$ since pseudo-Borel subgroups are precisely solvable pseudo-parabolic subgroups. $\square$

Proposition C.2.8. *Up to k-isomorphism, a pseudo-reductive k-group G has at most one quasi-split pseudo-inner form.*

Proof. We must show that if G and $\mathscr{G}$ are quasi-split with $\mathscr{G}$ a pseudo-inner form of G then $\mathscr{G} \simeq G$. Let $B \subset G$ and $\mathscr{B} \subset \mathscr{G}$ be pseudo-Borel k-subgroups, and let $C \subset B$ and $\mathscr{C} \subset \mathscr{B}$ be Cartan k-subgroups. By definition of "pseudo-inner form", there exists a k_s-isomorphism $f : \mathscr{G}_{k_s} \simeq G_{k_s}$ such that for all $\gamma \in \mathrm{Gal}(k_s/k)$ the k_s-automorphism $c(\gamma) = f \circ ({}^{\gamma}f)^{-1}$ of G_{k_s} lies in $\mathrm{PsInn}(G_{k_s})$. It is harmless to post-compose f with conjugation against an element of $G(k_s)$, so by the $G(k_s)$-conjugacy of pseudo-Borel k_s-subgroups of G_{k_s} we can arrange that $f(\mathscr{B}_{k_s}) = B_{k_s}$ and then further arrange via composing f with conjugation by an element of $B(k_s)$ that $f(\mathscr{C}_{k_s}) = C_{k_s}$.

Each $c(\gamma)$ belongs to $\mathrm{PsInn}(G_{k_s})$ and preserves the pair (C_{k_s}, B_{k_s}), so $c(\gamma) \in Z_{G,C}(k_s)$ by Lemma C.2.6. Thus, it suffices to show that $\mathrm{H}^1(k, Z_{G,C}) = 1$. The pseudo-parabolic k-subgroup $B' := B \cap \mathscr{D}(G)$ of $\mathscr{D}(G)$ inherits solvability from B and thus is a pseudo-Borel k-subgroup (so $\mathscr{D}(G)$ is quasi-split). The Cartan k-subgroup $C' = C \cap \mathscr{D}(G)$ of $\mathscr{D}(G)$ is contained in B', so C' is as described in Lemma C.2.7. Since $\mathscr{D}(G)$ has trivial anisotropic kernel (as it is quasi-split), so $Z_{\mathscr{D}(G),C'}$ is identified with the kernel considered in Proposition 6.3.12 (applied to the pseudo-semisimple $\mathscr{D}(G)$ and Cartan k-subgroup $C' \subset B'$), we have $Z_{\mathscr{D}(G),C'} \simeq \mathrm{R}_{F/k}(\mathrm{GL}_1)$ for a nonzero finite reduced k-algebra F. But $Z_{G,C} \simeq Z_{\mathscr{D}(G),C'}$ by Lemma 6.1.2(i), so we are done. $\square$

C.2.9. Now we address *existence* of a quasi-split pseudo-inner form of a pseudo-reductive k-group. Ultimately this will be reduced to a refined Galois descent problem for a quasi-split pseudo-reductive group over a finite Galois extension of k. Thus, we first describe the setup for this *refined descent problem.*

Let G' be a quasi-split pseudo-reductive group over a finite Galois extension k'/k, and let C' be a Cartan k'-subgroup contained in a pseudo-Borel k'-subgroup B'. Let $S' \subset C'$ be the maximal split k'-torus, so $C' = Z_{G'}(S')$ by Lemma C.2.7. (We do not assume this data is equipped with a k-structure!)

Suppose we are given k'-isomorphisms $f_\sigma : {}^\sigma G' \simeq G'$ for all $\sigma \in \mathrm{Gal}(k'/k)$, and moreover assume

$$f_{\sigma\tau} \in \mathrm{PsInn}(G') \cdot (f_\sigma \circ {}^\sigma f_\tau) \tag{C.2.9.1}$$

in $\mathrm{Isom}_{k'}({}^{\sigma\tau}G', G')$ for all $\sigma, \tau \in \mathrm{Gal}(k'/k)$. (One source of such $\{f_\sigma\}$ is a k'/k-descent datum on G', but this will not be the only case we need.)

The set of pairs (B', S') in G' is permuted transitively by $\mathscr{D}(G')(k')$, so we can adjust the choice of each f_σ via $\mathscr{D}(G')(k')$-conjugation such that $f_\sigma({}^\sigma C') = C'$ and $f_\sigma({}^\sigma B') = B'$ for all σ. After this adjustment, which retains (C.2.9.1), there exists $z(\sigma, \tau) \in \mathrm{PsInn}(G')$ preserving (B', C') such that

$$f_{\sigma\tau} = z(\sigma, \tau) \circ f_\sigma \circ {}^\sigma f_\tau. \tag{C.2.9.2}$$

By Lemma C.2.6 we have $z(\sigma, \tau) \in Z_{G',C'}(k')$, and the possible choices for such k'-isomorphisms ${}^\sigma G' \simeq G'$ are $f'_\sigma = z_\sigma \circ f_\sigma$ for varying $z_\sigma \in Z_{G',C'}(k')$.

The commutativity of $Z_{G',C'}$ implies that the associated k'-isomorphisms

$$h_\sigma : {}^\sigma(Z_{G',C'}) = Z_{{}^\sigma G', {}^\sigma C'} \simeq Z_{G',C'}$$

defined by $\zeta \mapsto f_\sigma \circ \zeta \circ f_\sigma^{-1}$ are *independent* of the choice of such $\{f_\sigma\}$. It follows from (C.2.9.2) and the commutativity of $Z_{G',C'}$ that $\{h_\sigma\}$ satisfies the cocycle condition $h_{\sigma\tau} = h_\sigma \circ {}^\sigma h_\tau$ for all $\sigma, \tau \in \mathrm{Gal}(k'/k)$; i.e., it is a Galois descent datum on the affine k'-group $Z_{G',C'}$ relative to k'/k. Let $\mathscr{Z}$ be the associated k-descent; this is a commutative pseudo-reductive k-group.

Since $B'(k') \cap N_{G'}(S')(k') = Z_{G'}(S')(k') = C'(k')$, any pseudo-Borel k'-subgroup of G' containing C' has the form $nB'n^{-1}$ for $n \in N_{G'}(S')(k')$ that is unique modulo $C'(k')$, say with image $w \in W(G', S')(k')$. The analogue of $\mathscr{Z}$ using $nB'n^{-1}$ is obtained from descent data defined by composing f_σ with n-conjugation on G' and $\sigma(n)^{-1}$-conjugation on ${}^\sigma G'$. Thus, via the natural $W(G', S')$-action on $Z_{G',C'}$, the w-action descends to a k-isomorphism between the associated $\mathscr{Z}$'s. In this sense, $\mathscr{Z}$ depends only on (G', C') and not on B'. By Lemma 6.1.2, $\mathscr{Z}$ only depends on $(\mathscr{D}(G'), \mathscr{D}(G') \cap C')$ and likewise $\mathscr{Z}$ is unaffected by replacing G' with a pseudo-reductive central quotient.

If we replace f_σ with $f'_\sigma := z_\sigma \circ f_\sigma$ for $z_\sigma \in Z_{G',C'}(k') = \mathscr{Z}(k')$ then the associated $z'(\sigma,\tau)$ is

$$z_{\sigma\tau} \cdot (z_\sigma \cdot (f_\sigma \circ {}^\sigma z_\tau \circ f_\sigma^{-1}))^{-1} \cdot z(\sigma,\tau) = (z_{\sigma\tau}/(z_\sigma \cdot \sigma(z_\tau))) \cdot z(\sigma,\tau),$$

where $x \mapsto \sigma(x)$ is the *canonical* action of $\mathrm{Gal}(k'/k)$ on $\mathscr{Z}(k')$. In other words, the function $(\sigma,\tau) \mapsto z(\sigma,\tau) \in Z_{G',C'}(k') = \mathscr{Z}(k')$ changes exactly by a 2-coboundary on $\mathrm{Gal}(k'/k)$ valued in $\mathscr{Z}(k')$.

By computing $f_{\sigma(\tau\rho)} = f_{(\sigma\tau)\rho}$ in two different ways, it is straightforward to see that $(\sigma,\tau) \mapsto z(\sigma,\tau) \in \mathscr{Z}(k')$ is a 2-cocycle. Thus, this 2-cocycle represents a class $c(G',C') \in \mathrm{H}^2(k'/k, \mathscr{Z}(k'))$ that is independent of the choice of $\{f_\sigma\}$ (required to satisfy $f_\sigma({}^\sigma B') = B'$ and $f_\sigma({}^\sigma C') = C'$) and independent of the choice of B' (in the same sense that $\mathscr{Z}$ is independent of the choice of B').

The construction shows that $c(G',C')$ vanishes if and only if we can find $\{f'_\sigma\}$ that is a k'/k-descent datum. For such $\{f'_\sigma\}$, the associated k'/k-descent of G' is quasi-split since $f'_\sigma({}^\sigma B') = B'$. Hence, the vanishing of $c(G',C') \in \mathrm{H}^2(k'/k, \mathscr{Z}(k'))$ is *exactly* the condition that G' admits a quasi-split k-descent whose k'/k-descent datum is related to the initial $\{f_\sigma\}$ through composition against *pseudo-inner* k'-automorphisms of G' (due to Proposition 6.2.4, Lemma C.2.6, and the equality $Z_{G',C'} = Z_{\mathscr{D}(G'),\mathscr{D}(G')\cap C'}$).

Theorem C.2.10. *A pseudo-reductive k-group G admits a quasi-split pseudo-inner form provided that if* $\mathrm{char}(k) = 2$ *then G is standard or* $[k:k^2] \leqslant 4$.

Proof. Pseudo-inner forms involve Galois-twisting against $(\mathrm{Aut}^{\mathrm{sm}}_{\mathscr{D}(G)/k})^0$, so we first reduce to the case where G is perfect. Clearly if G is standard then so is $\mathscr{D}(G)$. Suppose $\mathscr{D}(G)$ has a quasi-split k_s/k-form arising from Galois-twisting against a class $\xi \in \mathrm{H}^1(k, (\mathrm{Aut}^{\mathrm{sm}}_{\mathscr{D}(G)/k})^0)$, so the twist of G against ξ is a pseudo-inner k_s/k-form H of G such that $\mathscr{D}(H)$ is quasi-split. But $P \mapsto P \cap \mathscr{D}(H)$ is a bijection between the sets of pseudo-parabolic k-subgroups of H and $\mathscr{D}(H)$ and it is inclusion-preserving in both directions (see the discussion [CGP] immediately after [CGP, Rem. 11.4.2]), so H inherits the quasi-split property from $\mathscr{D}(H)$.

Now we may replace G with $\mathscr{D}(G)$ so that G is pseudo-semisimple. In particular, the quotient G/Z_G is pseudo-semisimple of minimal type with trivial center (see Proposition 4.1.3) and it is standard if G is standard. The formation of $(\mathrm{Aut}^{\mathrm{sm}}_{G/k})^0$ is insensitive to replacing G with a pseudo-reductive central quotient (Corollary 6.2.6), and a pseudo-inner form H of G is quasi-split if and only if the pseudo-inner form H/Z_H of G/Z_G is quasi-split, so we may replace G with G/Z_G to arrange that G is of minimal type. Now passing to the universal smooth k-tame central extension $\widetilde{G}$ (of minimal type by Proposition 5.3.3, and standard if G is by Lemma 3.2.8), we can assume $G_{\overline{k}}^{\rm ss}$ is simply connected.

Let $C \subset G$ be a Cartan k-subgroup, and let T be its maximal k-torus. Choose a finite Galois extension k'/k splitting T, so $G' := G_{k'}$ has split maximal k'-torus $T' := T_{k'}$ and admits a pseudo-Borel k'-subgroup B' containing $C' := C_{k'}$. Let $\Gamma = \mathrm{Gal}(k'/k)$ and define the k-descent $\mathscr{Z}$ of $Z_{G',C'}$ via the construction in C.2.9 applied to the k'/k-descent datum $\{f_\gamma\}_{\gamma\in\Gamma}$ on G' defined by G, thereby defining a class $c(G',C') \in \mathrm{H}^2(k'/k, \mathscr{Z}(k'))$. It suffices to show that $c(G',C') = 0$, as then G admits a quasi-split *pseudo-inner* k'/k-form H.

Let $\Delta \subset \Phi(G',T')$ be the basis associated to B', equipped with the $*$-action of Γ as in Definition 6.3.3. Note that if G is standard then so is G'_a for every $a \in \Phi(G',T')$. For each $a \in \Delta$, let $C'_a := Z_{G'_a}(a^\vee) = G'_a \cap C'$, so the natural map

$$\theta' : Z_{G',C'} \to \prod_{a\in\Delta} Z_{G'_a,C'_a}$$

is an isomorphism by Proposition 6.1.4. Let $k_a \subset k'$ be the subfield over k such that $\mathrm{Gal}(k'/k_a)$ is the stabilizer of a in $\mathrm{Gal}(k'/k)$ under the $*$-action.

For $\gamma \in \Gamma$, consider the unique $w_\gamma \in W(G',T')$ such that $w_\gamma(\gamma(\Delta)) = \Delta$. Define a k'-isomorphism $f'_\gamma : {}^\gamma G' \simeq G'$ carrying $({}^\gamma C', {}^\gamma B')$ onto (C',B') by composing the standard k'/k-descent isomorphism ${}^\gamma G' \simeq G'$ with conjugation against a choice of representative $n_\gamma \in N_{G'}(T')(k')$ of w_γ; note that f'_γ is only well-defined up to conjugation against $C'(k')$ (upon varying the choice of n_γ). The relation (C.2.9.1) is satisfied by $\{f'_\gamma\}_{\gamma\in\Gamma}$ due to Proposition 6.3.4(i) since the uniqueness of the w_γ's ensures that $w_{\sigma\tau} = w_\sigma \cdot \sigma(w_\tau)$. Note that the intervention of the n_γ's is a possible obstruction to $\{f'_\gamma\}_{\gamma\in\Gamma}$ being a k'/k-descent datum (so we needed to formulate (C.2.9.1) with the intervention of $\mathrm{PsInn}(G')$).

The k'-group G'_a is generally *not* equipped with a descent to a k_a-group, but it fits into the framework of (C.2.9.1) relative to k'/k_a. Indeed, if $\gamma \in \Gamma$ then conjugation by n_γ on G' carries $G'_{\gamma(a)}$ onto $G'_{\gamma * a}$ (see Definition 6.3.3), so f'_γ restricts to a composite isomorphism

$$f'_{\gamma,a} : {}^\gamma(G'_a) \simeq G'_{\gamma(a)} \simeq G'_{\gamma * a}.$$

For the pseudo-Borel k'-subgroup $B'_a := P_{G'_a}(a^\vee) = G'_a \cap B'$ of G'_a containing the Cartan k'-subgroup C'_a, $f'_{\gamma,a}$ carries ${}^\gamma(C'_a)$ onto $C'_{\gamma*a}$ and ${}^\gamma(B'_a)$ onto $B'_{\gamma*a}$. Hence, for $\gamma \in \Gamma_a := \mathrm{Gal}(k'/k_a)$ we get $f'_{\gamma,a} : {}^\gamma(G'_a) \simeq G'_a$ carrying the pair $({}^\gamma(B'_a), {}^\gamma(C'_a))$ over to (B'_a, C'_a) and satisfying (C.2.9.1) for G'_a relative to k'/k_a. As in C.2.9, the restrictions $h_{\gamma,a} : {}^\gamma(C'_a) \simeq C'_a$ of the $f'_{\gamma,a}$'s for $\gamma \in \Gamma_a$ are independent of the choice of n_γ's and constitute a k'/k_a-descent datum; we let

$\mathscr{Z}_a$ denote the associated k_a-descent.

For $\sigma, \tau \in \Gamma_a$ we have

$$f'_{\sigma\tau,a} = z_a(\sigma,\tau)\cdot(f'_{\sigma,a}\circ{}^{\sigma}(f'_{\tau,a})),$$

where the automorphism $z_a(\sigma,\tau) \in Z_{G'_a,C'_a}(k')$ is the restriction to $G'_a \hookrightarrow G'$ of $z(\sigma,\tau) \in Z_{G',C'}(k') = \mathscr{Z}(k')$. In other words, $\theta'(z(\sigma,\tau)) = (z_a(\sigma,\tau))_{a\in\Delta}$.

Let $\{a_i\}$ be a set of representatives in Δ for the orbits of the $*$-action, and define $k_i = k_{a_i} \subset k'$, $G'_i = G'_{a_i}$, $C'_i = Z_{G'_i}(a_i^\vee) = G'_i \cap C'$, $B'_i = P_{G'_i}(a_i^\vee) = G'_i \cap B'$, and $\mathscr{Z}_i = \mathscr{Z}_{a_i}$. Clearly B'_i is a pseudo-Borel k'-subgroup of G'_i and the k'-isomorphism θ' descends to a k-isomorphism

$$\theta : \mathscr{Z} \simeq \prod_i \mathrm{R}_{k_i/k}(\mathscr{Z}_i). \tag{C.2.10}$$

Note that the natural quotient map $\mathrm{R}_{k_i/k}(\mathscr{Z}_i)_{k_i} \twoheadrightarrow \mathscr{Z}_i$ is induced by the natural quotient map $k_i \otimes_k k_i \to k_i$ defined by multiplication. Thus, the canonical quotient k' of $k_i \otimes_k k'$ arising from the inclusion defining k_i inside k' over k identifies $\mathscr{Z}_i(k')$ as the quotient of $\mathscr{Z}(k') = Z_{G',C'}(k')$ induced by the natural projection $Z_{G',C'} \to Z_{G'_i,C'_i}$ arising from *restriction of automorphisms* due the definition of $\mathscr{Z}_i$ (using B'_i).

Under the isomorphism

$$\mathrm{H}^2(k'/k, \mathscr{Z}(k')) = \prod \mathrm{H}^2(k'/k, \mathscr{Z}_i(k_i \otimes_k k')) \simeq \prod \mathrm{H}^2(k_i, \mathscr{Z}_i(k'))$$

provided by (C.2.10) and Shapiro's Lemma, the ith projection is induced by restricting 2-cocycles along $\mathrm{Gal}(k'/k_i) \hookrightarrow \mathrm{Gal}(k'/k)$ and composing with the map $\mathscr{Z}(k') \twoheadrightarrow \mathscr{Z}_i(k')$ as just considered, so $c(G',C') = (c(G'_i,C'_i))_i$. Hence $c(G',C') = 0$ if $c(G'_i,C'_i) = 0$ for all i; we shall prove the latter.

For each i, the k'-group G'_i is *not* provided with a k-structure but for every $\gamma \in \Gamma_i := \mathrm{Gal}(k'/k_i)$ the γ-twist of G'_i admits the k'-isomorphism f'_{γ,a_i} onto G'_i. Moreover, G'_i is pseudo-split over k' and is absolutely pseudo-simple of minimal type with rank 1 and maximal geometric reductive quotient SL_2. Since A_1 and BC_1 have no nontrivial diagram automorphisms, by Proposition 6.3.4(i) (applied to G'_i) we have $\mathrm{PsInn}(G'_i) = \mathrm{Aut}(G'_i)$, so trivially the condition (C.2.9.1) is satisfied by $\{f'_{\gamma,a_i}\}_{\gamma\in\Gamma_i}$ for each i.

Lemma C.2.11 below is a rank-1 refinement of our original problem for pseudo-inner forms of k-groups (it concerns pseudo-split k'-groups not equipped with a k-structure), and it applies to each $(G'_i, k'/k_i)$ to prove that G'_i admits a pseudo-split k'/k_i-descent for all i, so every $c(G'_i,C'_i)$ vanishes. □

Lemma C.2.11. *Let k'/k be a finite Galois extension, and G' a pseudo-split and absolutely pseudo-simple k'-group of minimal type such that $G'^{\rm ss}_{\overline{k'}} \simeq \mathrm{SL}_2$. If G' is not standard and* $\mathrm{char}(k) = 2$ *then assume* $[k : k^2] \leqslant 4$.

If G' is k'-isomorphic to each of its $\mathrm{Gal}(k'/k)$*-twists then G' descends to a pseudo-split pseudo-semisimple k-group.*

Proof. The possibilities for such G' are classified up to k'-isomorphism (functorially in k'/k) using inseparable field extensions and some linear algebra data. The isomorphism hypothesis with Galois twists is expressable in terms of an equivalence relation on these fields and linear algebra data. We have to show that any such equivalence class of data over k' admits a representative that descends to k. Let Φ be the root system for G' (either A_1 or BC_1, the latter possible only if $\mathrm{char}(k) = 2$). Note that G is standard if $\mathrm{char}(k) \neq 2$, by Proposition 3.1.9.

Case (1) (Standard case, any characteristic): The k'-group G' is classified by the minimal field of definition K'/k' for the geometric unipotent radical, which can be any purely inseparable finite extension. Explicitly, $G' = \mathrm{R}_{K'/k'}(\mathrm{SL}_2)$. Choose a k'-isomorphism ${}^{\sigma}G' \simeq G'$ for each $\sigma \in \mathrm{Gal}(k'/k)$. This defines k'-isomorphisms between K' and its $\mathrm{Gal}(k'/k)$-twists; such isomorphisms are *unique* (in contrast with the k'-group isomorphisms) since K'/k' is purely inseparable, so we get a k'/k-descent datum on K'. Hence, there is an extension K of k such that $k' \otimes_k K \simeq K'$ as k'-algebras, so K/k is purely inseparable of finite degree and $\mathrm{R}_{K/k}(\mathrm{SL}_2)$ is a pseudo-split k-descent of G'.

Case (2) ($\mathrm{char}(k) = 2$, $[k : k^2] \leqslant 2$): The case of perfect k is trivial, so assume $[k : k^2] = 2$. The cases with $\Phi = \mathrm{A}_1$ are part of Case (1) (see Proposition 3.1.8), and by [CGP, Thm. 9.9.3(1)] the cases with $\Phi = \mathrm{BC}_1$ are classified up to isomorphism over k' by the minimal field of definition K'/k' of the geometric unipotent radical (which can be any *nontrivial* purely inseparable finite extension). Hence, the BC_1 cases proceed as in Case (1).

Case (3) ($\mathrm{char}(k) = 2$, $[k : k^2] = 4$, $\Phi = \mathrm{A}_1$): In these cases G' is classified by pairs $(K'/k', V')$ where K' is a purely inseparable finite extension of k' and V' is a nonzero $k'K'^2$-subspace of K' such that $k'\langle V'\rangle = K'$. Explicitly, by Proposition 3.1.8, $G' = H_{V',K'/k'}$ with K' the minimal field of definition over k' for the geometric unipotent radical of G', and k'-isomorphism classes correspond to $K'^{\times}$-scaling on V'. Hence, $\gamma(V')$ is a multiple of V' (by an element of $K'^{\times}$) for every $\gamma \in \mathrm{Gal}(k'/k)$.

As in Case (1) we have $K' = k' \otimes_k K$ for a purely inseparable finite extension K/k. We seek a multiple of V' that descends to a kK^2-subspace V of K. The case $V' = K'$ is trivial, so we may assume $V' \neq K'$. In particular, $[K' : k'K'^2] > 2$ (since $k'\langle V'\rangle = K'$), so clearly $\dim_{k'K'^2}(V') > 2$.

The degree $[K' : K'^2]$ is equal to 4 (as $[E : E^p]$ is invariant under finite extension of any field E of characteristic $p > 0$, as explained in Remark 8.3.1), so $k' \subset K'^2$ since otherwise $[K' : k'K'^2] \leqslant 2$. This forces $\dim_{K'^2}(V') = 3$, so V' is a K'^2-hyperplane in K'. Since $\mathrm{Hom}_{K'^2}(K', K'^2)$ is 1-dimensional over K', the descent after suitable $K'^{\times}$-scaling is obvious.

Case (4) ($\mathrm{char}(k) = 2$, $[k : k^2] = 4$, $\Phi = \mathrm{BC}_1$): It follows from [CGP, Thm. 9.8.6, Prop. 9.8.9] (with $n = 1$) that G' is classified up to k'-isomorphism by triples $(K'/k', V', W')$ consisting of a *nontrivial* purely inseparable finite extension K'/k', a nonzero $k'K'^2$-subspace $V' \subset K'$, and a K'^2-subspace $W' \subset K'$ containing V' such that W'/V' is nonzero with finite K'^2-dimension and $k'\langle W' \rangle = K'$. Such triples yield k'-isomorphic groups if and only if the field data coincide and the linear algebra data are related via scaling by a common element of $K'^{\times}$.

Once again there is a k-descent K of K' and $[K : K^2] = [k : k^2] = 4$. By hypothesis (V', W') is a $K'^{\times}$-multiple of each of its Galois twists, and we seek a pair (V, W) relative to K/k such that $(V_{k'}, W_{k'})$ is a $K'^{\times}$-multiple of (V', W').

First suppose the root field for V' is K'^2, so $k \subset K^2$ and $K' = K'^2\langle W' \rangle$. The nonzero proper K'^2-subspace V' of the 4-dimensional K'^2-vector space K' is either a line or a hyperplane over K'^2 since otherwise it is a line over a quadratic extension of K'^2 inside K', contradicting that its root field is K'^2. Since the sets of K'^2-lines and K'^2-hyperplanes in K' each constitute a single $K'^{\times}$-homothety class, some $K'^{\times}$-multiple of (V', W') descends if $W' = K'$. Thus, assume $W' \neq K'$, so W' is a K'^2-hyperplane in K' (any smaller K'^2-dimension for W' would contradict that $[K' : K'^2] = 4$ and $K'^2\langle W' \rangle = K'$). We may apply $K'^{\times}$-scaling so that W' descends to a K^2-hyperplane W in K. Thus, the only scaling factors in $K'^{\times}$ carrying the pair (V', W') to a Galois twist are elements of $(K'^2)^{\times}$, so V' is $\mathrm{Gal}(k'/k)$-stable and hence descends. This settles all cases with V' having root field K'^2.

Finally, assume V' is a subspace over a subfield $F' \subset K'$ strictly containing K'^2, so $[K' : F'] = 2$ and $\dim_{F'}(V') = 1$ (since $[K' : K'^2] = 4$ and $W'/V' \neq 0$). Necessarily $F' = K'^2\langle V' \rangle$, so F' descends to a quadratic extension F of K^2 inside K since V' is a $K'^{\times}$-multiple of its $\mathrm{Gal}(k'/k)$-twists. The quotient K'/V' is an F'-line (hence a K'^2-plane), and we can arrange $V' = F'$ via $K'^{\times}$-scaling. The nonzero K'^2-subspace W'/V' in the F'-line K'/V' is either the entire space or can be moved to any K'^2-line via suitable $F'^{\times}$-scaling, so some $F'^{\times}$-multiple of (V', W') descends. □

The following result provides a refinement to Proposition C.1.3, and examples in §C.3–§C.4 show that there is no room for further improvement concerning the general existence of pseudo-split k_s/k-forms.

Corollary C.2.12. *If k is imperfect with characteristic 2 and $[k : k^2] \leqslant 4$ then an absolutely pseudo-simple k-group G admits a pseudo-split k_s/k-form except possibly if G is standard with root system over k_s equal to D_{2n} for some $n \geqslant 2$.*

Proof. By Proposition C.1.3, we may assume that G is not standard and does not have root system F_4, so its irreducible root system is of type B, C, or BC with rank $n \geqslant 1$. Since $[k : k^2] \leqslant 4$, a quasi-split k_s/k-form is provided by Theorem C.2.10. By Lemma C.2.2, this k_s/k-form is pseudo-split. □

Remark C.2.13. Inspired by Proposition C.1.3 and Corollary C.2.12 (and Remark 6.3.16), a problem that naturally arises in connection with the Tits-style classification theorem (Theorem 6.3.11) is to determine when an absolutely pseudo-semisimple k_s-group $\mathscr{G}$ descends to a pseudo-split k-group G. This is not a matter of constructing k_s/k-forms since we are not given any k-group.

Let Φ be the irreducible root system of $\mathscr{G}$, and let Δ be a basis of Φ. A necessary condition is that for every $c \in \Delta$ the minimal field of definition over k_s for the geometric unipotent radical of $\mathscr{G}_c$ descends (necessarily uniquely up to unique isomorphism) to a purely inseparable finite extension K_c/k. Assuming such a K_c/k exists for every c, let K/k denote their compositum (inside the perfect closure of k). In the standard case we have $\mathscr{G} = \mathrm{R}_{K/k}(G')_{k_s}/Z$ for the split simply connected K-group G' with root system Φ and a closed k_s-subgroup $Z \subset \mathrm{R}_{K/k}(Z_{G'})_{k_s}$, so by [CGP, Prop. A.5.14] existence amounts to Z arising from a k-subgroup of $\mathrm{R}_{K/k}(Z_{G'})$. When $\mathscr{G}$ is not of minimal type, so $Z_{G'}$ is not K-étale, then such descent for Z appears to be difficult to control.

Now assume $\mathscr{G}$ is of minimal type (standard or not); this avoids the k_s-fiber of the type-D_{2n} construction in Proposition C.1.4. By Theorem 3.4.1 (when Φ is reduced) and [CGP, Thm. 9.8.6, Prop. 9.8.9] (when Φ is non-reduced), it is necessary and sufficient that $\mathscr{G}_c$ descends to a pseudo-split k-group for all $c \in \Delta$. However, we seek a more convenient criterion in terms of fields and linear algebra data.

By Proposition 3.1.8 and Theorem 3.4.1, under the above running assumptions the pseudo-split k-descent exists except possibly when k is imperfect of characteristic 2 and $\mathscr{G}_c$ is non-standard for some $c \in \Delta$, in which case Φ is of type B_n, C_n, or BC_n for some $n \geqslant 1$. In these remaining cases a further necessary condition is that the $K_s^\times$-homothety class of linear algebra data inside K_s classifying the rank-1 group $\mathscr{G}_c$ for non-divisible $c \in \Phi$ should be stable (as a homothety class!) under the action of $\mathrm{Gal}(K_s/K) = \mathrm{Gal}(k_s/k)$. This additional necessary condition is not sufficient when $[k : k^2] > 4$ and k admits a quadratic

Galois extension: counterexamples of every type B, C, or BC arise via the k_s-fiber of the k-groups (without a pseudo-split k_s/k-form) built in §C.3–§C.4 for such k. However, if $[k:k^2] \leqslant 4$ then sufficiency holds, as shown by the arguments in Cases (2)–(4) at the end of the proof of Lemma C.2.11.

C.3 Rank-1 cases

Let k be an imperfect field with $\mathrm{char}(k) = 2$, and consider absolutely pseudo-simple k-groups G of minimal type with root system A_1 over k_s such that $G^{\mathrm{ss}}_{\overline{k}} = \mathrm{SL}_2$. Let K/k be the minimal field of definition for $\mathscr{R}_u(G_{\overline{k}}) \subset G_{\overline{k}}$, and define $G' = G^{\mathrm{ss}}_K$, so $i_G : G \hookrightarrow \mathrm{R}_{K/k}(G')$ has trivial kernel and G' is simply connected of rank 1. If $[k:k^2] \geqslant 8$ then under some Galois-theoretic hypotheses on k we will now construct such a G having root field k and having no pseudo-split k_s/k-form. (In §C.4 this example is used to build analogues with root system over k_s equal to B_n, C_n, BC_n for any $n \geqslant 1$.)

Example C.3.1. Assume $[k:k^2] \geqslant 8$ and that k admits a quadratic Galois extension k'. (Eventually we will also assume that $\mathrm{Br}(k) \to \mathrm{Br}(k')$ has nontrivial kernel, as can be arranged in many cases, but we postpone the introduction of Brauer groups until later in the construction.) We may and do choose $K \subset k^{1/2}$ such that $[K:k] = 8$. Let $\{t_1, t_2, t_3\}$ be a 2-basis for K/k. Choose $a \in k' - k$ (so $k' = k(a)$). Let σ be the nontrivial k-automorphism of k'; σ will also denote the nontrivial K-automorphism of $K' := k' \otimes_k K$.

The 4-dimensional k'-subspace

$$V' = k' + k' \cdot t_1 + k' \cdot (t_2 + at_3) + k' \cdot t_1(t_2 + \sigma(a)t_3) \subset K'$$

satisfies $\sigma(V') = t_1 V'$ (since $t_1^2 \in k^\times$), and $k'[V'] = K'$. We claim that $F' := \{\lambda' \in K' \mid \lambda' \cdot V' \subset V'\}$ is equal to k'. Choose $x \in F'$. Since $1 \in V'$, so $x = x \cdot 1 \in V'$, we can assume

$$x = c_1 t_1 + c_2(t_2 + at_3) + c_3 t_1(t_2 + \sigma(a)t_3)$$

for $c_1, c_2, c_3 \in k'$ and we want $x = 0$. The k'-subspace V' contains xt_1, but

$$xt_1 \in W' := k' + k' \cdot (t_2 + \sigma(a)t_3) + k' \cdot t_1(t_2 + at_3).$$

Using the 2-basis property for $\{t_1, t_2, t_3\}$ relative to K' over k' we see that the intersection $V' \cap W'$ equals k', so $x \in k't_1$ yet $t_1 \notin F'$, so $x = 0$ as desired.

The k'-subgroup $H_{V',K'/k'} \subset \mathrm{R}_{K'/k'}(\mathrm{SL}_2)$ admits a k-descent G (necessarily of minimal type) via the isomorphism

$$\sigma^*(H_{V',K'/k'}) \simeq H_{\sigma(V'),K'/k'} = H_{t_1 V',K'/k'} \simeq H_{V',K'/k'}$$

defined in terms of conjugation on $\mathrm{R}_{K'/k'}(\mathrm{SL}_2) = \mathrm{R}_{K/k}(\mathrm{SL}_2)_{k'}$ by the k-point

$$\begin{pmatrix} 0 & t_1 \\ 1 & 0 \end{pmatrix} \in \mathrm{R}_{K/k}(\mathrm{PGL}_2)(k) = \mathrm{PGL}_2(K).$$

By Galois descent, the minimal field of definition over k for $\mathscr{R}_u(G_{\overline{k}}) \subset G_{\overline{k}}$ is K/k. Moreover, the maximal reductive quotient G_K^{ss} is the K-form of SL_2 arising from the class in $\mathrm{H}^1(K'/K, \mathrm{PGL}_2(K'))$ represented by the 1-cocycle $c : \sigma \mapsto \left(\begin{smallmatrix} 0 & t_1 \\ 1 & 0 \end{smallmatrix}\right) \in \mathrm{PGL}_2(K) \subset \mathrm{PGL}_2(K')$.

Suppose G admits a pseudo-split k_s/k-form. This form is k-isomorphic to $H_{V,K/k}$ for a nonzero k-subspace $V \subset K$ satisfying $k\langle V \rangle = K$, and Proposition C.1.1 applied over k' gives $H_{V_{k'},K'/k'} \simeq H_{V',K'/k'}$ with $V_{k'} := k' \otimes_k V$. Necessarily $V_{k'} = \mu V'$ for some $\mu \in K'^{\times}$, but $\sigma(V_{k'}) = V_{k'}$ and

$$\sigma(\mu V') = \sigma(\mu)\sigma(V') = \sigma(\mu) t_1 V',$$

so $(\sigma(\mu)/\mu)t_1 \in F' = k'$. Hence, $t_1 \in (\mu/\sigma(\mu))k'^{\times}$. Applying $\mathrm{N}_{K'/K}$ then gives that the element $t_1^2 \in k^{\times}$ lies in $\mathrm{N}_{k'/k}(k'^{\times})$.

In view of how K and $\{t_1, t_2, t_3\}$ were chosen, we can arrange for a contradiction (and hence conclude that G does not admit a pseudo-split k_s/k-form) provided that $\mathrm{N}_{k'/k}(k'^{\times}) \neq k^{\times}$ (for then we may choose K to contain $t_1 \in k^{1/2}$ such that $t_1^2 \notin \mathrm{N}_{k'/k}(k'^{\times})$, and $t_1 \notin k$ since $(k^{\times})^2 \subset \mathrm{N}_{k'/k}(k'^{\times})$); this is equivalent to the condition that $\ker(\mathrm{Br}(k) \to \mathrm{Br}(k'))$ is nontrivial. For example, if $k := \mathbf{F}(z_1, z_2, z_3)$ is a rational function field in 3 variables over a finite field $\mathbf{F}$ of characteristic 2 and $k' = L(z_2, z_3)$ for a separable quadratic extension $L/\mathbf{F}(z_1)$ then the kernel of $\mathrm{Br}(\mathbf{F}(z_1)) \to \mathrm{Br}(L)$ is nontrivial by global class field theory and it injects into $\ker(\mathrm{Br}(k) \to \mathrm{Br}(k'))$.

The description of G_K^{ss} in terms of $\mathrm{H}^1(K'/K, \mathrm{PGL}_2(K'))$ shows that G_K^{ss} is the algebraic group of norm-1 units in the quaternion algebra over K associated to the class $t_1 \in K^{\times}/\mathrm{N}_{K'/K}(K'^{\times}) \subset \mathrm{Br}(K)[2]$. But t_1 is not a norm from K' since (by design) the element $t_1^2 \in k^{\times}$ is not a norm from k', so G_K^{ss} is K-anisotropic.

C.4 Higher-rank and non-reduced cases

Now we adapt the construction in Example C.3.1 to give additional examples of absolutely pseudo-simple G over k of minimal type which do not admit a pseudo-split k_s/k-form and whose root system (over k_s) is of type B, C, or BC of any rank. No such G has a quasi-split k_s/k-form (by Lemma C.2.2, since the Dynkin diagrams of these root systems have no nontrivial automorphism).

Example C.4.1 (B and C). Let k be an imperfect field of characteristic 2 such that $[k : k^2] \geqslant 8$ and k admits a quadratic Galois extension k'/k (with nontrivial automorphism denoted as σ). Also assume $\mathrm{Br}(k) \to \mathrm{Br}(k')$ has nontrivial kernel. Choose K/k as in Example C.3.1 with a 2-basis $\{t_1, t_2, t_3\}$ such that the element $t_1^2 \in k^\times$ is not a norm from k', so we obtain a 4-dimensional k'-subspace $V' \subset K' := k' \otimes_k K$ such that $H_{V',K'/k'}$ has a k-descent with no pseudo-split k_s/k-form and root field k.

The k-descent is defined by using a descent datum on the k'-subgroup $H_{V',K'/k'} \subset \mathrm{R}_{K'/k'}(\mathrm{SL}_2)$ via $\nu = \left(\begin{smallmatrix} 0 & t_1 \\ 1 & 0 \end{smallmatrix}\right) \in \mathrm{PGL}_2(K')$. That is, we use the Galois descent datum on the k'-group $H_{V',K'/k'}$ obtained by composing the canonical k'-isomorphism $\sigma^*(H_{V',K'/k'}) \simeq H_{\sigma(V'),K'/k'} = H_{t_1 V',K'/k'}$ with the natural action of ν on $\mathrm{R}_{K'/k'}(\mathrm{SL}_2)$ (which carries $H_{t_1 V',K'/k'}$ onto $H_{V',K'/k'}$, restricting to inversion on the common diagonal Cartan k'-subgroup).

For $n \geqslant 2$, consider generalized basic exotic pseudo-split absolutely pseudo-simple k'-groups (G', T') with root system $\Phi = \Phi(G', T')$ equal to B_n or C_n. By Proposition 8.2.5, up to k'-isomorphism there is a unique G' such that $G'_c \simeq H_{V',K'/k'}$ for $c \in \Phi_<$ when $\Phi = \mathrm{B}_n$ with $n \geqslant 2$ and for $c \in \Phi_>$ when $\Phi = \mathrm{C}_n$ with $n \geqslant 3$. Let $L' \subset G'$ be a Levi k'-subgroup containing T', Δ a basis of Φ, and c_0 the unique short root in Δ when $\Phi = \mathrm{B}_n$ with $n \geqslant 2$ and the unique long root in Δ when $\Phi = \mathrm{C}_n$ with $n \geqslant 3$. Choose a pinning of (L', T', Δ) so that the Levi k'-subgroup $\mathrm{SL}_2 \simeq L'_{c_0} \hookrightarrow G'_{c_0}$ identifies G'_{c_0} with $H_{V',K'/k'}$ via $i_{G'_{c_0}}$.

By Theorem 6.1.1, there is a unique Galois descent datum on G' relative to k'/k whose effect on $G'_{c_0} = H_{V',K'/k'}$ is as above and whose effect on $G'_c \subset \mathrm{R}_{K'/k'}(\mathrm{SL}_2)$ for $c \in \Delta - \{c_0\}$ is the composition of the canonical descent datum with transpose-inverse (having the effect of inversion on the diagonal Cartan k'-subgroup). This descent datum preserves the split k'-torus T' and defines a descent of (G', T') to a pair (G, T) over k where G is of minimal type and T splits over k' with the nontrivial element of $\mathrm{Gal}(k'/k)$ acting on $\mathrm{X}^*(T_{k'})$ via inversion. In particular, for every $c \in \Phi(G_{k'}, T_{k'}) = \Phi$ the absolutely pseudo-simple k'-subgroup $G'_c \subset G' = G_{k'}$ of rank 1 generated by the $\pm c$-root groups is preserved by the k'/k-descent datum and so descends to an absolutely pseudo-simple k-subgroup $G_c \subset G$ of rank 1.

Suppose there is a pseudo-split k_s/k-form H of G, say with a pseudo-split maximal k-torus S, so $H_{k'}$ and $G_{k'}$ are pseudo-split and k'_s/k'-forms of each other. By Proposition C.1.1, there is a k'-isomorphism $H_{k'} \simeq G_{k'} = G'$. This isomorphism can be chosen to carry $S_{k'}$ over to $T_{k'} = T'$, so we may identify $\Phi(H, S)$ with Φ. This defines H_c for all $c \in \Phi$, and $(H_{c_0})_{k'} \simeq G'_{c_0} = H_{V',K'/k'}$, so H_{c_0} is a pseudo-split k-descent of $H_{V',K'/k'}$. But by construction of $(V', K'/k')$ in Example C.3.1, $H_{V',K'/k'}$ has no such k-descent. Thus, there is no such H; i.e., G does not admit a pseudo-split k_s/k-form. As $G_{k'} = G'$, the minimal field of definition over k for the geometric unipotent radical of G is K/k and $G_K^{\rm ss}$ is a simply connected semisimple K-group that splits over K' (with root system Φ).

Example C.4.2 (BC). Let $(K/k, V', \sigma)$ be as in Example C.3.1, and let W' be a k'-linear complement of V' in K', so $\dim_{k'} W' = 4$. Define

$$V^{(2)} = K'^2 w' \oplus K'^2 t_1\sigma(w') \subset k'w' \oplus k't_1\sigma(w')$$

for a nonzero $w' \in W'$. To justify the description of $V^{(2)} \subset K'$ inside a direct sum over k', note that if $t_1\sigma(w') = \lambda' w'$ for some $\lambda' \in k'^{\times}$ then $\sigma(w')/w' = \lambda'/t_1$, so applying $\mathrm{N}_{K'/K}$ yields $t_1^2 = \mathrm{N}_{k'/k}(\lambda')$, an absurdity.

Observe that $\sigma(V^{(2)}) = t_1 V^{(2)}$ since $t_1^2 \in k^{\times}$, and $t_1\sigma(w') \notin V'$ (since $\sigma(V') = t_1 V' = t_1^{-1} V'$ and $W' \cap V' = 0$), so $V^{(2)} \cap V'$ is a proper K'^2-subspace of the K'^2-plane $V^{(2)}$. Thus, if $V^{(2)} \cap V' \neq 0$ then $V^{(2)} \cap V'$ is spanned over $K'^2 \subset k'$ by $w' + \alpha^2 t_1\sigma(w')$ for some $\alpha \in K'^{\times}$.

We may and do choose W' to contain $w' := t_1t_2$. Since $\{t_1, t_2, t_3\}$ is a 2-basis for K' over k', it follows that $V^{(2)} \cap V' = 0$. Note that $V^{(2)} \oplus V'$ contains 1 and generates K' as a k'-algebra since the same holds for V'. Hence, for arbitrary $n \geqslant 1$ there exists a pseudo-split pseudo-semisimple k'-group $G' = \mathscr{D}(G_{K'/k',V',V^{(2)},n})$ of minimal type with root system BC_n as in [CGP, Thm. 9.8.1(1)].

The k'-group G' is equipped with a standard split maximal k'-torus D', and if Δ is a basis of $\Phi(G', D')$ with unique multipliable element c_0 then $G'_{2c_0} = H_{V',K'/k'}$. Thus, by using [CGP, Prop. 9.8.9(i)] with $n = 1$ (applied to G'_{c_0} and its σ-twist) and Theorem 6.1.1, the Galois-twisting construction on G'_{2c_0} in Example C.3.1 extends to a descent datum on (G', D') that is the composition of transpose-inverse with standard descent datum on $G'_c = \mathrm{R}_{K'/k'}(\mathrm{SL}_2)$ for all $c \in \Delta - \{c_0\}$. By the same reasoning as in Example C.4.1, the k-descent G of G' does not have a pseudo-split k_s/k-form.

Appendix D

Basic exotic groups of type F_4 of relative rank 2

Let k be an imperfect field of characteristic 2, and consider a basic exotic k-group H such that H_{k_s} has root system F_4. The minimal field of definition K/k for the geometric unipotent radical of H is a nontrivial purely inseparable finite extension of k satisfying $K^2 \subset k$, and $H(k') = \mathrm{Aut}_{k'}(H_{k'})$ for all separable extensions k'/k by [CGP, Cor. 8.2.5].

Fix a nontrivial purely inseparable finite extension K/k inside $k^{1/2}$, so there exists a *pseudo-split* basic exotic k-group $\mathscr{G}$ with root system F_4 such that the minimal field of definition for $\mathscr{R}_u(\mathscr{G}_{\overline{k}}) \subset \mathscr{G}_{\overline{k}}$ is K/k; $\mathscr{G}$ is unique up to k-isomorphism by Proposition C.1.1. The possibilities up to k-isomorphism for G with the associated K/k are exactly the k_s/k-forms of $\mathscr{G}$, and the non-pseudo-split forms are those G with k-rank less than 4.

It is classical over an arbitrary field (see [Spr, 17.5.1–17.5.2]) that a connected *semisimple* group of type F_4 which is neither anisotropic nor split has relative rank 1. In contrast, we will see below that for suitable k there do exist G as above with k-rank 2. Such k-groups give rise to Moufang quadrangles "of type F_4" as defined in (16.7) of [TW] (see also [TW, Ch. 14, (17.4)] as well as [MV]).

In this appendix we use techniques developed in this monograph and [CGP] to show that k-rank 3 cannot occur and to classify the isomorphism classes of k_s/k-forms G of k-rank 2 in terms of conformal isometry classes of certain anisotropic regular degenerate quadratic spaces over k.

D.1 General preparations

D.1.1. Let S be a maximal k-split torus in G and let P be a minimal pseudo-parabolic k-subgroup of G containing S. Since $\mathscr{G}$ is pseudo-split, every pseudo-parabolic k_s-subgroup of $\mathscr{G}_{k_s}$ is $\mathscr{G}(k_s)$-conjugate to one that descends to a pseudo-parabolic k-subgroup of $\mathscr{G}$. Thus, we can choose a pseudo-parabolic k-subgroup $\mathscr{P}$ of $\mathscr{G}$ such that there is a k_s-isomorphism $(G_{k_s}, P_{k_s}) \simeq (\mathscr{G}_{k_s}, \mathscr{P}_{k_s})$. Since $\mathscr{G}(k_s) = \mathrm{Aut}_{k_s}(\mathscr{G}_{k_s})$ and $\mathscr{P}$ is its own normalizer in $\mathscr{G}$ [CGP, Thm. 3.5.7], it follows that the pair (G, P) is obtained from $(\mathscr{G}, \mathscr{P})$ via twisting by a continuous $\mathscr{P}(k_s)$-valued 1-cocycle on $\mathrm{Gal}(k_s/k)$.

Let $\mathscr{U} = \mathscr{R}_{u,k}(\mathscr{P})$, so $\mathscr{U}$ is k-split, and let $\mathscr{M}$ be a smooth closed k-subgroup of $\mathscr{P}$ such that $\mathscr{U} \rtimes \mathscr{M} \to \mathscr{P}$ is an isomorphism. (For example, we can use $\mathscr{M} = Z_{\mathscr{G}}(\lambda)$ for a 1-parameter k-subgroup λ of G such that $\mathscr{P} = P_{\mathscr{G}}(\lambda)$.) Let $\mathscr{M}'$ be the derived group of $\mathscr{M}$.

Every k_s/k-form of $\mathscr{U}$ is k-split by [CGP, Thm. B.3.4], so the twist of $\mathscr{U}$ by any continuous $\mathscr{M}(k_s)$-valued 1-cocycle on $\mathrm{Gal}(k_s/k)$ is k-split. Hence, the natural injective map $\mathrm{H}^1(k, \mathscr{M}) \to \mathrm{H}^1(k, \mathscr{P})$ is bijective. In particular, we can find a continuous 1-cocycle $c : \mathrm{Gal}(k_s/k) \to \mathscr{M}(k_s)$ that twists $(\mathscr{G}, \mathscr{P})$ into (G, P), and using that to define the initial isomorphism $(G_{k_s}, P_{k_s}) \simeq (\mathscr{G}_{k_s}, \mathscr{P}_{k_s})$ above ensures that this isomorphism carries $\mathscr{M}_{k_s}$ over to M_{k_s} for a smooth closed k-subgroup M of P. For $U := \mathscr{R}_{u,k}(P)$, clearly $P = U \rtimes M$ and so the natural map $M \to P/U$ is an isomorphism. We now have an isomorphism of triples

$$(G_{k_s}, P_{k_s}, M_{k_s}) \simeq (\mathscr{G}_{k_s}, \mathscr{P}_{k_s}, \mathscr{M}_{k_s}). \tag{D.1.1}$$

Remark D.1.2. The only relevant property of the Galois extension k_s/k above is that G_{k_s} is pseudo-split. Later we will encounter situations in which G_F is pseudo-split for a quadratic Galois extension F/k, so we can use F in place of k_s when building (D.1.1).

By [CGP, Prop. C.2.4], minimality of P in G implies $P = U \rtimes Z_G(S)$ for any maximal split k-torus S of G lying in P. The isomorphism $M \simeq P/U$ implies that such an S can be found inside M, so we make such a choice of S. Choose a maximal k-torus T in $M = Z_G(S)$, so $S \subset T$ and (D.1.1) carries T_{k_s} over to a maximal k_s-torus in $\mathscr{M}_{k_s}$. Since T_{k_s} is k_s-split and $\mathscr{M}$ contains a split maximal k-torus $\mathscr{T}$, the $\mathscr{M}(k_s)$-conjugacy of split maximal k_s-tori in $\mathscr{M}_{k_s}$ allows us to choose (D.1.1) so that it also carries T_{k_s} over to $\mathscr{T}_{k_s}$.

Fix a basis Δ of $\Phi(G_{k_s}, T_{k_s})$ contained in the parabolic subset $\Phi(P_{k_s}, T_{k_s})$. We enumerate the basis Δ as $\{a_1, a_2, a_3, a_4\}$ corresponding to the Dynkin diagram

Let Δ_0 be the set of elements of Δ that are trivial on the subtorus $S_{k_s} \subset T_{k_s}$, and call an element of Δ *distinguished* if it lies in $\Delta - \Delta_0$. As shown in the proof of [CGP, Thm. C.2.15], the restrictions to S of the distinguished roots in Δ constitute the basis ${}_k\Delta$ for the positive system of roots $\Phi(P, S)$ of ${}_k\Phi =$

$\Phi(G, S)$, and the fibers of $\Delta - \Delta_0 \twoheadrightarrow {}_k\Delta$ are the orbits under the $*$-action of $\mathrm{Gal}(k_s/k)$ on $\Delta - \Delta_0$. The F_4-diagram has no nontrivial automorphisms, so the $*$-action is trivial and hence $\Delta - \Delta_0 \twoheadrightarrow {}_k\Delta$ is bijective. Since (D.1.1) carries T_{k_s} over to $\mathscr{T}_{k_s}$, it identifies Δ with a basis of $\Phi := \Phi(\mathscr{G}, \mathscr{T})$.

The root system $\Psi = \Phi(\mathscr{M}, \mathscr{T})$ consists of the roots in Φ which are $\mathbf{Z}$-linear combination of roots in Δ_0, so the Dynkin diagram of Ψ is the subdiagram of Δ with vertices given by Δ_0. Let $\{\Psi_i\}$ be the set of irreducible components of Ψ. By [CGP, Thm. 3.4.6] there exists a Levi k-subgroup $\mathscr{L} \subset \mathscr{G}$ containing the k-split maximal torus $\mathscr{T}$, and since $\mathscr{L}$ is simply connected we see that its k-subgroup $\mathscr{L}_i$ generated by the root groups corresponding to the roots in Ψ_i is simply connected and $\Phi(\mathscr{L}_i \cdot \mathscr{T}, \mathscr{T}) = \Psi_i$. The set $\Delta^\vee$ is a basis of $X_*(\mathscr{T})$, so the k-subgroup of $\mathscr{L}$ generated by $\mathscr{T}$ and the $\mathscr{L}_i$'s is a semi-direct product of $\prod_{c \in \Delta - \Delta_0} c^\vee(\mathrm{GL}_1)$ against $\prod_i \mathscr{L}_i$. The analogue then holds over K for the pair $(\mathscr{L}_K, \mathscr{T}_K)$, so the inclusion

$$i_{\mathscr{G}} : \mathscr{G} \hookrightarrow \mathrm{R}_{K/k}(\mathscr{G}_K^{\mathrm{ss}}) = \mathrm{R}_{K/k}(\mathscr{L}_K)$$

shows via inspection of the open cell of the basic exotic $\mathscr{G}$ of type F_4 (see [CGP, Rem. 7.2.8]) that

$$\mathscr{M} = \mathscr{M}' \cdot Z_{\mathscr{G}}(\mathscr{T}) = \prod_i \mathscr{M}'_i \rtimes \prod_{c \in \Delta - \Delta_0} \mathrm{R}_{K_c/k}(\mathrm{GL}_1) \tag{D.1.2}$$

where $K_c = K$ for short $c \in \Delta$, $K_c = k$ for long $c \in \Delta$, and $\mathscr{M}'_i \subset \mathrm{R}_{K/k}((\mathscr{L}_i)_K)$ is the k-subgroup of $\mathscr{M}$ generated by the root groups U_a for $a \in \Psi_i$.

We will now show that G cannot be of k-rank 3. More generally:

Proposition D.1.3. *If a_2 or a_3 is distinguished then G is pseudo-split.*

Proof. Assume G is not pseudo-split, so $\mathrm{H}^1(k, \mathscr{M}) \neq 1$ and Ψ is not empty (cf. Lemma C.2.2). The omission of a_2 or a_3 from Δ_0 implies that each Ψ_i is of type A_1 or A_2. Thus, each $\mathscr{L}_i$ is either SL_2 or SL_3. Define $K_i = K$ if Ψ_i consists of short roots in $\Phi(\mathscr{G}, \mathscr{T})$ and $K_i = k$ if Ψ_i consists of long roots in $\Phi(\mathscr{G}, \mathscr{T})$, so $\mathscr{M}'_i \simeq \mathrm{R}_{K_i/k}((\mathscr{L}_i)_{K_i})$. Hence, (D.1.2) gives

$$\mathscr{M} = \prod_i \mathrm{R}_{K_i/k}((\mathscr{L}_i)_{K_i}) \rtimes \prod_{c \in \Delta - \Delta_0} \mathrm{R}_{K_c/k}(\mathrm{GL}_1),$$

so the non-abelian Shapiro's Lemma yields an exact sequence of pointed sets

$$\prod_i \mathrm{H}^1(K_i, (\mathscr{L}_i)_{K_i}) \to \mathrm{H}^1(k, \mathscr{M}) \to \prod_c \mathrm{H}^1(K_c, \mathrm{GL}_1).$$

The outer terms vanish, contradicting that $\mathrm{H}^1(k, \mathscr{M}) \neq 1$. □

Remark D.1.4. By the same method we see that if k is imperfect of characteristic 3 and G is a basic exotic k-group of type G_2 that is k-isotropic (i.e., some root is distinguished) then G is pseudo-split. This is analogous to the classical fact [Spr, 17.4.2] that isotropic connected *semisimple* groups of type G_2 over a field are always split.

D.2 Forms of k-rank 2

D.2.1. Assume G is k-isotropic but not pseudo-split, so by Proposition D.1.3 the roots a_2 and a_3 are non-distinguished. The k-rank of G is therefore equal to 1 or 2. We now focus on the case of k-rank 2, so the distinguished vertices are precisely a_1 and a_4. The minimal pseudo-parabolic k-subgroup P is equal to $U \rtimes Z_G(S)$ with k-split $U = \mathscr{R}_{u,k}(P)$, and the derived group M' of $M := Z_G(S)$ is k-anisotropic and of type B_2, corresponding to the subdiagram of Δ with vertex set $\Delta_0 = \{a_2, a_3\}$. We denote the split maximal k-torus $\mathscr{T} \cap \mathscr{M}'$ of $\mathscr{M}'$ by $\mathscr{S}'$.

The structure of open cells of pseudo-split basic exotic groups of type F_4 (such as $\mathscr{G}$) and the Isomorphism Theorem for pseudo-split pseudo-reductive groups (Theorem 6.1.1) imply that $\mathscr{M}' \simeq \mathrm{Spin}(\mathfrak{q})$ for the orthogonal sum

$$(\mathscr{V}, \mathfrak{q}) = \mathscr{H}_1 \perp \mathscr{H}_2 \perp (K, x^2) \qquad \text{(D.2.1)}$$

over k where $\mathscr{H}_i$ is a hyperbolic plane and K is equipped with the quadratic form given by the squaring map into k (making K the defect space $\mathscr{V}^{\perp}$). By the $\mathscr{M}'(k)$-conjugacy of split maximal k-tori in $\mathscr{M}'$ and the natural bijection (respecting k-rank) between the sets of maximal k-tori in $\mathrm{Spin}(\mathfrak{q})$ and its maximal central quotient $\mathrm{SO}(\mathfrak{q})$, the isomorphism $\mathscr{M}' \simeq \mathrm{Spin}(\mathfrak{q})$ can be chosen so that $\mathscr{S}'$ corresponds to the split maximal k-torus $\overline{\mathscr{S}'}$ in $\mathrm{SO}(\mathfrak{q})$ that keeps stable each of the two hyperbolic planes $\mathscr{H}_1$ and $\mathscr{H}_2$ under its action on $\mathscr{V}$. We identify $\mathscr{M}'$ with $\mathrm{Spin}(\mathfrak{q})$ using such an isomorphism.

The nontrivial weights of $\mathscr{S}'$ in the standard representation of $\mathrm{Spin}(\mathfrak{q})$ (via its image in $\mathrm{SO}(\mathfrak{q})$) on $\mathscr{V}$ are the short roots. Swapping $\mathscr{H}_1$ and $\mathscr{H}_2$ if necessary, we may assume that the weights on $\mathscr{H}_1$ are $\pm a_3$ and on $\mathscr{H}_2$ are $\pm(a_2 + a_3)$. Let $e_1 \in \mathscr{H}_1$ be a weight vector for a_3, let $e_2 \in \mathscr{H}_2$ be a weight vector for $a_2 + a_3$, and let $f_i \in \mathscr{H}_i$ be the unique weight vector for the opposite weight to that of e_i such that the value $B_{\mathfrak{q}}(e_i, f_i) \in k^{\times}$ is equal to 1.

Lemma D.2.2. *There exists an involution $\iota \in N_{\mathscr{M}'}(\mathscr{S}')(k)$ inducing inversion on $\mathscr{S}'$ such that $f_i = \iota(e_i)$.*

Proof. The $\mathbf{F}_2$-span V_0 of the e_i's, f_i's, and $h = 1 \in K = \mathscr{V}^{\perp}$ defines a non-degenerate quadratic space $(V_0, \mathfrak{q}_0)$ over $\mathbf{F}_2$ such that $\mathrm{Spin}(\mathfrak{q}_0)$ is naturally an $\mathbf{F}_2$-descent of the Levi k-subgroup $\mathscr{L}' = \mathrm{Spin}(\mathfrak{q}|_{\mathscr{H}_1 \oplus \mathscr{H}_2 \oplus kh})$ of $\mathscr{M}' = \mathrm{Spin}(\mathfrak{q})$ that contains $\mathscr{S}'$ as a split maximal k-torus. Under the resulting identification of the adjoint central quotient $\mathrm{SO}(\mathfrak{q}_0)$ as an $\mathbf{F}_2$-descent of the Levi k-subgroup $\overline{\mathscr{L}}' = \mathrm{SO}(\mathfrak{q}|_{\mathscr{H}_1 \oplus \mathscr{H}_2 \oplus kh})$ of $\mathrm{SO}(\mathfrak{q})$, it is clear that $\overline{\mathscr{S}}'$ descends to the split maximal $\mathbf{F}_2$-torus $\overline{S}_0 \subset \mathrm{SO}(\mathfrak{q}_0)$ consisting of elements preserving the lines spanned by each of e_1, f_1, e_2, f_2, h (with trivial action on h). The corresponding split maximal $\mathbf{F}_2$-torus $S_0 \subset \mathrm{Spin}(\mathfrak{q}_0)$ is an $\mathbf{F}_2$-descent of $\mathscr{S}' \subset \mathscr{L}'$.

It suffices to find an involution in $N_{\mathscr{L}'}(\mathscr{S}')(k)$ inducing inversion on $\mathscr{S}'$ and carrying e_i to f_i for $i = 1, 2$, so it is even sufficient to find such an involution in $N_{\mathrm{Spin}(\mathfrak{q}_0)}(S_0)(\mathbf{F}_2)$. Since $\mathbf{F}_2^{\times} = 1$, the natural map $N_{\mathrm{Spin}(\mathfrak{q}_0)}(S_0)(\mathbf{F}_2) \to N_{\mathrm{SO}(\mathfrak{q}_0)}(\overline{S}_0)(\mathbf{F}_2)$ is identified with the map of Weyl groups

$$W(\mathrm{Spin}(\mathfrak{q}_0), S_0) \to W(\mathrm{SO}(\mathfrak{q}_0), \overline{S}_0),$$

and this is an isomorphism since $\mathrm{Spin}(\mathfrak{q}_0) \to \mathrm{SO}(\mathfrak{q}_0)$ is a central isogeny. Thus, we just need to find an involution $\overline{\iota}$ in $N_{\mathrm{SO}(\mathfrak{q}_0)}(\overline{S}_0)(\mathbf{F}_2)$ carrying e_i to f_i for $i = 1, 2$, and acting on the isogenous quotient $\overline{S}_0$ of S_0 via inversion. Define $\overline{\iota} \in \mathrm{GL}(V_0)$ to be the automorphism that fixes h and swaps e_i and f_i for $i = 1, 2$. Clearly $\overline{\iota}$ is an involution that lies in $\mathrm{O}(\mathfrak{q}_0)(\mathbf{F}_2)$, and $\mathrm{O}(\mathfrak{q}_0) = \mathrm{SO}(\mathfrak{q}_0) \times \mu_2$ since $\dim V_0$ is odd, so $\mathrm{O}(\mathfrak{q}_0)$ and $\mathrm{SO}(\mathfrak{q}_0)$ have the same points valued in any field. By design we see that $\overline{\iota}$-conjugation restricts to inversion on $\overline{S}_0$, so we are done. □

We begin our analysis of the possibilities for G of k-rank 2 by showing that M' is the spin group of an anisotropic k_s/k-form of $(\mathscr{V}, \mathfrak{q})$; an explicit description of the possibilities for such quadratic spaces will be given later.

Lemma D.2.3. *There exists a regular degenerate anisotropic quadratic space (V, q) over k such that $M' \simeq \mathrm{Spin}(q)$, and (V, q) is unique up to conformal isometry. Moreover, there exists a separable quadratic extension F/k such that M_F is pseudo-split (so $M_F \simeq \mathscr{M}_F$ and the derived group M'_F is pseudo-split).*

Since S is unique up to $G(k)$-conjugacy, the conformal isometry class of (V, q) is uniquely determined by the k-isomorphism class of G.

Proof. Recall that we have a continuous 1-cocycle $c : \mathrm{Gal}(k_s/k) \to \mathscr{M}(k_s)$ that twists $(\mathscr{G}, \mathscr{P}, \mathscr{M})$ into (G, P, M). This compatibly twists the derived group

$\mathscr{M}' = \mathrm{Spin}(\mathfrak{q})$ into the derived group M'. As in (D.1.2), we have

$$\mathscr{M} = \mathscr{M}' \rtimes \prod_{c \in \Delta - \Delta_0} \mathrm{R}_{K_c/k}(\mathrm{GL}_1), \tag{D.2.3}$$

so $\mathrm{H}^1(k, \mathscr{M}') \to \mathrm{H}^1(k, \mathscr{M})$ is surjective and hence we can replace c with a cohomologous 1-cocycle to arrange that it is valued in $\mathscr{M}'(k_s)$.

Via the maximal central quotient map $\mathscr{M}' = \mathrm{Spin}(\mathfrak{q}) \to \mathrm{SO}(\mathfrak{q})$, c is carried onto a 1-cocycle $\overline{c}$ valued in $\mathrm{SO}(\mathfrak{q})(k_s)$, so $M'/Z_{M'} \simeq \mathrm{SO}(q)$ where q is the $\overline{c}$-twist of $\mathfrak{q}$. This latter k-isomorphism uniquely lifts to a k-isomorphism $M' \simeq \mathrm{Spin}(q)$ between universal smooth k-tame central extensions, and q is anisotropic since M' is k-anisotropic (see Proposition 7.1.6). The uniqueness of (V, q) up to conformal isometry follows from Proposition 7.2.2(i).

Since $\mathrm{char}(k) = 2$ and q written in terms of linear coordinates must have a nonzero cross-term (as its defect space is a proper subspace), we can find a plane on which the restriction of the anisotropic q is not additive and hence is nondegenerate. Thus, there exists a separable quadratic extension F/k such that q_F is isotropic. Clearly $(M'/Z_{M'})_F$ is F-isotropic, so M'_F is also F-isotropic. But M' commutes with S and the intersection $M' \cap S$ is finite, so G_F has F-rank at least 3 and hence is pseudo-split by Proposition D.1.3. In particular, the F-group $M_F = Z_G(S)_F$ is pseudo-split. □

D.2.4. We fix F/k as in Lemma D.2.3, so G_F is pseudo-split. By Remark D.1.2, after possibly changing M inside P we obtain an isomorphism of triples

$$(G_F, P_F, M_F) \simeq (\mathscr{G}_F, \mathscr{P}_F, \mathscr{M}_F) \tag{D.2.4}$$

arising from a 1-cocycle $\mathfrak{c} : \mathrm{Gal}(F/k) \to \mathscr{M}(F)$. (This retains the property that M_F is pseudo-split, and every $M \subset P$ has the form $Z_G(S)$ for some $S \subset P$. The maximal k-torus T will no longer play any role, so it does not matter whether or not the above F-isomorphism carries $\mathscr{T}_F$ to an F-torus of M_F that descends to a k-torus of M.) Via (D.1.2), Hilbert's Theorem 90 gives the surjectivity of

$$\mathrm{H}^1(F/k, \mathscr{M}'(F)) \to \mathrm{H}^1(F/k, \mathscr{M}(F)),$$

so replacing $\mathfrak{c}$ with a cohomologous 1-cocycle makes it valued in $\mathscr{M}'(F)$.

Next, we seek a maximal k-torus S' of M' that splits over F. For this purpose, fix a minimal pseudo-parabolic F-subgroup Q of M'_F, and let γ be the nontrivial k-automorphism of F. By [CGP, Prop. 3.5.12(2), Prop. C.2.7], the intersection $Q \cap \gamma^*(Q)$ is a smooth connected F-subgroup that contains a split maximal F-torus T' of M'_F. The F-group $Q \cap \gamma^*(Q)$ visibly descends to a k-subgroup H' of M', so

$$Q \cap \gamma^*(Q) = H'_F.$$

Lemma D.2.5. *The split maximal F-torus T' in H'_F is central. In particular, T' is unique and so descends to a maximal k-torus in H' that splits over F.*

Proof. Suppose to the contrary, so T' has a nontrivial weight a' on $\mathrm{Lie}(H'_F)$. Consider the associated root group $U_{a'}$ in H'_F as in [CGP, Def. 2.3.4]. This is a nontrivial F-split smooth connected unipotent F-subgroup of H'_F $(\subset Q)$. As $U_{a'}$ is F-split and Q is a minimal pseudo-parabolic F-subgroup of the pseudo-split pseudo-semisimple F-group M'_F, it follows from [CGP, Cor. C.3.9] that $U_{a'} \subset \mathscr{R}_{us,F}(Q)$. The subgroup U' of H'_{k_s} generated by the k_s-split unipotent groups $h(U_{a'})_{k_s}h^{-1}$ for varying $h \in H'(k_s)$ is contained in $\mathscr{R}_{us,k_s}(Q_{k_s})$, so it is a smooth connected unipotent normal k_s-subgroup of H'_{k_s}.

By [CGP, Thm. B.3.4] a smooth connected unipotent k_s-group is k_s-split if it is generated by k_s-split subgroups. Thus, U' is k_s-split and hence $U' \subset \mathscr{R}_{us,k_s}(H'_{k_s})$. Since $\mathscr{R}_{us,k}(H')_{k_s} = \mathscr{R}_{us,k_s}(H'_{k_s})$, we see that $\mathscr{R}_{us,k}(H') \neq 1$. But $H' \subset M'$, so M' contains the nontrivial split smooth connected unipotent k-subgroup $\mathscr{R}_{us,k}(H')$. This contradicts the k-anisotropicity of the pseudo-reductive M', by [CGP, Cor. C.3.9]. □

D.2.6. Let S' be a maximal k-torus of M' that splits over F. Any two F-split maximal tori of $\mathscr{M}'_F$ are conjugate under $\mathscr{M}'(F)$, so by applying this to $\mathscr{S}'_F$ and $S'_F \subset M'_F \simeq \mathscr{M}'_F$ (using (D.2.4)) we can replace $\mathfrak{c}$ with a cohomologous 1-cocycle so that it takes values in $N_{\mathscr{M}'}(\mathscr{S}')(F)$ and the associated k-isomorphism $M' \simeq {}_{\mathfrak{c}}\mathscr{M}'$ carries S' over to ${}_{\mathfrak{c}}\mathscr{S}'$.

As S' is anisotropic over k and splits over the separable quadratic extension F, the effect of conjugation by $\mathfrak{c}(\gamma)$ on $\mathscr{S}'_F$ is inversion. Likewise, the image $\overline{\mathfrak{c}}(\gamma) \in \mathrm{SO}(\mathfrak{q})(F) = \mathrm{O}(\mathfrak{q}_F)$ of $\mathfrak{c}(\gamma) \in \mathscr{M}'(F) = \mathrm{Spin}(\mathfrak{q})(F)$ fixes the defect space $K_F := F \otimes_k K$ pointwise (due to Proposition 7.1.2) and must move each weight space for the action of the split rank-2 torus $\mathscr{S}'_F$ on $\mathscr{V}_F := F \otimes_k \mathscr{V}$ to the weight space for the opposite weight (since $\mathfrak{c}(\gamma)$ acts on $\mathscr{S}'_F$ through inversion).

The involution $\iota \in N_{\mathscr{M}'}(\mathscr{S}')(k)$ from Lemma D.2.2 acts on $\mathscr{S}'$ via inversion, so $\mathfrak{c}(\gamma)$ is a point of

$$Z_{\mathscr{M}'}(\mathscr{S}')(F) \cdot \iota = ((a_2)^\vee(\mathrm{GL}_1)(F) \times \mathrm{R}_{K/k}((a_3)^\vee_K(\mathrm{GL}_1))(F)) \cdot \iota.$$

Thus, $\mathfrak{c}(\gamma) = a_2^\vee(s)a_3^\vee(st)\cdot\iota$ for unique $s \in F^\times$ and $t \in (F\otimes_k K)^\times$. As e_1 has weight a_3 and e_2 has weight a_2+a_3, the identities $\langle a_2, a_3^\vee\rangle = -2$ and $\langle a_3, a_2^\vee\rangle = -1$ imply

$$a_3(a_2^\vee(s)a_3^\vee(st)) = st^2,\ (a_2+a_3)(a_2^\vee(s)a_3^\vee(st)) = s.$$

Since $\iota^2 = 1$, so $\iota(f_i) = e_i$ for each i (as $\iota(e_i) = f_i$ for each i), we obtain

$$\overline{\mathfrak{c}}(\gamma)(e_1) = (st^2)^{-1}f_1,\ \overline{\mathfrak{c}}(\gamma)(f_1) = st^2e_1,\ \overline{\mathfrak{c}}(\gamma)(e_2) = s^{-1}f_2,\ \overline{\mathfrak{c}}(\gamma)(f_2) = se_2.$$

Hence, the cocycle condition $\overline{\mathfrak{c}}(\gamma)\cdot\gamma(\overline{\mathfrak{c}}(\gamma)) = 1$ is equivalent to the combined conditions $s\in k^\times$ and $t\in K^\times$, which is to say

$$\mathfrak{c}(\gamma) = a_2^\vee(s)a_3^\vee(st)\cdot\iota \in N_{\mathscr{M}'}(\mathscr{S}')(k). \tag{D.2.6.1}$$

Conversely, for $s\in k^\times$, $t\in K^\times$, the element $a_2^\vee(s)a_3^\vee(st)\cdot\iota \in N_{\mathscr{M}'}(\mathscr{S}')(k)$ has order 2, so we can define a 1-cocycle $\mathfrak{c}: \mathrm{Gal}(F/k)\to N_{\mathscr{M}'}(\mathscr{S}')(k)$ such that $\mathfrak{c}(\gamma)$ is this element. The action of $\overline{\mathfrak{c}}(\gamma)$ on e_i and f_i is as given above. The quadratic space (V,q) obtained by twisting $(\mathscr{V},\mathfrak{q})$ by $\overline{\mathfrak{c}}$ is the restriction of $\mathfrak{q}_F$ to the k-subspace V of $\mathscr{V}_F$ consisting of v such that $\overline{\mathfrak{c}}(\gamma)(\gamma\cdot v) = v$. Let $H_i = (F\otimes_k\mathscr{H}_i)\cap V$, so $V = H_1\oplus H_2\oplus K$. The elements of H_2 are the F-linear combinations ye_2+zf_2 such that

$$ye_2+zf_2 = s\gamma(z)e_2+s^{-1}\gamma(y)f_2,$$

which is to say that the elements of H_2 are $s\gamma(z)e_2+zf_2$ with $z\in F$. But

$$\mathfrak{q}(s\gamma(z)e_2+zf_2) = s\mathrm{N}_{F/k}(z)$$

since $B_{\mathfrak{q}}(e_2,f_2) = 1$, and likewise $\mathfrak{q}_F|_{H_1} = st^2\mathrm{N}_{F/k}$. Hence, (V,q) is isomorphic to

$$st^2\mathrm{N}_{F/k}\perp s\mathrm{N}_{F/k}\perp (K,x^2) \tag{D.2.6.2}$$

for a separable quadratic extension field F/k and scalars $s\in k^\times$ and $t\in K^\times$.

Assume (V,q) is anisotropic, so the twist ${}_{\mathfrak{c}}\mathscr{M}'\simeq \mathrm{Spin}(q)$ is k-anisotropic (by Proposition 7.1.6). Since $P := {}_{\mathfrak{c}}\mathscr{P} = {}_{\mathfrak{c}}\mathscr{U}\rtimes{}_{\mathfrak{c}}\mathscr{M}$ is a pseudo-parabolic k-subgroup of $G := {}_{\mathfrak{c}}\mathscr{G}$ and ${}_{\mathfrak{c}}\mathscr{M}'$ is k-anisotropic, we see that P must be minimal as a pseudo-parabolic k-subgroup of G. The maximal tori of $\mathscr{M}'$ are 2-dimensional, so G has k-rank equal to $4-2=2$. We have shown:

Proposition D.2.7. *For every separable quadratic extension field F/k and elements $s\in k^\times$ and $t\in K^\times$ such that* (D.2.6.2) *is anisotropic, the homomorphism*

$\mathfrak{c} : \mathrm{Gal}(F/k) \to N_{\mathscr{M}'}(\mathscr{S}')(k)$ defined by $\gamma \mapsto a_2^\vee(s) a_3^\vee(st) \cdot \iota$ yields a twist $G = {}_{\mathfrak{c}}\mathscr{G}$ that has k-rank equal to 2. Moreover, all basic exotic pseudo-reductive k-groups of type F_4 *with k-rank 2 and minimal field of definition K/k for the geometric unipotent radical arise in this manner.*

Example 7.1.7 provides k, K, and F such that for some $s \in k^\times$ and $t \in K^\times$ the associated quadratic form (D.2.6.2) is k-anisotropic. Hence, for such k and K there exist basic exotic k-groups G of type F_4 whose k-rank is 2.

By Lemma D.2.3, a necessary condition on such triples $(F/k, s, t)$ and $(F'/k, s', t')$ for the corresponding k_s/k-forms of $\mathscr{G}$ to be k-isomorphic is that the respective associated quadratic spaces (V, q) and (V', q') as in (D.2.6.2) are conformal. Such conformality is also sufficient. Indeed, the "anisotropic kernel" $M' = \mathscr{D}(Z_G(S))$ of the k_s/k-form G arising from $(F/k, s, t)$ satisfies $M'/Z_{M'} = \mathrm{SO}(q)$, so by Proposition 7.2.2(i) specifying the conformal isometry class of (V, q) is the same as specifying the k-isomorphism class of $M'/Z_{M'}$. But by Theorem 6.3.11, the k-isomorphism class of G is determined by K/k and the k-isomorphism class of $M'/Z_{M'}$ (since the k_s-isomorphism class of G_{k_s} is determined by the extension $K_s = k_s \otimes_k K$ of k_s, and the Dynkin diagrams of G_{k_s} and M'_{k_s} do not admit nontrivial automorphisms).

Bibliography

[Ar] M. Artin, "Brauer–Severi varieties" in *Brauer groups in ring theory and algebraic geometry*, LNM **917**, Springer-Verlag, New York (1982), 194–210.

[Bo] A. Borel, *Linear algebraic groups* (2nd ed.), Springer-Verlag, New York, 1991, 288pp.

[BLR] S. Bosch, W. Lütkebohmert, M. Raynaud, *Néron models*, Springer-Verlag, New York, 1990, 325pp.

[Bou] N. Bourbaki, *Lie groups and Lie algebras* (Ch. 4–6), Springer-Verlag, New York, 2002.

[C1] B. Conrad, *Finiteness theorems for algebraic groups over function fields*, Compositio Math. **148** (2012), 555–639.

[C2] B. Conrad, "Reductive group schemes" in *Autour des schémas en groupes I*, Panoramas et Synthèses 42–43, Société Mathématique de France, 2014.

[CGP] B. Conrad, O. Gabber, G. Prasad, *Pseudo-reductive groups* (2nd ed.), Cambridge Univ. Press, 2015.

[SGA3] M. Demazure, A. Grothendieck, *Schémas en groupes* I, II, III, Lecture Notes in Math **151, 152, 153**, Springer-Verlag, New York, 1970.

[EKM] R. Elman, N. Karpenko, A. Merkurjev, *The algebraic and geometric theory of quadratic forms*, AMS Colloq. Publ. **56**, 2008.

[EGA] A. Grothendieck, *Eléments de Géométrie Algébrique*, Publ. Math. IHES **24, 28, 32**, 1966–7.

[SGA2] A. Grothendieck, *Cohomologie Locale des Faisceaux Cohérents et Théorèmes de Lefschetz Locaux et Globaux*, North-Holland, Amsterdam, 1968.

[Hum] J. Humphreys, *Linear algebraic groups* (2nd ed.), Springer-Verlag, New York, 1987.

[KMRT] M-A. Knus, A. Merkurjev, M. Rost, J-P. Tignol, *The book of involutions*, AMS Colloq. Publ. **44**, Providence, 1998.

[Mat] H. Matsumura, *Commutative ring theory*, Cambridge Univ. Press, 1990.

[MV] B. Mühlherr, H. Van Maldeghem, *Exceptional Moufang quadrangles of type* F_4, Canad. J. Math. **51** (1999), 347–371.

[PY] G. Prasad, J-K Yu, *On quasi-reductive group schemes*, Journal of algebraic geometry **15** (2006), 507–549.

[R] B. Rémy, *Groupes algébriques pseudo-réductifs et applications*, Séminaire Bourbaki, Exp. 1021, Astérisque 339 (2011), 259–304.

[Spr] T. A. Springer, *Linear algebraic groups* (2nd ed.), Birkhäuser, New York, 1998.

[Tits] J. Tits, "Classification of algebraic semisimple groups" in *Algebraic groups and discontinuous groups*, Proc. Symp. Pure Math.,vol. 9, AMS, 1966.

[TW] J. Tits, R. Weiss, *Moufang polygons*, Springer Monographs in Mathematics, Springer-Verlag, New York, 2002.

[T] B. Totaro, *Pseudo-abelian varieties*, Ann. Sci. Ecole Norm. Sup. **46** (5), 2013, 693–721.

Index

Milton Keynes UK
Ingram Content Group UK Ltd.
UKHW020239110624
443885UK00006BA/218

9 780691 167930